Water Distribution

Grades 1 & 2

American Water Works Association

Association of Boards of Certification

Copyright © 1979, 1995, 2003, 2010, 2016 American Water Works Association.
All rights reserved.
Printed in the United States of America.

No part of this publication may be reproduced or transmitted in any form or by any means, electronic or mechanical, including photocopy, recording, or any information or retrieval system, except in the form of brief excerpts or quotations for review purposes, without the written permission of the publisher.

Disclaimer
Many of the photographs and illustrative drawings that appear in this book have been furnished through the courtesy of various product distributors and manufacturers. Any mention of trade names, commercial products, or services does not constitute endorsement or recommendation for use by the American Water Works Association or the US Environmental Protection Agency. In no event will AWWA be liable for direct, indirect, special, incidental, or consequential damages arising out of the use of information presented in this book. In particular, AWWA will not be responsible for any costs, including, but not limited to, those incurred as a result of lost revenue. In no event shall AWWA's liability exceed the amount paid for the purchase of this book.

Library of Congress Cataloging-in-Publication Data
CIP data have been applied for.

ISBN: 9781625761262

000200010272023379

6666 West Quincy Avenue
Denver, CO 80235-3098
303.794.7711

Contents

Foreword vii
Acknowledgments viii
How to Use This Book ix

Chapter 1 USEPA Drinking Water Regulations — 1
Federal Regulations — 1
State Regulations — 9
Requirements of Special Interest to Distribution System Operators — 10

Chapter 2 Operator Math — 21
Volume Measurements — 21
Conversions — 28
Per Capita Water Use — 52
Average Daily Flow — 53
Basic Pipeline Disinfection Calculations — 57
Basic Storage Facility Disinfection Calculations — 59
Calculating Heads — 61
Instantaneous Flow Rate Calculations — 63

Chapter 3 Water Use and System Design — 69
Water Use — 69
Water Rights — 73
Distribution System Purpose and Planning — 77
System Layout — 82
Mapping — 83
Valving — 83
Sizing Mains — 84

Chapter 4 Hydraulics — 89
Fluids at Rest and in Motion — 89
Measuring Pressure — 90
Head — 95

Chapter 5 Pipe — 99
Pipe Material Selection — 99
Types of Pipe Materials — 105

Chapter 6 Water Main Installation and Rehabilitation — 123
- Pipe Shipment — 123
- Pipe Handling — 124
- Excavation — 127
- Laying Pipe — 138

Chapter 7 Backfilling, Main Testing, and Installation Safety — 153
- Backfilling — 153
- Pressure and Leak Testing — 156
- Flushing and Disinfection — 158
- Final Inspection — 162
- Site Restoration — 162
- Water Main Installation Safety — 164

Chapter 8 Water Services — 169
- Meter Locations — 169
- Service Line Sizes, Materials, and Equipment — 171
- Water Service Taps — 175
- Leaks and Breaks — 179
- Thawing — 179
- Service Line Responsibility — 180
- Service Line Records — 181

Chapter 9 Valves — 183
- Uses of Water Utility Valves — 183
- Classification of Water Utility Valves — 186
- Valve Operation — 195
- Valve Storage — 196
- Valve Joints — 196
- Valve Boxes and Vaults — 197
- Valve Records — 198

Chapter 10 Fire Hydrants — 201
- Fire Hydrant Uses — 201
- System Problems Caused by Hydrant Operation — 203
- Types of Fire Hydrants — 203
- Hydrant Parts — 207
- Inspection and Installation — 209
- Operation and Maintenance — 212
- Hydrant Records — 214
- Hydrant Safety — 214

Chapter 11 Water Storage — 217
- Water Storage Requirements — 217
- Types of Treated-Water Storage Facilities — 220
- Location of Distribution Storage — 225
- Water Storage Facility Equipment — 227
- Operation and Maintenance of Water Storage Facilities — 233
- Water Storage Facility Safety — 236

Chapter 12 Electrical and Instrumentation-and-Control
Systems 239
Electricity and Magnetism 239
Electrical Measurements and Equipment 241
Instrumentation-and-Control Systems 243
Supervisory Control and Data Acquisition 251

Chapter 13 Motors and Engines 255
Motors 255
Motor Control Equipment 257
Internal-Combustion Engines 258
Pump, Motor, and Engine Records 259

Chapter 14 Pumps and Pumping Stations 261
Types of Pumps 261
Operation of Centrifugal Pumps 268
Mechanical Details of Centrifugal Pumps 271
Centrifugal Pump Maintenance 278

Chapter 15 Meters 281
Customer Water Meters 281
Mainline Metering 284

Chapter 16 Basic Chlorination 289
Basics of Chemical Disinfection 289
Chlorine Feed Equipment 294
Gas Chlorination Facilities 294
Hypochlorination Facilities 307
$C \times T$ Values 309
Disinfection By-products 309
Booster Disinfection 310

Chapter 17 System Operations 313
Maintaining Water Quality 313

Chapter 18 Water Quality Testing 321
Sampling 321
Monitoring for Chemical Contaminants 332
Laboratory Certification 333
Record Keeping and Sample Labeling 334
Sample Preservation, Storage, and Transportation 334
Common Water Quality Tests 337

Chapter 19 Backflow Prevention and
Cross-Connection Control 341
Terminology 341
Cross-Connections and Locations 342
Types of Cross-Connections 342

Backflow Control Methods and Devices	347
Cross-Connection Control Programs	357
Records and Reports	360

Chapter 20 Information Management and System Mapping — 363

Distribution System Maps	363
Equipment Records	367
Geographic Information Management Systems	370

Chapter 21 Safety, Security, and Emergency Response — 373

Personal Safety Considerations	373
Equipment Safety	374
Water Supply System Security	379

Chapter 22 Public Relations — 383

The Importance of Public Relations	383
The Role of Public Relations	383
The Role of Water Distribution Personnel	384

Appendix A Conversion Tables — 389

Appendix B Specifications and Approval of Treatment Chemicals and System Components — 411

Drinking Water Additives	411
Coatings and Equipment in Contact with Water	411
NSF International Standards and Approval	412
AWWA Standards	413

Appendix C Other Sources of Information — 415

Appendix D Sample Material Safety Data for Chlorine — 419

Study Question Answers   427
References   439
Glossary   441
Index   455

Foreword

This book is part of the *Water System Operations* (WSO) series. This water operator education series was designed by the American Water Works Association (AWWA) to address core test content on certification exams by operator certification type (treatment or distribution) and certification grade level.

The current books in the series are:

WSO Water Treatment, Grade 1
WSO Water Treatment, Grade 2
WSO Water Treatment, Grades 3 & 4
WSO Water Distribution, Grades 1 & 2
WSO Water Distribution, Grades 3 & 4

Acknowledgments

The WSO series was developed by AWWA with the help of a volunteer steering committee of subject matter experts.

We would like to extend our thanks to the following individuals for their invaluable help and expertise:

- William Lauer, Project Manager, QualQuest, LLC (Lakewood, CO)
- Zac Bertz, Joint Water Commission (Hillsboro, OR)
- Mary Howell, Backflow Management, Inc. (Portland, OR)
- Bob Hoyt, City of Worcester Department of Public Works and Parks (Worcester, MA)
- Ted Kenney, New England Water Works Association (Holliston, MA)
- Darin LaFalam, City of Worcester Department of Public Works and Parks (Worcester, MA)
- Kenneth C. Morgan, KCM Consulting Services, LLC (Phoenix, AZ)
- Ray Olson, Distribution System Resources (Littleton, CO)
- Paul Riendeau, New England Water Works Association (Holliston, MA)
- Ben Wright, City of Cayce Water and Sewer Department (Cayce, SC)

Additionally, we would like to thank Barbara Martin, AWWA/Partnership for Safe Water (Denver, CO), for her technical assistance and review.

How to Use This Book

Thank you for purchasing this volume in the Water System Operations series. AWWA's WSO series conforms to the latest Association of Boards of Certification (ABC) Need-to-Know criteria. ABC administers water operator certification testing for most of North America. Some states and provinces utilize different certification testing authorities, but this book covers all of the fundamentals every water operator needs to effectively pass their certification test and do their job effectively.

To help you advance in your career as a water operator, the WSO series is divided by subject areas (water treatment and water distribution) and certification grades (1, 2, 3, and 4, following the most common practice among states and provinces). Reference material, including basic science and mathematics concepts and information on water sources and quality, is included throughout all volumes in the series.

New features have been added to help you get the most from this book:

Key words   Important terms are highlighted in blue when they are first introduced, with a corresponding definition box in the margin of the page. You can also look up key words in the glossary at the end of the book.

The first pretreatment provided in most surface water treatment systems is screening. Coarse screens located on an intake structure are usually called *trash racks* or *debris racks*. Their function is to prevent clogging of the intake by removing sticks, logs, and other large debris in a river, lake, or reservoir. Finer screens may then be used at the point where the water enters the treatment system to remove smaller debris that has passed the trash racks.

The two basic types of screens used by water systems are bar screens and wire-mesh screens. Both types are available in models that are manually cleaned or automatically cleaned by mechanical equipment.

Bar Screens

Bar screens are made of straight steel bars, welded at both ends to two horizontal steel members. The screens are usually ranked by the open distance between bars as follows:

- Fine: spacing of 1/16 to 1/2 in. (1.5–13 mm)
- Medium: spacing of 1/2 to 1 in. (13–25 mm)
- Coarse: spacing of 1 1/4 to 4 in. (32–100 mm)

> **screening**
> A pretreatment method that uses coarse screens to remove large debris from the water to prevent clogging of pipes or channels to the treatment plant.
>
> **bar screens**
> A series of straight steel bars, welded at their ends to horizontal steel beams, forming a grid. Bar screens are placed on intakes or in waterways to remove large debris.

Video clips   A number of subjects in this book are linked to helpful video clips available on the AWWA WSO website (www.awwa.org/wsovideoclips). The videos present visual, hands-on information to supplement the descriptions in the text.

WATCH THE VIDEO
Sedimentation and Clarifiers (www.awwa.org/wsovideoclips)

End-of-chapter questions To test your understanding of the core concepts and applications in the book, questions are provided at the end of each chapter. The answers are provided at the end of the book, with the steps worked-out for math problems.

End-of-book questions In the higher-level WSO books, questions are included at the end of the book to test your understanding of topics covered in previous grades. Keep in mind that certification tests will cover material from both lower-level and higher-level books.

Visit our website at www.awwa.org/wso to check out the additional books in the series and to access the free online resources associated with this book.

Chapter 1
USEPA Drinking Water Regulations

Drinking water regulations have undergone major and dramatic changes during the past two decades, and trends indicate that they will continue to become more stringent and complicated. It is important that all water system operators understand the basic reasons for having regulations, how they are administered, and why compliance with them is essential. The reader should recognize that regulatory requirements are constantly changing. It is the operator's responsibility to keep current on all regulatory requirements.

Federal Regulations

Although the regulations required by the Safe Drinking Water Act (SDWA) are of prime interest in the operation and administration of water distribution systems, operators must also adhere to regulations required by several other federal environmental and safety acts.

Safe Drinking Water Act Requirements

Requirements under the SDWA are quite extensive, and complete details can be found through the USEPA website. The SDWA includes a number of current and proposed rules, including the following:

- Surface Water Treatment Rule (SWTR)
- Revised Total Coliform Rule (RTCR)
- Interim Enhanced Surface Water Treatment Rule (IESWTR)
- Long-Term 1 Enhanced Surface Water Treatment Rule (LT1ESWTR)
- Long-Term 2 Enhanced Surface Water Treatment Rule (LT2ESWTR)
- Ground Water Rule (GWR)
- Stage 1 Disinfectants and Disinfection By-products Rule (Stage 1 DBPR)
- Stage 2 Disinfectants and Disinfection By-products Rule (Stage 2 DBPR)
- Lead and Copper Rule (LCR)
- Public Notification Rule
- Filter Backwash Recycle Rule (FBRR)
- Unregulated Contaminant Monitoring Rule (UCMR)

The following discussion will primarily center on requirements that affect the operation of water distribution systems.

Prior to 1975, review of public water supplies was done by each state, usually by the state health department. The SDWA was passed by Congress in 1975 for a combination of reasons. One of the primary purposes was to create uniform national standards for drinking water quality to ensure that every public water supply in the country would meet minimum health standards. Another was that scientists and public health officials had recently discovered many previously unrecognized disease organisms and chemicals that could contaminate drinking water and might pose a health threat to the public. It was considered beyond the capability of the individual states to deal with these problems.

The SDWA delegates responsibility for administering the provisions of the act to the US Environmental Protection Agency (USEPA). The agency is headquartered in Washington DC and has 10 regional offices in major cities of the United States. Some principal duties of the agency are as follows:

- Set maximum allowable concentrations for contaminants that might present a health threat in drinking water. These limits are called **maximum contaminant levels (MCLs)**.
- Delegate primary enforcement responsibility for local administration of the requirements to state agencies.
- Provide grant funds to the states to assist them in operating the greatly expanded program mandated by the federal requirements.
- Monitor state activities to ensure that all water systems are being required to meet the federal requirements.
- Provide continued research on drinking water contaminants and improvement of treatment methods.

State Primacy

The intent of the SDWA is for each state to accept primary enforcement responsibility (primacy) for the operation of the state's drinking water program. Under the provisions of the delegation, the state must establish requirements for public water systems that are at least as stringent as those set by USEPA. The primacy agency in each state was designated by the state governor. In some states the primacy agency is the state health department, and in others it is the state environmental protection agency, department of natural resources, or pollution control agency. USEPA has primacy in any state that has not accepted this role (e.g., Wyoming).

Classes of Public Water Systems

The basic definition of a public water system in the SDWA is, in essence, a system that supplies piped water for human consumption and has at least 15 service connections or serves 25 or more persons for 60 or more days of the year. Examples of water systems that would not fall under the federal definition are private homes, groups of fewer than 15 homes using the same well, and summer camps that operate for fewer than 60 days per year. These systems are, however, generally under some degree of supervision by a local, area, or state health department.

USEPA has further divided public water systems into three classifications (Figure 1-1):

1. *Community public water systems* serve 15 or more homes. Besides municipal water utilities, this classification also covers mobile home parks and small homeowner associations that have their own water supply and serve more than 15 homes.

maximum contaminant level (MCL)
The maximum permissible level of a contaminant in water as specified in the regulations of the Safe Drinking Water Act.

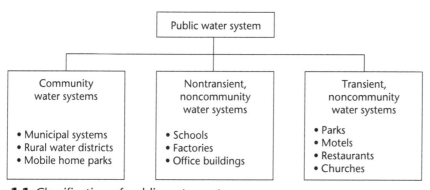

Figure 1-1 Classification of public water systems
Source: *Drinking Water Handbook for Public Officials* (1993).

2. *Nontransient, noncommunity public water systems* are establishments that have their own private water systems, serving an average of at least 25 persons who do not live at the location, but the same people use the water for more than 6 months per year. Examples are schools and factories.

3. *Transient, noncommunity public water systems* are establishments such as parks and motels that have their own water systems and serve an average of at least 25 persons per day, but these persons use the water only occasionally and for short periods of time.

The monitoring requirements for community and nontransient, noncommunity systems include all contaminants that are considered a public health threat. Transient, noncommunity systems are required to monitor only for nitrate, nitrite, and microbiological contamination.

Regulation of Contaminants

The National Primary Drinking Water Regulations (NPDWRs) specify MCLs or a treatment technique requirement for contaminants that may be found in drinking water that could have an adverse health effect on humans.

Specific concentration limits for the chemicals are listed, and all community and nontransient, noncommunity systems must test for their presence. If a water system is found to have concentrations of chemicals present that are above the MCL, the system must either change its water source or treat the water to reduce the chemical concentration. Primary regulations are mandatory and must be complied with by all water systems to which they apply.

The National Secondary Drinking Water Regulations apply to drinking water contaminants that may adversely affect the aesthetic qualities of water, such as taste, odor, or color. These qualities have no known adverse health effects, but they seriously affect public acceptance of the water. Secondary regulations are not mandatory but are strongly urged by USEPA. Some state regulatory agencies have made some of the secondary limits mandatory in their states.

Public Notification

The SDWA mandates that the public be kept informed of noncompliance with federal requirements by requiring that noncomplying systems provide public notification. If public water systems violate any of the operating, monitoring, or reporting requirements, or if the water quality exceeds an MCL, the system must inform the public of the problems. Even though the problem may have already been corrected, an explanation must be provided in the news media describing the public health significance of the violation.

The language and methods of providing public notification are mandated by USEPA to ensure that the public is fully informed. If a system is required to provide public notification, the state primacy agency will provide full instructions.

Water distribution operators should understand that although public notification is intended to keep the public informed, if it is necessitated by a simple mistake such as forgetting to send in the monthly samples, it can cause some embarrassment for the system's staff. To avoid this situation, careful attention must be given to state requirements. Any problem in meeting any of the requirements should be discussed with the state agency's representative.

If an operator is required to provide public notification, it should be made as positive as possible. Although the basic wording is mandatory, other wording can be added to keep the announcement from sounding completely negative to the public. Such wording can be discussed with the primacy agency's representative.

Monitoring and Reporting

To ensure that the drinking water supplied by all public water systems meets federal and state requirements, system operators are required to regularly collect samples and have the water tested. The regulations specify minimum sampling frequencies, sampling locations, testing procedures, methods of keeping records, and frequency of reporting to the state. The regulations also mandate special reporting procedures to be followed if a contaminant exceeds an MCL.

All systems must provide periodic monitoring for microbiological contaminants and some chemical contaminants. The frequency of sampling and the chemicals that must be tested for depend on the size of the water system, the source of water, and the history of analyses.

State policies vary on providing laboratory services. Some states have the laboratory facilities available to perform all required analyses or, in some cases, a certain number of the required analyses for a system. Most states charge for all or some of the laboratory services. Sample analyses that are required and cannot be performed by a state laboratory must be taken or sent to a state-certified private laboratory.

If the analysis of a sample exceeds an MCL, resampling is required, and the state should be contacted immediately for special instructions. There is always a possibility that such a sample was caused by a sampling or laboratory error, but it must be handled as though it was caused by contamination of the water supply.

The results of all water analyses must be periodically sent to the state. Failure to have the required analyses performed or to report the results to the state will usually result in the system having to provide public notification. States typically have special forms for submitting the data and specify a number of days following the end of the monitoring period by which the form must be submitted. The minimum information that must be provided in the form is listed in Table 1-1. State regulators may require other information for their own records and documentation.

There are also requirements specifying the length of time a water system must retain records. Table 1-2 lists the record-keeping requirements mandated by USEPA.

Water Quality Monitoring

Although most water quality monitoring is related to ensuring proper quality of the source water or treatment processes, many of the samples are collected from the distribution system. Thus, sample collection often becomes a duty of

Table 1-1 Laboratory report summary requirements

Type of Information	Summary Requirement
Sampling information	Date, place, and time of sampling
	Name of sample collector
	Identification of sample
	• Routine or check sample
	• Raw or treated water
Analysis information	Date of analysis
	Laboratory conducting analysis
	Name of person responsible for analysis
	Analytical method used
	Analysis results

Table 1-2 Typical record-keeping requirements

Type of Records	Time Period
Bacteriological and turbidity analyses	5 years
Chemical analyses	10 years
Actions taken to correct violations	3 years
Sanitary survey reports	10 years
Exemptions	5 years following expiration

distribution system personnel. The reason for collecting samples from the distribution system is that there are some opportunities for water quality to change after it enters the distribution system, and under the requirements of the SDWA, it is the duty of the water purveyor to deliver water of proper quality to the consumer's tap.

Methods of Collecting Samples

Two basic methods of collecting samples are grab sampling and composite sampling. A *grab sample* is a single volume of water collected at one time from a single place. To sample water in the distribution system, a faucet is used to fill a bottle. This sample represents the quality of the water only at the time the sample was collected. If the quality of the water is relatively uniform, the sample will be quite representative. If the quality varies, the sample may not be representative.

A *composite sample* consists of a series of grab samples collected from the same point at different times and mixed together. The composite is then analyzed to obtain the average value. The two types of composite samples commonly used are time composite samples and flow-proportional samples. Time composite samples are made up of equal-volume samples collected at regular intervals. Flow-proportional composite samples are also collected at regular time intervals, but the size of each grab sample is proportional to the flow at the time of sampling.

Although composite sampling appears to be a good idea because it provides an average of water quality, it cannot be used for most analyses of drinking water quality because a majority of parameters are not stable over a period of time.

time composite sample
A composite sample consisting of several equal-volume samples taken at specified times.

flow-proportional composite sample
A composite sample in which individual sample volumes are proportional to the flow rate at the time of sampling.

Sample Storage and Shipment

Care must always be taken to use the exact sample containers specified or provided by the laboratory that will be doing the analyses. Most sample containers are now plastic to avoid the possibility of glass breaking during shipment. Some samples for organic chemical analysis must be collected in special glass containers because some of the chemical might permeate the walls of a plastic container.

Sample holding time before analysis is quite critical for some parameters. If a laboratory receives a sample that has passed the specified holding time, it is supposed to declare the sample invalid and request resampling. Some samples can be refrigerated or treated once they arrive at the laboratory to extend the holding time, allowing the laboratory a few more days before the analyses must be completed.

Many laboratories do not work on weekends, so this factor should be taken into consideration when sending samples. Bacteriological analyses must, for example, be performed immediately by the laboratory. The best time to collect and send these samples is on a Monday or Tuesday so they will reach the laboratory by mid-week. Samples should be sent to the laboratory by the fastest means available, such as a next-day delivery service through the US Post Office or special carrier.

Sample Point Selection

Samples are collected from various points in the distribution system to determine the quality of water delivered to consumers. In some cases, distribution system samples may be significantly different from samples collected as the water enters the system. For example, corrosion in pipelines, bacterial growth, or algae growth in the pipes can cause increases in color, odor, turbidity, and chemical content (e.g., lead and copper). More seriously, a cross-connection between the distribution system and a source of contamination can result in chemical or biological contamination of the water.

Most samples collected from the distribution system will be used to test for coliform bacteria and chlorine residual. There are two primary considerations in determining the number and location of sampling points:

1. They should be representative of each different source of water entering the system (i.e., if there are several wells that pump directly into the system, samples should be obtained that are representative of the water from each one).
2. They should be representative of the various conditions within the system (such as dead ends, loops, storage facilities, and each pressure zone).

The required number of samples that must be collected and the frequency of sampling depend on the number of customers served, the water source, and other factors. Specific sampling instructions must be obtained from the state primacy agency. See *Standard Methods for the Examination of Water and Wastewater* for more information.

Sample Faucets

After representative sample points have been located on the distribution system, specific locations having suitable faucets for sampling must be identified. If suitably located, public buildings and the homes of utility employees are convenient places to collect samples. Otherwise, arrangements must be made to collect samples from businesses or private homes.

The following types of sampling faucets should not be used:

- Any faucet located close to the bottom of a sink, because containers may touch the faucet
- Any leaking faucet with water running out from around the handle and down the outside
- Any faucet with threads, such as a sill cock, because water generally does not flow smoothly from these faucets and may drip contamination from the threads
- Any faucet connected to a home water-treatment unit, such as a water softener or carbon filter
- Drinking fountain

It is also best to try to find a faucet without an aerator. If faucets with aerators must be used, follow the state recommendations on whether or not the aerator should be removed for sampling.

Each sample point must be described in detail on the sample report form—not just the house address, but which faucet in which room. If resampling is necessary, the same faucet used for the first sample must be used.

When it is necessary to establish a sampling point at a location on the water system where no public building or home gives access for regular sampling, a permanent sampling station can be installed (Figure 1-2).

Sample Collection

For collection of bacteriological and most other samples, the procedure is to open the faucet so that it will produce a steady, moderate flow. Opening the faucet to full flow for flushing is not usually desirable because the flow may not be smooth and water will splash up onto the outside of the spout, which may introduce contamination not representative of the water. If a steady flow cannot be obtained, the faucet should not be used.

The water should be allowed to run long enough to flush any stagnant water from the house plumbing, which usually takes 2–5 minutes. The line is usually clear when the water temperature drops and stabilizes. The sample is then collected without changing the flow setting. The sample container lid should be held (not set down on the counter) with the threads down during sample collection and replaced immediately. The sample container should then be labeled.

Figure 1-2 Example of a permanent sampling station
Courtesy of Gil Industries, Inc.

The exception to the above-mentioned procedure is sampling for lead and copper analysis. These are to be first-draw samples and require special procedures.

Bottles to be used for collection of bacteriological samples should not be rinsed before they are filled. These bottles are usually prepared with a small quantity of thiosulfate at the bottom to immediately stop the action of the residual chlorine in the water.

WATCH THE VIDEO
Sampling Techniques (www.awwa.org/wsovideoclips)

Special-Purpose Samples

It is occasionally necessary to collect special samples, particularly in response to customer complaints, such as taste and odor issues. To check on this type of complaint, one sample should be collected immediately as the tap is opened to be representative of water that has been in the plumbing system, then a second sample should be collected after the line has been flushed. It is sometimes helpful to collect both hot- and cold-water samples in this manner. These samples can be used to identify whether the problem is in the customer's plumbing system or coming from the water distribution system. Many customer complaints of taste, odor, or color are found to be from their own water heaters, water softeners, or home water-treatment devices.

Laboratory Certification

It is imperative that the monitoring of all water systems be consistent; therefore, all laboratory analyses must be performed by experienced technicians under carefully controlled conditions. For this reason, compliance sample analyses are acceptable to the state only if they have been performed by a certified laboratory. The only exceptions are measurements for turbidity, chlorine residual, temperature, and pH, which may be performed by a person acceptable to the state, using approved equipment and methods.

Most states operate certified laboratories that can accept some or all of the samples from water systems. The states also certify private laboratories that may be used for performing water analyses. Most large water utilities have their own certified laboratories because of the great number of samples that must be processed.

Consumer Confidence Reports

One of the very significant provisions of the 1996 SDWA Amendments is the consumer confidence report (CCR) requirement. The purpose of the CCR is to provide all water customers with basic facts regarding their drinking water so that individuals can make decisions about water consumption based on their personal health. This directive has been likened to the requirement that packaged food companies disclose what is in their food products.

The reports must be prepared yearly by every community water system. Water systems serving more than 10,000 people must mail the report to customers. Smaller systems must notify customers as directed by the state primacy agency.

A water system that distributes only purchased water (satellite system) must prepare the report for their consumers. Information on the source water and chemical analyses must be provided to the satellite system by the system selling the water (parent system).

Some states are preparing much of the information for their water systems, but the system operator must still add local information. Water system operators should keep in mind that CCRs provide an opportunity to educate consumers about the sources and quality of their drinking water. Educated consumers are more likely to help protect drinking water sources and be more understanding of the need to upgrade the water system to make their drinking water safe.

USEPA Regulation Information

Current information on USEPA regulations can be obtained by contacting the Safe Drinking Water Hotline at 800-426-4791. Also see the Office of Ground Water and Drinking Water webpage at http://water.epa.gov/drink.

State Regulations

Under the provisions of primacy delegation, each state must have requirements applying to public water systems that are at least as stringent as those set by USEPA. States occasionally establish requirements that are more stringent. Federal requirements are only for factors that USEPA considers directly related to public health. So, in addition to the federal requirements, each state establishes other requirements to ensure proper water system operation.

Operator Certification

One requirement of the 1996 SDWA Amendments is that USEPA must establish minimum standards for state operator certification programs. Most states have had some form of certification for water system operators, but, unfortunately, each state has its own idea of how operators should be classified, so there has been little national consistency.

Among the more important requirements are that each water system must at all times be under the direct supervision of a certified operator, operators must have a high school or equivalent education and pass an examination to receive certification, and the state must establish training requirements for certification renewal. Most states have a separate certification class for distribution system operators.

Cross-Connection Control

The states also generally promote cross-connection control programs for all water systems. Many states have their own cross-connection control manuals and assist water systems in setting up local programs. Cross-connection control is covered in detail in Chapter 19.

Construction Approval

The SDWA requires states to review plans for water system construction and improvements. In general, plans and specifications for the proposed work must be prepared by a professional engineer and submitted for approval before work begins. State engineers review the plans for suitability of materials, conformance with state regulations, and other factors.

Some states allow small distribution system additions without approval or allow approval after construction. State regulations should be reviewed to ensure compliance with requirements.

Technical Assistance

One of the staff functions of the state drinking water program is to provide technical assistance to water system operators. Field staff with training and experience are usually available to provide advice and assistance. If possible, they will provide advice over the phone, but if the problem is of sufficient magnitude, they will arrange personal visits. Staff may also, on some occasions, suggest other sources of information or assistance.

Enforcement

Because of the direct relationship between drinking water quality and public health, it is rare for anyone to purposely disregard state and federal regulations. Most violations of regulations are caused by not understanding requirements or forgetting something that must be done.

The SDWA requires states to use enforcement actions when federal requirements are violated. Then if the state does not take appropriate action, USEPA is prepared to step in and do it. Minor infractions are handled by public notification, but intentional disregard for requirements can result in substantial monetary fines.

Requirements of Special Interest to Distribution System Operators

Distribution system regulations address three main areas of concern: microbiological safety, disinfection by-products (DBPs), and lead. The microbiological safety of the water reaching customers' taps is of primary concern, and this was the initial focus of the distribution system regulatory requirements.

DBPs, such as total trihalomethanes (TTHM), are created by chemical reactions between disinfectants (e.g., chlorine) and other substances in the water. High levels in water may increase the risk of cancer for some individuals over a lifetime. Therefore, MCLs and monitoring requirements are included in the appropriate rules. These requirements are changing as more is learned about the levels of concern.

Lead is hazardous if consumed in high amounts, particularly for children. Water with certain characteristics may dissolve lead from solder or plumbing fixtures (or lead service lines) and may pose a risk to consumers. Therefore, special tap sampling requirements are mandated to determine the need to stabilize the water or perhaps replace lead water services. The applicable regulatory rules are discussed in more detail in the following sections.

Total Coliform Rule (TCR) and Revised Total Coliform Rule (RTCR)

The objective of the TCR is to promote routine surveillance of distribution system water quality to search for fecal matter and/or disease-causing bacteria. All points in a distribution system cannot be monitored, and complete absence of fecal matter and disease-causing bacteria cannot be guaranteed. The TCR is an attempt to persuade water utilities to implement monitoring programs sufficient to verify that public health is being protected as much as possible, as well as allowing utilities to identify any potential contamination problems in their distribution system. The rule requires monthly sampling at each distribution sampling point.

The TCR, and the RTCR that was finalized in 2013, impact all public water sources. The RTCR requires public water sources that are vulnerable to microbial contamination to identify and fix problems. The RTCR also established criteria for systems to qualify for and stay on reduced monitoring, thereby providing incentives for improved water system operation.

The RTCR rule established an **maximum contaminant level goal (MCLG)** and an MCL for *Escherichia coli* (*E. coli*) and eliminated the MCLG and MCL for total coliform, replacing it with a treatment technique for coliform that requires assessment and corrective action. The rule establishes an MCLG and an MCL of zero for *E. coli*, a more specific indicator of fecal contamination and potentially harmful pathogens than total coliform. USEPA removed the MCLG and MCL of zero for total coliform. Many of the organisms detected by total coliform methods are not of fecal origin and do not have any direct public health implication.

Under the treatment technique for coliform, total coliform serves as an indicator of a potential pathway of contamination into the distribution system. A public water source that exceeds a specified frequency of total coliform occurrence must conduct an assessment to determine if any sanitary defects exist and, if found, correct them. In addition, a public water source that incurs an *E. coli* MCL violation must conduct an assessment and correct any sanitary defects found.

The RTCR also changed monitoring frequencies. It links monitoring frequency to water quality and system performance and provides criteria that well-operated small systems must meet to qualify and stay on reduced monitoring. It also requires increased monitoring for high-risk small systems with unacceptable compliance history and establishes some new monitoring requirements for seasonal systems such as state and national parks.

The revised rule eliminated monthly public notification requirements based only on the presence of total coliforms. Total coliforms in the distribution system may indicate a potential pathway for contamination but in and of themselves do not indicate a health threat. Instead, the rule requires public notification when an *E. coli* MCL violation occurs, indicating a potential health threat, or when a public water source fails to conduct the required assessment and corrective action.

The rule requires that public water sources collect total coliform samples at sites representative of water quality throughout the distribution system according to a written plan approved by the state or primacy agency. Samples are collected at regular intervals monthly. Positive total coliform samples must be tested for *E. coli*. If any positive total coliform sample is also positive for *E. coli* the state must be notified by the end of the day on which the result was received. Repeat samples are required within 24 hours of any total coliform–positive routine sample. Three repeat samples are required, one at the site of the positive sample and one within five service taps both upstream and downstream of the positive site. Any positive total coliform samples must be tested for *E. coli*. Any positive *E. coli* (EC+) samples must be reported by the end of the day. Any positive total coliform (TC+) samples require another set of repeat samples.

A Level 1 or Level 2 sanitary assessment and corrective action is triggered to occur within 30 days if there is indication of coliform contamination. A Level 1 assessment by the public water source is triggered if more than 5 percent of the routine/repeat monthly samples (if at least 40 are required) are total coliform positive or a repeat sample is not taken for a total coliform positive result. A Level 2 assessment conducted by the state or its representative is triggered if the public water source has an *E. coli* violation or repeated Level 1 assessment triggers.

maximum contaminant level goal (MCLG)
Nonenforceable health-based goals published along with the promulgation of an MCL. Originally called recommended maximum contaminant levels (RMCLs).

Major violations of the RTCR are MCL violations and treatment technique violations. A public water source will receive an *E. coli* MCL violation when there is any combination of an EC+ sample result with a routine/repeat TC+ or EC+ sample result, as follows:

E. coli MCL Violation Occurs With the
Following Sample Result Combination

Routine	Repeat
EC+	TC+
EC+	Any missing sample
EC+	EC+
TC+	EC+
TC+	TC+ (but no *E. coli* analysis)

A public water source will receive a treatment technique violation given any of the following conditions:

- Failure to conduct a Level 1 or Level 2 assessment within 30 days of a trigger
- Failure to correct all sanitary defects from a Level 1 or Level 2 assessment within 30 days of a trigger or in accordance with the state-approved time frame
- Failure of a seasonal system to complete state-approved start-up procedures prior to serving water to the public

Disinfectants and Disinfection By-product Rules

There are several rules that, together, address the issues created by the formation of various potentially harmful compounds by the addition of some disinfectants. Chlorine, for example, can form trihalomethanes (THMs) if certain organic substances are present. The concentration of some by-products can increase in the distribution system. Therefore, the rules require testing samples collected at sites throughout the system. Some important aspects of these rules for distribution system operators are given in the following sections.

Stage 1 Disinfectants and Disinfection By-products Rule (Stage 1 DBPR)

The Stage 1 DBPR applies to community water systems and nontransient, noncommunity systems, including those serving fewer than 10,000 people, that add a disinfectant to the drinking water during any part of the treatment process.

The rule includes the following key provisions:

- Maximum residual disinfectant levels (MRDLs) for three disinfectants—chlorine (4.0 mg/L), chloramines (4.0 mg/L), and chlorine dioxide (0.8 mg/L)
- MCLs for TTHM—0.080 mg/L; haloacetic acids (HAA5)—0.060 mg/L; and two inorganic DBPs—chlorite (1.0 mg/L) and bromate (0.010 mg/L)
- A treatment technique for removal of DBP precursor material (*enhanced coagulation*)

Stage 2 Disinfectants and Disinfection By-products Rule (Stage 2 DBPR)

The rule tightened requirements for DBPs, but compliance is not achieved by modifying the numerical value of the MCLs or by requiring monitoring of new constituents. Instead, the rule makes compliance more difficult than under the Stage 1 DBPR by (1) changing the way the compliance value is calculated and (2) changing the compliance monitoring locations to sites representative of the greatest potential for THM and HAA formation. These changes were incorporated to attempt to account for peak spatial occurrence in the system. This change in focus reflects concerns of utilities and regulators caused by the potential for reproductive and developmental health effects associated with repeated exposure over a 12-month period at peak locations within the system.

The compliance value in the Stage 2 DBPR is called the locational running annual average (LRAA), and it is calculated by separately averaging the four quarterly samples at each monitoring location. Compliance is based on the maximum LRAA value (see Table 1-1). Furthermore, the Stage 2 DBPR includes several interim steps that led to the replacement of many existing Stage 1 DBPR monitoring locations with new locations representative of the greatest potential for consumer exposure to high levels of TTHM and HAA5.

The Stage 2 DBPR requires that facilities maintain compliance with the Stage 1 DBPR using the existing monitoring locations during the first three years after the final version of the Stage 2 DBPR was published. In the time period between the third and sixth year after the Stage 2 DBPR was published, compliance continued to be based on maintaining 80/60 (TTHM and HAA5) or lower for the running annual average; it also included a requirement for maximum LRAA at existing Stage 1 monitoring locations. These time periods during the Stage 2 DBPR transition were called "Stage 2A" and "Stage 2B."

The long-term goal of the Stage 2 DBPR is to identify locations within the distribution system with the greatest potential for either TTHM or HAA5 formation and then base compliance on maintenance of LRAA at or below 80/60 for each of these locations. Many of these locations were identified during the initial distribution system evaluation (IDSE). Consequently, the IDSE and the Stage 2A were actually just transition phases between the Stage 1 DBPR and the eventual long-term requirements of Stage 2B.

The IDSE included monitoring, modeling, and/or other evaluations of drinking water distribution systems to identify locations representative of the greatest potential for consumer exposure to high levels of TTHM and HAA5. The goal of the IDSE was to evaluate a number of potential monitoring locations to justify selection of monitoring locations for long-term compliance (i.e., Stage 2B) with the Stage 2 DBPR.

One item to note regarding the Stage 2 DBPR as it applies to TTHM and HAA5 is that the goal is to find the locations in the distribution system where average annual levels of these DBPs are highest. TTHM formation increases as contact time with free or combined chlorine increases, although formation in the presence of combined chlorine is limited. Therefore, establishing points in the distribution system with highest potential for TTHM formation is related to points with maximum water age. Utilities that have not performed a tracer study in the distribution system to determine water age should consider doing so.

By contrast, peak locations for HAA5 are more complicated because microorganisms in biofilm attached to distribution system pipe surfaces can biodegrade HAA5. Consequently, increasing formation of HAA5 over time is offset by biodegradation, eventually reaching a point where HAA5 levels decrease over time, even to the point where they drop to zero.

In chloramination systems, HAA5 formation is limited. In fact, ammonium chloride is added as a quenching agent in HAA5 compliance samples in order to halt HAA5 formation prior to analysis (see *Standard Methods for the Examination of Water and Wastewater*, latest edition). Therefore, little additional HAA5 formation occurs after chloramination to offset HAA5 biodegradation occurring in the distribution system.

WATCH THE VIDEO
Disinfection By-products (www.awwa.org/wsovideoclips)

Surface Water Treatment Rule (SWTR)

This rule is primarily directed at the treatment of water from surface water sources. It was originally intended to protect the public from exposure to *Giardia lamblia*. The rule was expanded by the Interim Enhanced Surface Water Treatment Rule to include *Cryptosporidium*. The Long-Term 1 and 2 Enhanced Surface Water Treatment Rules strengthen the requirements for microbial protection of all sizes of water systems. Portions of these rules affect distribution systems, so it is important to describe the rules and to highlight these requirements.

Interim Enhanced Surface Water Treatment Rule (IESWTR)

The IESWTR applies to systems using surface water, or groundwater under the direct influence of surface water, that serve 10,000 or more persons. The rule also includes provisions for states to conduct sanitary surveys for surface water systems regardless of system size. The rule builds on the treatment technique requirements of the SWTR, with the following key additions and modifications of importance in distribution systems:

- Disinfection profiles must be prepared by systems with TTHM or HAA5 annual distribution system levels of 0.064 mg/L or 0.048 mg/L, respectively, or higher. The disinfection profiles will consist of daily *G. lamblia* log inactivation over a period of 1–3 years. These will be used to establish benchmarks for microbial protection to ensure that there are no significant reductions as systems modify disinfection practices to meet the Stage 1 DBPR.
- Systems using groundwater under the direct influence of surface water are subject to the new rules dealing with *Cryptosporidium*.
- *Cryptosporidium* is included in the watershed control requirements for unfiltered public water systems.
- Covers are required on new finished water reservoirs.
- Sanitary surveys, conducted by states, are required for all surface water systems regardless of size.
- The rule includes disinfection benchmark provisions to ensure continued levels of microbial protection while facilities take the necessary steps to comply with new DBP standards.

Sanitary Surveys

Sanitary surveys are a requirement of the IESWTR. A sanitary survey is "an onsite review of the water source, facilities, equipment, operation, and maintenance of the public water system for the purpose of evaluating the adequacy of such source, facilities, equipment, operation, and maintenance for producing and distributing safe drinking water" (USEPA, 1999). These surveys are usually performed by the state primacy agency and are required of all surface water systems and groundwater systems under the direct influence of surface water.

Sanitary surveys are typically divided into eight main sections, although some state primacy groups may have more:

1. Water sources
2. Water treatment process
3. Water supply pumps and pumping facilities
4. Storage facilities
5. Distribution systems
6. Monitoring, reporting, and data verification
7. Water system management and operations
8. Operator compliance with state requirements

Sanitary surveys are required on a periodic basis usually every 3 years. Surveys may be comprehensive or focused according to the regulatory agency requirements.

Long-Term 1 Enhanced Surface Water Treatment Rule (LT1ESWTR)

The LT1ESWTR strengthened microbial controls for small systems (i.e., those systems serving fewer than 10,000 people). The rule also prevents significant increase in microbial risk where small systems take steps to implement the Stage 1 DBPR. The rule also addresses disinfection profiling and benchmarking.

Long-Term 2 Enhanced Surface Water Treatment Rule (LT2ESWTR)

The update to the SWTR is called the LT2ESWTR, and it supplements SWTR requirements contained in the IESWTR for large surface water systems (>10,000 persons) and the LT1ESWTR for small systems (<10,000 persons).

One of the key elements of the LT2ESWTR was the use of *Cryptosporidium* monitoring results to classify surface water sources into one of four USEPA-defined risk levels called "bins." Facilities in the lowest bin (bin 1) are required to maintain compliance with the current IESWTR. Facilities in higher bins (bins 2–4) are required to either (1) provide additional *Cryptosporidium* protection from new facilities or programs not currently in use at a facility or (2) demonstrate greater *Cryptosporidium* protection capabilities of existing facilities and programs using a group of USEPA-approved treatment technologies, watershed programs, and demonstration studies, collectively referred to as the "Microbial Toolbox."

Implementation of the LT2ESWTR was phased over many years according to system size. Four size categories were established (schedule 1–4, with 4 being the smallest at <10,000 population) for implementing the rule. The rule for schedule 4 systems allows filtered supplies to perform initial monitoring for fecal coliform to determine if *Cryptosporidium* monitoring is required.

One of the most potentially useful and cost-effective tools for utilities that was used to comply with the LT2ESWTR and demonstrate the true *Cryptosporidium* removal capability of an existing system is the demonstration of performance (DOP) credit. It was especially advantageous for facilities in bin 2. The DOP study can be conducted on an entire treatment process or a specific segment of the process. It can include monitoring of ambient aerobic spores in full-scale treatment processes or in pilot-scale spiking studies using *Cryptosporidium*, aerobic spores, or some other suitable microbial surrogate. The *Long-Term 2 Enhanced Surface*

Water Treatment Rule: Toolbox Guidance Manual (USEPA, 2003) describes cases where the DOP credit is likely the most cost-effective solution if the facility is assigned to bin 2, and the DOP credit can also be useful as a low-cost safety factor if the facility is assigned to bins 3 or 4.

Lead and Copper Rule (LCR)

The LCR (promulgated in 1991 and revised in 2007) seeks to minimize lead and copper at users' taps. The rule establishes action levels for lead (0.015 mg/L) and copper (1.30 mg/L) for the 90th percentile of the samples measured at customer taps. Monitoring for a variety of water quality parameters is required. In addition to monitoring, all large systems are required to conduct corrosion studies to determine optimal lead and copper corrosion control strategies.

If the action triggers are exceeded, the system is required to evaluate several approaches: public education, source water treatment, corrosion control practices, and possibly lead pipe replacement. Corrosion control can include pH/alkalinity adjustment, corrosion inhibitor addition, and calcium adjustment.

This rule can affect disinfection strategies because some of the control measures for lead and copper involve water chemistry adjustments (specifically pH control). These adjustments can affect the formation of DBPs and disinfection effectiveness. Therefore, corrosion control measures employed to comply with the LCR must also be considered in the selection of an overall disinfection strategy.

The objective of the LCR is to control corrosiveness of the finished water in drinking water distribution systems to limit the amount of lead and copper that may be leached from certain metal pipes and fittings in the distribution system. Of particular concern are pipes and fittings connecting the household tap to the distribution system service line at individual homes or businesses, especially because water can remain stagnant in these service lines for long periods of time, increasing the potential to leach lead, copper, and other metals. Although the utility is not responsible for maintaining and/or replacing these household connections, they are responsible for controlling pH and corrosiveness of the water delivered to consumers.

Details of the LCR include the following:

- The LCR became effective December 7, 1992.
- The action level for lead is 0.015 mg/L and for copper is 1.3 mg/L.
- A utility is in compliance at each sampling event (frequency discussed in the following paragraphs) when <10 percent of the distribution system samples are above the action level for lead and copper (i.e., 90th percentile value for the sampling event must be below the action level).
- Utilities found not to be in compliance must modify water treatment until they are in compliance. The term *action level* is used rather than MCL because noncompliance (i.e., exceeding an action level) triggers a need for modifications in treatment.
- The utility must sample each entry point into the distribution system during each sampling event.
- Additional revisions to the LCR are under evaluation; check with your local regulatory agency for additional clarifications.

After identifying sampling locations and determining initial tap water lead and copper levels at each of these locations, utilities must also monitor other water quality parameters (WQPs) at these same locations as needed to monitor and evaluate corrosion control characteristics of treated water. The only exemptions from analysis of these WQPs are systems serving less than 50,000 people for which lead and copper levels in initial samples are below action levels. Lead, copper, and WQPs are initially collected at 6-month intervals, and then this frequency can be reduced if action levels are not exceeded and optimal water treatment is maintained. Systems that are in noncompliance and are performing additional corrosion-control activities must continue to monitor at 6-month intervals, plus they must collect WQPs from distribution system entry points every 2 weeks.

Each utility must complete a survey and evaluate materials that comprise their distribution system, in addition to using other available information, to target homes that are at high risk for lead or copper contamination.

Revisions to the LCR were enacted in 2007. These clarifications to the existing rule were made in seven areas:

1. Minimum number of samples required
2. Definitions for compliance and monitoring periods
3. Reduced monitoring criteria
4. Consumer notice of lead tap water monitoring results (Within 30 days of learning the results, all systems must provide individual lead tap results to people who receive water from sites that were sampled, *regardless of whether the results exceed the lead action level.*)
5. Advanced notification and approval of long-term treatment changes
6. Public education requirements (Community water systems must deliver materials to bill-paying customers and post lead information on water bills, work in concert with local health agencies to reach at-risk populations [children, pregnant woman], deliver to other organizations serving "at-risk" populations, provide press releases, and include new outreach activities.)
7. Reevaluation of lead service lines (A sample is required from any lead service lines not completely replaced to determine impact on lead levels.)

The local regulatory agency can be consulted for those revisions that are applicable to a particular system.

Ground Water Rule (GWR)

USEPA promulgated the final GWR in October 2006 to reduce the risk of exposure to fecal contamination that may be present in public water systems that use groundwater sources.

The GWR establishes a risk-targeted strategy to identify groundwater systems that are at high risk for fecal contamination. The rule also specifies when corrective action (which may include disinfection) is required to protect consumers who receive water from groundwater systems from bacteria and viruses.

A sanitary survey is required, by the state primacy agency, at regular intervals depending on the condition of the water system as determined in the initial survey. Systems found to be at high risk for fecal contamination are required to provide 4-log inactivation of viruses. Increased monitoring for fecal contamination indicators may be required by the regulatory authority. Federal regulations do not currently require disinfection of groundwater unless the well has been designated by the state as vulnerable to contamination by surface water (termed "groundwater under the direct influence of surface water"). These are generally relatively shallow wells. Many states, though, have their own requirements for required disinfection of various sizes, types, or classes of well systems.

National Secondary Drinking Water Regulations

A National Secondary Drinking Water Regulation is a nonenforceable guideline regarding contaminants that may cause aesthetic effects such as taste, odor, and color. Some states choose to adopt them as enforceable standards. Table 1-3 lists the secondary MCLs, and Table 1-4 lists the adverse effects of secondary contaminants.

Table 1-3 National Secondary Drinking Water Regulations

Contaminant	Secondary Standard
Aluminum	0.05–0.2 mg/L
Chloride	250 mg/L
Color	15 color units
Copper	1.0 mg/L
Corrosivity	Noncorrosive
Fluoride	2.0 mg/L
Foaming agents	0.5 mg/L
Iron	0.3 mg/L
Manganese	0.05 mg/L
Odor	3 threshold odor number
pH	6.5–8.5
Silver	0.10 mg/L
Sulfate	250 mg/L
Total dissolved solids	500 mg/L
Zinc	5 mg/L

Note: For more information, read *Secondary Drinking Water Regulations: Guidance for Nuisance Chemicals.*

Table 1-4 Adverse Effects of Secondary Contaminants

Contaminant	Adverse Effect
Chloride	Causes taste
	Adds to total dissolved solids and scale
	Indicates contamination
	Can accelerate the corrosion of some metals
Color	Indicates dissolved organics may be present, which may lead to trihalomethane formation
	Unappealing appearance
Copper	Undesirable metallic taste
Corrosivity	Corrosion products unappealing to consumers.
	Corrosion causes tastes and odors.
	Corrosion products can affect health.
	Corrosion causes costly deterioration of water system.
Fluoride	Dental fluorosis (mottling or discoloration of teeth)
Foaming agents	Unappealing appearance
	Indicates possible contamination

Table 1-4 Adverse Effects of Secondary Contaminants (continued)

Contaminant	Adverse Effect
Hydrogen sulfide	Offensive odor
	Causes black stains on contact with iron
	Can accumulate to deadly concentration in poorly ventilated areas
	Flammable and explosive
Iron	Discolors laundry brown
	Changes taste of water, tea, coffee, and other beverages
Manganese	Discolors laundry
	Changes taste of water, tea, coffee, and other beverages
Odor	Unappealing to drink
	May indicate contamination
pH	Below 6.5, water is corrosive.
	Above 8.5, water will form scale, taste bitter.
Sulfate	Has a laxative effect
Total dissolved solids	Associated with taste, scale, corrosion, and hardness
Zinc	Undesirable taste
	Milky appearance

Note: For more information, read *Secondary Drinking Water Regulations: Guidance for Nuisance Chemicals*.

Study Questions

1. Which agency sets legal limits on the concentration levels of harmful contaminants in potable water distributed to customers?
   a. National Primary Drinking Water Regulations Agency
   b. US Environmental Protection Agency
   c. US Public Health Service
   d. Occupational Health and Safety Administration

2. The number of monthly distribution system bacteriological samples required is
   a. based on the water withdrawal permit limit.
   b. based on system size.
   c. based on population served.
   d. different for each state.

3. What is the maximum contaminant level for total trihalomethanes (TTHM) in the United States?
   a. 0.040 mg/L
   b. 0.060 mg/L
   c. 0.080 mg/L
   d. 0.100 mg/L

4. Under the Surface Water Treatment Rule, disinfection residuals must be collected at the same location in the distribution system as
   a. coliform samples.
   b. total trihalomethanes.
   c. disinfection by-products.
   d. alkalinity, conductivity, and pH for corrosion studies.

5. Iron can cause "red water" and thus customer complaints when its concentration is above its secondary maximum contaminant level of
   a. 0.01 mg/L.
   b. 0.05 mg/L.
   c. 0.10 mg/L.
   d. 0.30 mg/L.

6. The goal of the Surface Drinking Water Act is for each _____ to accept primary enforcement responsibility (primacy) for the operation of the state's drinking water program.
   a. state
   b. water treatment plant
   c. operator
   d. municipality

7. What is the objective of the Total Coliform Rule?
   a. To provide *Cryptosporidium* protection in new facilities
   b. To ensure safe levels of lead and copper in drinking water
   c. To protect the public from exposure to *Giardia lamblia* and *Cryptosporidium*
   d. To promote routine surveillance of distribution system water quality to search for fecal matter and/or disease-causing bacteria

8. Which agency sets standards on the concentration levels of harmful contaminants in drinking water?

9. Explain the difference between the different tiers of violations. Which one is the most serious?

10. A positive *E. coli* test must be reported to the primacy health agency within what time period?

11. How is the required number of monthly bacteriological samples determined?

12. What is the action level for lead in first-draw samples taken from customer taps?

13. A water system is designated a community public water system if it serves how many homes?

14. What authority provides specific water sampling instructions?

15. According to the Lead and Copper Rule, what is the action level for lead?

Chapter 2
Operator Math

Volume Measurements

A volume measurement defines the amount of space that an object occupies. The basis of this measurement is the *cube*, a square-sided box with all edges of equal length, as shown in the diagram below. The customary units commonly used in volume measurements are cubic inches, cubic feet, cubic yards, gallons, and acre-feet. The metric units commonly used to express volume are cubic centimeters, cubic meters, and liters.

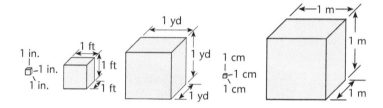

The calculations of surface area and volume are closely related. For example, to calculate the surface area of one of the cubes above, you would multiply two of the dimensions (length and width) together. To calculate the volume of that cube, however, a *third dimension* (depth) is used in the multiplication. The concept of volume can be simplified as

$$\text{volume} = (\text{area of surface})(\text{third dimension})$$

The *area of surface* to be used in the volume calculation is the *representative surface* area, the side that gives the object its basic shape. For example, suppose you begin with a rectangular area as shown on the next page. Notice the shape that would be created by stacking a number of those same rectangles one on top of the other:

21

Because the rectangle gives the object its basic shape in this example, it is considered the representative area. Note that the same volume could have been created by stacking a number of smaller rectangles one behind the other:

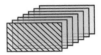

Although an object may have more than one representative surface area, as illustrated in the two preceding diagrams, sometimes only one surface is the representative area. Consider, for instance, the following shape:

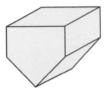

Let's compare two different sides of this shape—the top and the front—to determine if they are representative areas. In the first case, a number of the top shapes (rectangles) stacked together does not result in the same shape volume. Therefore, this rectangular area is not a representative area. In the second case, however, a number of front shapes (pentagons) stacked one behind the other results in the same shape volume as the original object. Therefore, the pentagonal area may be considered a representative surface area.

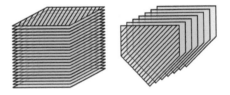

Rectangles, Triangles, and Circles

For treatment plant calculations, representative surface areas are most often rectangles, triangles, circles, or a combination of these. The following diagrams illustrate the three basic shapes for which volume calculations are made.

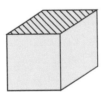

In the first diagram, the rectangle defines the shape of the object; in the second diagram, the triangle, rather than the rectangle, gives the trough its basic shape; in the third, the surface that defines the shape of the cylinder is the circle.

The formulas for calculating the volume of each of these three shapes are given below. Note that they are closely associated with the area formulas given previously.

> **Volume Formulas**
>
> Rectangle tank volume = $\left(\begin{array}{c}\text{area of}\\\text{rectangle}\end{array}\right)\left(\begin{array}{c}\text{third}\\\text{dimension}\end{array}\right)$
>
> $\qquad\qquad\qquad = lw \left(\begin{array}{c}\text{third}\\\text{dimension}\end{array}\right)$
>
> Trough volume = $\left(\begin{array}{c}\text{area of}\\\text{triangle}\end{array}\right)\left(\begin{array}{c}\text{third}\\\text{dimension}\end{array}\right)$
>
> $\qquad\qquad\quad = \left(\dfrac{bh}{2}\right)\left(\begin{array}{c}\text{third}\\\text{dimension}\end{array}\right)$
>
> Cylinder volume = $\left(\begin{array}{c}\text{area of}\\\text{circle}\end{array}\right)\left(\begin{array}{c}\text{third}\\\text{dimension}\end{array}\right)$
>
> $\qquad\qquad\quad = (0.785\ D^2)\left(\begin{array}{c}\text{third}\\\text{dimension}\end{array}\right)$
>
> Cone volume = 1/3(volume of a cylinder)
>
> Sphere volume = $\left(\dfrac{\pi}{6}\right)$ (diameter)^3

Use of the volume formulas is demonstrated in the examples that follow.

Example 1

Calculate the volume of water contained in the tank illustrated below if the depth of water (called *side water depth*, or SWD) in the tank is 10 ft.

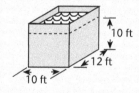

First, recall the volume formula:

volume = (area of surface)(third dimension)

In this example, the *rectangle* is the representative surface, and the area of a rectangle can be calculated as length times width. The dimension not used in the area calculation is the *depth*. Thus, the volume calculation is as follows:

$$\text{volume} = (\text{length})(\text{width})(\text{depth})$$
$$= (12 \text{ ft})(10 \text{ ft})(10 \text{ ft})$$
$$= 1{,}200 \text{ ft}^3$$

Example 2

What is the volume (in cubic inches) of water in the trough shown below if the depth of water is 8 in.?

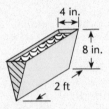

First, all dimensions must be expressed in the same terms. Because the answer is desired in cubic inches, the 2-ft dimension should be converted to inches:

$$(2 \text{ ft})(12 \text{ in./ft}) = 24 \text{ in.}$$

Then we apply the volume formula:

$$\text{volume} = (\text{area of surface})(\text{third dimension})$$

The triangle is the representative surface. The area of a triangle is calculated as follows:

$$\frac{(\text{base})(\text{height})}{2}$$

The third dimension may be considered length or width—a difference in terminology. Thus, the volume calculation is as follows:

$$\text{volume} = \left(\frac{bh}{2}\right)(\text{length})$$
$$= \frac{(4 \text{ in.})(8 \text{ in.})}{2}(24 \text{ in.})$$
$$= \frac{(4 \text{ in.})(8 \text{ in.})(24 \text{ in.})}{2}$$
$$= 384 \text{ in.}^3$$

Example 3

What is the volume of water contained in the tank shown below if the depth of water (SDW) is 28 ft?

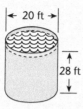

Recall the volume formula:

volume = (area of surface)(third dimension)

The circle is the representative surface, and the area of a circle is calculated as follows:

$(0.785)(D^2)$

The third dimension is depth. Thus, the volume calculation is as follows:

volume = $(0.785)(D^2)$(depth)

= (0.785)(20 ft)²(28 ft)

= (0.785)(20 ft)(20 ft)(28 ft)

= 8,792 ft³

Example 4

A tank with a cylindrical bottom has dimensions as shown below. What is the capacity of the tank? (Assume that the cross section of the bottom of the tank is a half circle.)

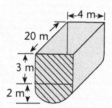

In problems involving a representative surface area that is a combination of shapes, it is often easier to calculate the representative surface area first, then calculate the volume:

representative surface area = area of rectangle + area of half-circle

$$= (4 \text{ m})(3 \text{ m}) + \frac{(0.785)(4 \text{ m})(4 \text{ m})}{2}$$

= 12 m² + 6.28 m²

= 18.28 m²

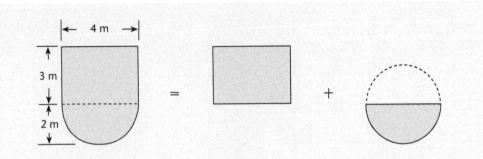

And now calculate the volume of the tank:

$$\text{volume} = (\text{area of surface})(\text{third dimension})$$
$$= (18.28 \text{ m}^2)(20 \text{ m})$$
$$= 365.6 \text{ m}^3$$

Cones and Spheres

There are many shapes (though very few in water treatment calculations) for which the concept of a "representative surface" does not apply. The cone and sphere are notable examples of this. That is, we cannot "stack" areas of the same size on top of one another to obtain a cone or sphere.

Calculating the volume of a cylinder was discussed earlier. The volume of a cone represents ⅓ of that volume:

$$\text{volume of a cone} = \tfrac{1}{3}(\text{volume of a cylinder})$$

or

$$= \frac{(0.785)(D^2)(\text{depth})}{3}$$

The volume of a sphere is more difficult to relate to the other calculations or even to a diagram. In this case, the formula should be memorized. To express the volume of a sphere mathematically, use the following formula:

$$\text{volume of a sphere} = \left(\frac{\pi}{6}\right)(\text{diameter})^3$$

The symbol π, or pi, represents the relationship between the circumference and diameter of a circle. The number 3.14 is used for pi. Thus, the equation may be expressed as follows:

$$\text{volume of a sphere} = \left(\frac{3.14}{6}\right)(\text{diameter})^3$$

Example 5

Calculate the volume of a cone that is 3 m tall and has a base diameter of 2 m.

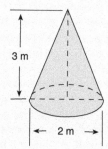

To begin, recall the formula used for calculating the volume of a cone:

$$\text{volume of a cone} = \tfrac{1}{3}(\text{volume of a cylinder})$$

Thus, the volume is calculated as follows:

$$\text{volume} = \frac{(0.785)(D^2)(\text{third dimension})}{3}$$

$$= \frac{(0.785)(2 \text{ m})(2 \text{ m})(3 \text{ m})}{3}$$

$$= 3.14 \text{ m}^3$$

Example 6

If a spherical tank is 30 ft in diameter, what is its capacity?

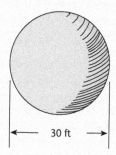

Apply the formula used for calculating the volume of a sphere:

$$\text{volume of a sphere} = \left(\frac{\pi}{6}\right)(\text{diameter})^3$$

The volume is calculated as follows:

$$\text{volume of a sphere} = \left(\frac{\pi}{6}\right)(\text{diameter})^3$$

$$= \left(\frac{3.14}{6}\right)(30\text{ ft})(30\text{ ft})(30\text{ ft})$$

$$= 14{,}130\text{ ft}^3$$

Occasionally it is necessary to calculate the volume of a tank that consists of two distinct shapes. In other words, there is no representative surface area for the entire shape. In this case, the volumes should be calculated separately, then the two volumes added. The diagrams below illustrate this method:

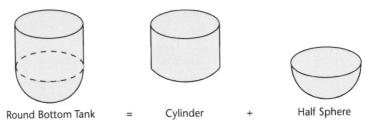

Round Bottom Tank = Cylinder + Half Sphere

 WATCH THE VIDEO
Volume Measurements (www.awwa.org/wsovideoclips)

Conversions

In making the conversion from one unit to another, you must know (1) the number that relates the two units and (2) whether to multiply or divide by that number. For example, in converting from feet to inches, you must know that in 1 ft there are 12 in., and you must know whether to multiply or divide the number of feet by 12.

Although the number that relates the two units of a conversion is usually known or can be looked up, there is often confusion about whether to multiply or divide. *Dimensional analysis*, discussed previously, is one method to help decide whether to multiply or divide for a particular conversion. Another, usually faster, method is to use conversion tables.

Conversion Tables

A conversion table, such as the one included in Appendix A, allows you to convert units simply by following the instructions indicated in the table headings. For example, if you want to convert from feet to inches, look in the *Conversion* column of the table for *From* "feet" *To* "inches." Read across this line and perform the operation indicated by the headings of the other columns; that is, multiply the number of feet by 12 to get the number of inches.

Suppose, however, that you want to convert inches to feet. Look in the *Conversion* column for *From* "inches" *To* "feet," and read across this line. The headings tell you to multiply the number of inches by 0.08333 (which is the decimal equivalent of 1/12) to get the number of feet. Multiplying by either 1/12 or 0.08333 is the same as dividing by 12.

The instruction to *multiply* by certain numbers (called conversion factors) is used throughout the conversion table. There is no column headed *Divide by* because the fractions representing division (such as 1/12) were converted to decimal numbers (such as 0.08333) when the table was prepared.

To use the conversion table, remember the following three steps:

1. In the *Conversion* column, find the units you want to change *From* and *To*. (Go *From* what you have *To* what you want.)
2. Multiply the *From* number you have by the conversion factor given.
3. Read the answer in *To* units.

Example 7

Convert 288 in. to feet.

In the *Conversion* column of the table, find *From* "inches" *To* "feet." Reading across the line, perform the multiplication indicated; that is, multiply the number of inches (288) by 0.08333 to get the number of feet:

$$(288 \text{ in.})(0.08333) = 24 \text{ ft}$$

Example 8

A tank holds 50 gal of water. How many cubic feet of water is this, and what does it weigh?

First, convert gallons to cubic feet. Using the table, you find that to convert *From* "gallons" *To* "cubic feet," you must multiply by 0.1337 to get the number of cubic feet:

$$(50 \text{ gal})(0.1337) = 6.69 \text{ ft}^3$$

Note that this number of cubic feet is actually a rounded value (6.685 is the actual calculated number). Rounding helps simplify calculations.

Next, convert gallons to pounds of water. Using the table, you find that to convert *From* "gallons" *To* "pounds of water," you must multiply by 8.34 to get the number of pounds of water:

$$(50 \text{ gal})(8.34) = 417 \text{ lb of water}$$

Notice that you could have arrived at approximately the same weight by converting 6.69 ft³ to pounds of water. Using the table, we get

$$(6.69 \text{ ft}^3)(62.4) = 417.46 \text{ lb of water}$$

This slight difference in the two answers is due to rounding numbers both when the conversion table was prepared and when the numbers are used in solving the problem. You may notice the same sort of slight difference in answers if you have to convert from one kind of units to two or three other units, depending on whether you round intermediate steps in the conversions.

Box Method

Another method that may be used to determine whether multiplication or division is required for a particular conversion is called the *box method*. This method is based on the relative sizes of different squares ("boxes"). The box method can be used when a conversion table is not available (such as during a certification exam). This method of conversion is often slower than using a conversion table, but many people find it simpler.

Because multiplication is usually associated with an *increase* in size, moving from a smaller box to a larger box corresponds to using multiplication in the conversion:

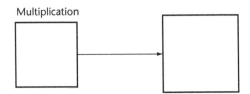

Division, on the other hand, is usually associated with a *decrease* in size. Therefore, moving from a larger box to a smaller box corresponds to using division in the conversion:

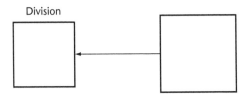

To use the box method to determine whether to multiply or divide in making a conversion, set up and label the boxes according to the following procedure:

1. Write the equation that relates the two types of units involved in the conversion. One of the two numbers in the equation must be a 1 (e.g., 1 ft = 12 in. or 1 ft = 0.305 m).
2. Draw a small box on the left, a large one on the right, and connect them with a line.
3. In the *smaller* box, write the name of the units associated with the 1 (for example, 1 *ft* = 12 in.—*ft* should be written in the smaller box). Note that the name of the units next to the 1 must be written in the *smaller* box; otherwise, the box method will give incorrect results.
4. In the larger box, write the name of the remaining units. Those units will also have a number next to them, a number that is not 1. Write that number over the line between the boxes.

Suppose, for example, that you want to make a box diagram for feet-to-inches conversions. First, write the equation that relates feet to inches:

$$1 \text{ ft} = 12 \text{ in.}$$

Next, draw the conversion boxes (smaller box on the left) and the connecting line:

Now label the diagram. Because the number 1 is next to the units of feet (1 ft), write *ft* in the smaller box. Write the name of the other units, inches (*in.*), in the larger box. And write the number that is next to inches, 12, over the line between the boxes.

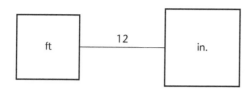

To convert from feet to inches, then *multiply by 12* because you are moving from a smaller box to a larger box. And to convert from inches to feet, *divide by 12* because you are moving from a larger box to a smaller box.

Let's look at another example of making and using the box diagram. Suppose you want to convert cubic feet to gallons. First write down the equation that relates these two units:

$$1 \text{ ft}^3 = 7.48 \text{ gal}$$

Then draw the smaller and larger boxes and the connecting line; label the boxes, and write in the conversion number:

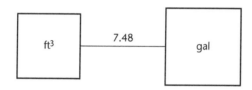

The smaller box corresponds to cubic feet and the larger box to gallons. To convert from cubic feet to gallons according to this box diagram, *multiply by 7.48* because you are moving from a smaller to a larger box. And to convert from gallons to cubic feet, *divide by 7.48* because you are moving from a larger to a smaller box.

Conversions of US Customary Units

This section discusses important conversions between terms expressed in US customary units (based on the box method).

Conversions From Cubic Feet to Gallons to Pounds

In making the conversion from cubic feet to gallons to pounds of water, you must know the following relationships:

$$1 \text{ ft}^3 = 7.48 \text{ gal}$$

$$1 \text{ gal} = 8.34 \text{ lb}$$

You must also know whether to multiply or divide, and which of the above numbers are used in the conversion. The following box diagram should assist in making these decisions:

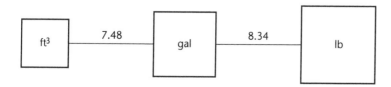

Example 9

Convert 1 ft³ to pounds.

First write down the diagram to aid in the conversion:

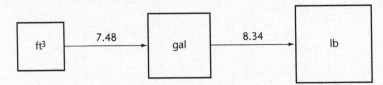

When you are converting from cubic feet to pounds, you are moving from smaller to larger boxes. Therefore, *multiplication* is indicated in both conversions:

$$(1 \text{ ft}^3)(7.48 \text{ gal/ft}^3)(8.34 \text{ lb/gal}) = 62.38 \text{ lb}$$

This total can be rounded to 62.4 lb, the number commonly used for water treatment calculations.

Example 10

A tank has a capacity of 60,000 ft³. What is the capacity of the tank in gallons?

First write down the diagram to aid in the conversion:

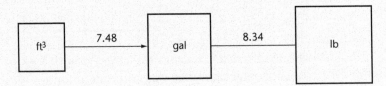

When converting from cubic feet to gallons, you are moving from a smaller to a larger box. Therefore, *multiplication* by 7.48 is indicated:

$$(60,000 \text{ ft}^3)(7.48 \text{ gal/ft}^3) = 448,800 \text{ gal}$$

Example 11

If a tank will hold 1,500,000 lb of water, how many cubic feet of water will it hold?

First write down the diagram to aid in the conversion:

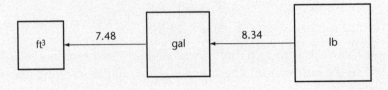

A move from larger to smaller boxes indicates *division* in both conversions:

$$\frac{1,500,000 \text{ lb}}{(8.34 \text{ lb/gal})(7.48 \text{ gal/ft}^3)} = 24,045 \text{ ft}^3$$

Flow Conversions

The relationships among the various US customary flow units are shown by the following diagram. Note the abbreviations commonly used in discussing flow:

- gps = gallons per second
- gpm = gallons per minute
- gpd = gallons per day

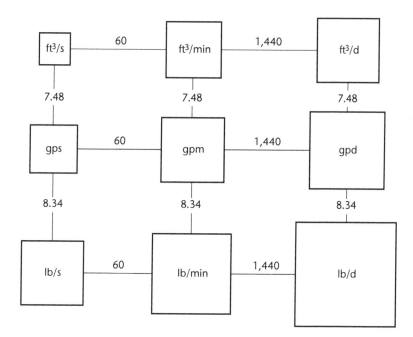

Because flows of cubic feet per hour, gallons per hour, and pounds per hour are less frequently used, they have not been included in the diagram. However, some chemical feed rate calculations require converting from or to these units.

The lines that connect the boxes and the numbers associated with them can be thought of as *bridges* that relate two units directly and all other units in the diagram indirectly. The relative sizes of the boxes are an aid in deciding whether multiplication or division is appropriate for the desired conversion.

The relationship among the boxes should be understood, not merely memorized. The principle is basically the same as that described in the preceding section. For example, looking at just part of the diagram, notice how every box in a single vertical column has the same *time* units; a conversion in this direction corresponds to a change in volume units. Every box in a single horizontal row has the same *volume* units; a conversion in this direction corresponds to a change in time units.

Although you need not draw the nine boxes each time you make a flow conversion, it is useful to have a mental image of these boxes to make the calculations. For the examples that follow, however, the boxes are used in analyzing the conversions.

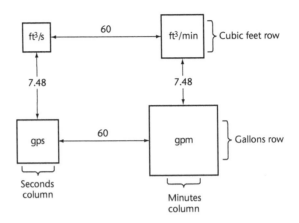

Example 12

If the flow rate in a water line is 2.3 ft³/s, what is this rate expressed as gallons per minute? (Assume the flow is steady and continuous.)

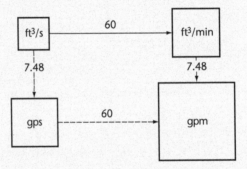

There are two possible paths from cubic feet per second to gallons per minute. Either will give the correct answer. Notice that each path has factors of 60 and 7.48, with only a difference in order. In each case, you are moving from a smaller to a larger box, and thus *multiplication* by both 60 and 7.48 is indicated:

$$(2.3 \text{ ft}^3/\text{s})(60 \text{ s/min})(7.48 \text{ gal/ft}^3) = 1{,}032 \text{ gpm}$$

Notice that you can write both multiplication factors into the same equation; you do not need to write one equation for converting cubic feet per second to cubic feet per minute and another for converting cubic feet per minute to gallons per minute.

Example 13

The flow rate to a sedimentation basin is 2,450,000 gpd. At this rate, what is the average flow in cubic feet per second?

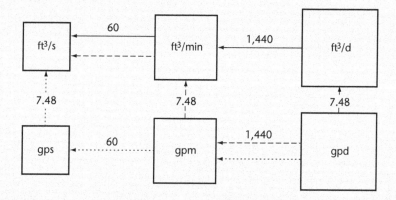

There are three possible paths from gallons per day to cubic feet per second. In each case, you would be moving from a larger to a smaller box, thus indicating *division* by 7.48, 1,440, and 60 (in any order):

$$\frac{2,450,000 \text{ gpd}}{(7.48 \text{ gal/ft}^3)(1,440 \text{ min/d})(60 \text{ s/min})} = 3.79 \text{ ft}^3/\text{s}$$

Again, the divisions are all written into one equation.

Example 14

If a flow rate is 200,000 gpd, what is this flow expressed as pounds per minute?

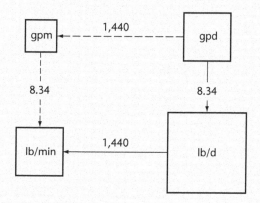

There are two possible paths from gallons per day to pounds per minute. The only difference in these paths is the order in which the numbers appear. The answer is the same in either case. In the following explanation, the solid-line path of the diagram is used.

Converting from gallons per day to pounds per day, you are moving from a smaller box to a larger box. Therefore, *multiplication* by 8.34 is indicated. Then from pounds per day to pounds per minute, you are moving from a larger to a smaller box, which indicates *division* by 1,440. These multiplication and division steps are combined into one equation:

$$\frac{(200{,}000 \text{ gpd})(8.34 \text{ lb/gal})}{1{,}440 \text{ min/d}} = 1{,}158 \text{ lb/min}$$

Linear Measurement Conversions

Linear measurement defines the distance along a line; it is the measurement between two points. The US customary units of linear measurement include the inch, foot, yard, and mile. In most treatment plant calculations, however, the mile is not used. Therefore, this section discusses conversions of inches, feet, and yards only. Here is the box diagram associated with these conversions:

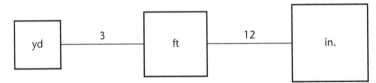

Example 15

The maximum depth of sludge drying beds is 14 in. How many feet is this?

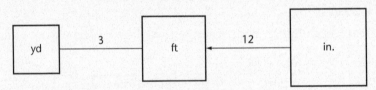

In converting from inches to feet, you are moving from a larger to a smaller box. Therefore, *division* by 12 is indicated.

$$\frac{14 \text{ in.}}{12 \text{ in./ft}} = 1.17 \text{ ft}$$

Example 16

During backwashing, the water level drops 0.6 yd during a given time interval. How many feet has it dropped?

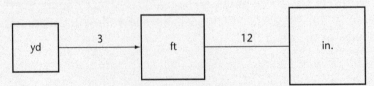

Moving from a smaller to a larger box indicates *multiplication* by 3:

$$(0.6 \text{ yd})(3 \text{ ft/yd}) = 1.8 \text{ ft}$$

Area Measurement Conversions

To make area conversions in US customary units, you work with units such as square yards, square feet, or square inches. These units are derived from the following multiplications:

$$(\text{yards})(\text{yards}) = \text{square yards}$$
$$(\text{feet})(\text{feet}) = \text{square feet}$$
$$(\text{inches})(\text{inches}) = \text{square inches}$$

By examining the relationship of yards, feet, and inches in linear terms, you can recognize the relationship between yards, feet, and inches in square terms. For example,

$$1 \text{ yd} = 3 \text{ ft}$$
$$(1 \text{ yd})(1 \text{ yd}) = (3 \text{ ft})(3 \text{ ft})$$
$$1 \text{ yd}^2 = 9 \text{ ft}^2$$

This method of comparison may be used whenever you wish to compare linear terms with square terms. Compare the diagram used for linear conversions with that used for square measurement conversions:

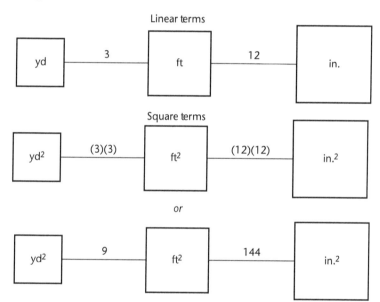

Example 17

The surface area of a sedimentation basin is 170 yd². How many square feet is this?

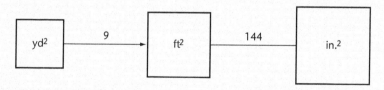

When converting from square yards to square feet, you are moving from a smaller to a larger box. Therefore, *multiplication* by 9 is indicated:

$$(170 \text{ yd}^2)(9 \text{ ft}^2/\text{yd}^2) = 1{,}530 \text{ ft}^2$$

Example 18

The cross-sectional area of a pipe is 64 in². How many square feet is this?

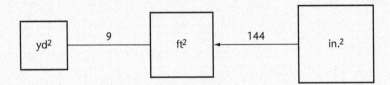

Converting from square inches to square feet, you are moving from a larger to a smaller box. *Division* by 144 is indicated:

$$\frac{64 \text{ in.}^2}{144 \text{ in.}^2/\text{ft}^2} = 0.44 \text{ ft}^2$$

One other area conversion important in treatment plant calculations is that between square feet and acres. This relationship is expressed mathematically as follows:

$$1 \text{ acre} = 43{,}560 \text{ ft}^2$$

A box diagram can be devised for this relationship. However, the diagram should be separate from the diagram relating square yards, square feet, and square inches because you usually wish to relate directly with square feet. As in the other diagrams, the relative sizes of the boxes are important:

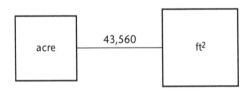

Example 19

A treatment plant requires 0.2 acre for drying beds. How many square feet are required?

Converting acres to square feet, you are moving from a smaller to a larger box. *Multiplication* by 43,560 is therefore indicated:

$$(0.2 \text{ acre})(43{,}560 \text{ ft}^2/\text{acre}) = 8{,}712 \text{ ft}^2$$

Volume Measurement Conversions

To make volume conversions in US customary unit terms, you work with such units as cubic yards, cubic feet, and cubic inches. These units are derived from the following multiplications:

$$(\text{yards})(\text{yards})(\text{yards}) = \text{cubic yards, or yd}^3$$

$$(\text{feet})(\text{feet})(\text{feet}) = \text{cubic feet, or ft}^3$$

$$(\text{inches})(\text{inches})(\text{inches}) = \text{cubic inches, or in.}^3$$

By examining the relationship of yards, feet, and inches in linear terms, you can recognize the relationship between yards, feet, and inches in cubic terms. For example,

$$1 \text{ yd} = 3 \text{ ft}$$
$$(1 \text{ yd})(1 \text{ yd})(1 \text{ yd}) = (3 \text{ ft})(3 \text{ ft})(3 \text{ ft})$$
$$1 \text{ yd}^3 = 27 \text{ ft}^3$$

Here is the box diagram associated with these cubic conversions:

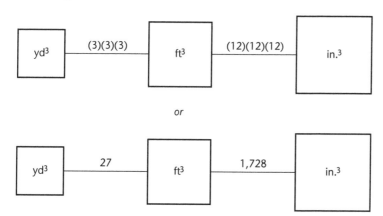

or

Example 20

Convert 15 yd³ to cubic inches.

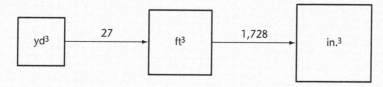

In converting from cubic yards to cubic feet and from cubic feet to cubic inches, you are moving from smaller to larger boxes. Thus, *multiplication* by 27 and 1,728 is indicated:

$$(15 \text{ yd}^3)(27 \text{ ft}^3/\text{yd}^3)(1{,}728 \text{ in.}^3/\text{ft}^3) = 699{,}840 \text{ in.}^3$$

Example 21

The required volume for a chemical is 325 ft³. What is this volume expressed as cubic yards?

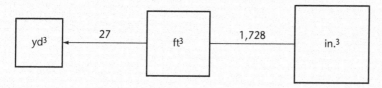

When you move from a larger to a smaller box, *division* is indicated:

$$\frac{325 \text{ ft}^3}{27 \text{ ft}^3/\text{yd}^3} = 12.04 \text{ yd}^3$$

Another volume measurement important in treatment plant calculations is that of acre-feet. A reservoir with a surface area of 1 acre and a depth of 1 ft holds exactly 1 acre-ft:

$$1 \text{ acre-ft} = 43{,}560 \text{ ft}^3$$

The relative sizes of the boxes are again important:

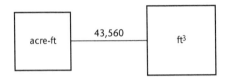

Example 22

The available capacity of a reservoir is 220,000 ft³. What is this volume expressed in terms of acre-feet?

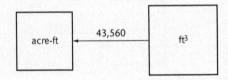

When you move from the larger to the smaller box, *division* by 43,560 is indicated:

$$\frac{220{,}000 \text{ ft}^3}{43{,}560 \text{ ft}^3/\text{acre-ft}} = 5.05 \text{ acre-ft}$$

Concentration Conversions

A milligrams-per-liter (mg/L) concentration can also be expressed in terms of grains per gallon (gpg) or parts per million (ppm). Of the three, the preferred unit of concentration is milligrams per liter.

Milligrams per Liter to Grains per Gallon

Conversions between milligrams per liter and grains per gallon are based on the following relationship:

$$1 \text{ gpg} = 17.12 \text{ mg/L}$$

As with any other conversion, often the greatest difficulty in converting from one term to another is deciding whether to multiply or divide by the number given. The following box diagram should help in making this decision:

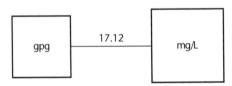

Example 23

Convert 25 mg/L to grains per gallon.

In this example, you are converting from milligrams per liter to grains per gallon. Therefore, you are moving from the larger to the smaller box:

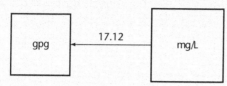

Larger to smaller indicates *division* by 17.12:

$$\frac{25 \text{ mg/L}}{17.12 \text{ mg/L/gpg}} = 1.46 \text{ gpg}$$

Example 24

Express a 20-gpg concentration in terms of milligrams per liter.

In this example, you are converting from grains per gallon to milligrams per liter. Therefore, you are moving from the smaller to the larger box:

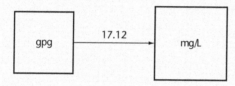

Smaller to larger indicates multiplication by 17.12:

$$(20 \text{ gpg})(17.12 \text{ mg/L/gpg}) = 342.4 \text{ mg/L}$$

Example 25

If the dosage of alum is 1.5 gpg, what is the dosage rate expressed in milligrams per liter?

The desired conversion is from grains per gallon to milligrams per liter:

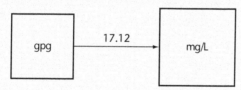

Smaller to larger indicates *multiplication* by 17.12:

$$(1.5 \text{ gpg})(17.12 \text{ mg/L/gpg}) = 25.68 \text{ mg/L}$$

Milligrams per Liter to Parts per Million

The concentration of impurities in water is usually so small that it is measured in milligrams per liter. This means the impurities in a standard volume (a liter) of water are measured by weight (milligrams). Concentrations in the range of 0–2,000 mg/L are roughly equivalent to concentrations expressed as the same number of parts per million (ppm). For example, "12 mg/L of calcium in water" expresses roughly the same concentration as "12 ppm calcium in water." However, milligrams per liter are the preferred units of concentration.

Metric System Conversions

To convert from the system of US customary units to the metric system, or vice versa, you must understand how to convert within the metric system. This requires a knowledge of the common metric prefixes. These prefixes should be learned before any conversions are attempted.

As shown in Table 2-1, the metric system is based on *powers*, or multiples of 10, just like the decimal system is. These prefixes may be associated with positions in the place value system.

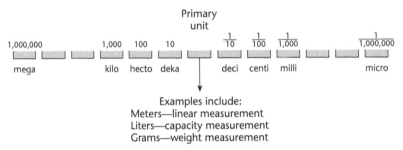

Understanding the position of these prefixes in the place value system is important because the method discussed next for metric-to-metric conversions is based on this system.

It is also important to understand the abbreviations used for metric terms. The basic measurement terms and their abbreviations are meters (m), liters (L), and grams (g). The prefixes added to the basic measurement terms may also be abbreviated (as shown in Table 2-1), as in the following examples:

$$1 \text{ megaliter} = 1 \text{ ML}$$

$$1 \text{ millimeter} = 1 \text{ mm}$$

$$1 \text{ kilogram} = 1 \text{ kg}$$

Use of these abbreviations greatly simplifies expressions of measurement.

Table 2-1 Metric system notations

Prefix	Abbreviation	Mathematical Value	Power Notation
giga	G	1,000,000,000	10^9
mega	M	1,000,000	10^6
kilo	k	1,000	10^3
hecto*	h	100	10^2
deka*	da	10	10^1
(none†)	(none)	1	10^0
deci*	d	1/10 or 0.1	10^{-1}
centi*	c	1/100 or 0.01	10^{-2}
milli	m	1/1,000 or 0.001	10^{-3}
micro	m	1/1,000,000 or 0.000001	10^{-6}
nano	n	1/1,000,000,000 or 0.000000001	10^{-9}

*Use of these units should be avoided when possible.
†Primary units, such as meters, liters, grams.

Metric-to-Metric Conversions

When conversions are being made for linear measurement (meters), volume measurement (liters), and weight measurement (grams), each change in prefix place value represents one decimal point move. This system of conversion is demonstrated by the following examples.

Example 26

Convert 1 m to decimeters (dm).

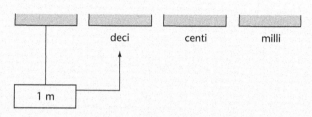

Converting from meters to decimeters requires moving one place to the right. Therefore, move the decimal point from its present position to one place to the right:

$$1.0 = 10 \text{ decimeters}$$

Example 27

Convert 1 g to (a) decigrams, (b) centigrams, and (c) milligrams.

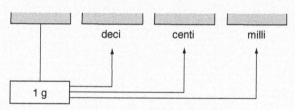

(a) Move the decimal point one place to the right:

$$1.0 = 10 \text{ decigrams}$$

(b) Move the decimal point two places to the right:

$$1.00 = 100 \text{ centigrams}$$

(c) Move the decimal point three places to the right:

$$1.000 = 1,000 \text{ milligrams}$$

In the preceding examples, the units are being converted from a primary unit (meter and gram) to a smaller unit (decimeter, decigram, etc.). Note that this system of conversion applies regardless of the initial type of unit and regardless of whether the number is increasing or decreasing, as shown in the following examples.

Example 28
Convert 1 dL to milliliters.

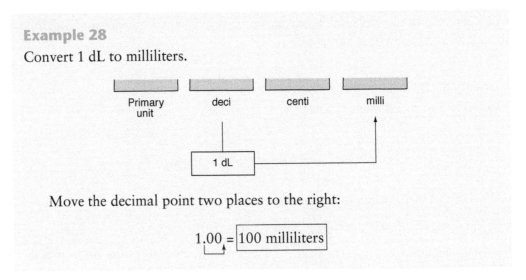

Move the decimal point two places to the right:

1.00 = 100 milliliters

This system of conversion also applies whether the number you are converting is a whole number—such as in the preceding examples, which use the number 1—or any other number.

Example 29
Convert 3.5 kg to grams.

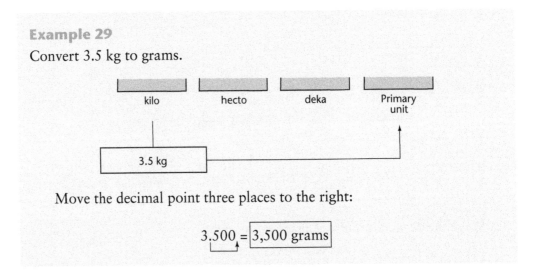

Move the decimal point three places to the right:

3.500 = 3,500 grams

Example 30
Convert 0.28 cm to meters.

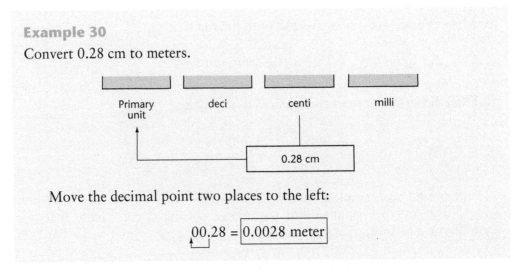

Move the decimal point two places to the left:

00.28 = 0.0028 meter

Most metric conversion errors are made in moving the decimal point to the left. You must be very careful in moving the decimal point from its present position, counting every number (including zeros) to the left as a decimal point move.

Example 31

Convert 1,750 L to kiloliters.

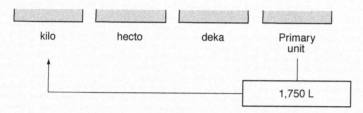

Move the decimal point three places to the left:

$$1750. = 1.75 \text{ kiloliters}$$

In the examples just given, there were no conversions of square or cubic terms. However, area and volume measurements can be expressed as square and cubic meters, centimeters, kilometers, and so on. The following discussion shows the special techniques needed for converting between these units.

Square meters indicates the mathematical operation meters times meters:

$$(\text{meter})(\text{meter}) = \text{square meters}$$

Square meters may also be written as m^2. The exponent of 2 indicates that *meter* appears twice in the multiplication. In conversions, the term *square* (or exponent of 2) indicates that *each prefix place value move requires two decimal point moves*. All other aspects of the conversions are similar to the preceding examples.

Example 32

Convert 1 m^2 to square decimeters.

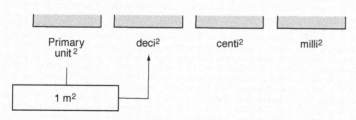

Converting from square meters to square decimeters requires moving one place value to the right. In square terms, *each prefix place move requires two decimal point moves*. Making the move in groups of two may be easier:

$$1.00 = 100 \text{ square decimeters}$$

Now check this conversion. From Example 26, you know that 1 m = 10 dm. Squaring both sides of the equation, we get the following:

$$(1\ m)(1\ m) = (10\ dm)(10\ dm)$$

$$1\ m^2 = 100\ dm^2$$

Example 33
Convert 32,000 m² to square kilometers.

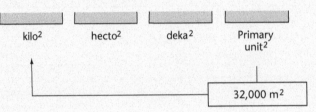

Three place value moves to the left correspond to six total decimal point moves to the left:

$$\underset{\leftarrow}{032,000.} = \boxed{0.032\ \text{square kilometer}}$$

Cubic meters indicates the following mathematical operation:

$$(\text{meters})(\text{meters})(\text{meters}) = \text{cubic meters}$$

Cubic meters may also be written as m³. The exponent of 3 indicates that *meter* appears three times in the multiplication. When you are converting cubic terms, *each prefix place value move requires three decimal point moves.* Again, it may be easier to make the decimal point moves in groups—groups of three for cubic-term conversions as opposed to groups of two for square-term conversions.

Example 34
Convert 1 m³ to cubic decimeters.

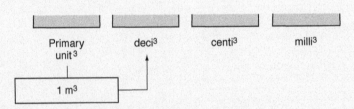

Converting from cubic meters to cubic decimeters requires moving one place value to the right. In cubic terms, *each prefix place value move requires three decimal point moves:*

$$1.000 = \boxed{1,000\ \text{cubic decimeters}}$$

Now check this conversion. From Example 26, you know that 1 m = 10 dm. Cubing both sides of the equation, we get the following:

$$(1 \text{ m})(1 \text{ m})(1 \text{ m}) = (10 \text{ dm})(10 \text{ dm})(10 \text{ dm})$$

$$1 \text{ m}^3 = 1{,}000 \text{ dm}^3$$

Example 35

Convert 155,000 mm³ to cubic meters.

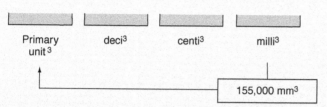

Three place value moves to the left indicate three *groups of three* decimal point moves to the left:

$$\underset{\leftarrow}{000115{,}000.} = \boxed{0.000155 \text{ cubic meter}}$$

Cross-System Conversions

For conversions from the US customary unit system to the metric system, or vice versa, the conversion table appearing in Appendix A is useful.

Example 36

Convert 20 gal to liters.

The factor given in the table to convert *From* "gallons" *To* "liters" is 3.785. This means that there are 3.785 L in 1 gal. Therefore, the conversion is

$$(20 \text{ gal})(3.785) = 75.7 \text{ L}$$

Example 37

Convert 3.7 acres to square meters.

The factor given in the table for converting *From* "acres" *To* "square meters" is 4,047. The conversion is therefore

$$(3.7 \text{ acre})(4{,}047) = 14{,}974 \text{ m}^2$$

Example 38

Convert 0.8 m/s to ft/min.

The factor given in the table for converting *From* "meters per second" *To* "feet per minute" is 196.8. Therefore, the conversion is

$$(0.8 \text{ m/s})(196.8) = 157.44 \text{ ft/min}$$

Occasionally, when making cross-system conversions, you may not find the factor in the table for the two terms of interest. For example, suppose you wish to convert from inches to decimeters, but the table gives factors only for converting from inches to centimeters or inches to millimeters. Or suppose you wish to convert cubic centimeters to gallons, but the table gives factors only for converting cubic meters to gallons. In such situations, it is usually easiest to make sure the US customary system unit is in the desired form and then make any necessary changes to the metric unit (e.g., changing inches to centimeters, then centimeters to decimeters). As shown in the example problems for metric-to-metric conversions, changing units in the metric system requires only a decimal point move.

The following two examples illustrate how the cross-system conversion may be made when the table does not give the precise units you need.

Example 39

The water depth in a channel is 1.2 ft. How many decimeters is this?

First check the conversion table in Appendix A to see if a factor is given for converting feet to decimeters. The conversion factor is given only for feet to kilometers, meters, centimeters, or millimeters.

To make the conversion from feet to decimeters then, first convert from feet to the closest metric unit to decimeters given in the table (centimeters); then convert that answer to decimeters. The conversion from feet to centimeters is

$$(1.2 \text{ ft})(30.48) = 36.58 \text{ cm}$$

Converting from centimeters to decimeters, we get

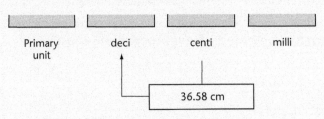

Move the decimal point one place to the left:

$$36.58 \text{ cm} = 3.658 \text{ dm}$$

Example 40

If you use 0.12 kg of a chemical to make up a particular solution, how many ounces of that chemical are used?

First, check the conversion table in Appendix A to determine if a factor is given for converting kilograms to ounces. No such conversion factor is given.

Try to find a conversion in the table from some other metric unit (such as milligrams or grams) to ounces. The conversion from grams to ounces is given in the table. Therefore, first convert the 0.12 kg to grams, then convert grams to ounces:

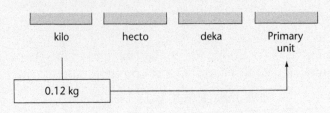

Move the decimal point three places to the right:

$$0.12 \text{ kg} = 120 \text{ g}$$

Then

$$(120 \text{ g})(0.03527) = 4.23 \text{ oz}$$

Temperature Conversions

The formulas used for Fahrenheit and Celsius temperature conversions are as follows:

$$°C = 5/9(°F - 32)$$
$$°F = (9/5)°C + 32$$

These formulas are difficult to remember unless used frequently. There is, however, another method of conversion that is perhaps easier to remember because the following three steps are used for both Fahrenheit and Celsius conversions:

1. Add 40°.
2. Multiply by the appropriate fraction (5/9 or 9/5).
3. Subtract 40°.

The only variable in this method is the choice of 5/9 or 9/5 in the multiplication step. To make this choice, you must know something about the two scales. As shown in Figure 2-1, on the Fahrenheit scale the freezing point of water is 32°, whereas it is 0° on the Celsius scale. The boiling point of water is 212° on the Fahrenheit scale and 100° on the Celsius scale. Thus for the same temperature, higher numbers are associated with the Fahrenheit scale and lower numbers with the Celsius scale. This information helps you decide whether to multiply by 5/9 or 9/5. Let's look at a few conversion problems to see how the three-step process works.

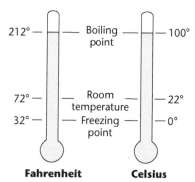

Figure 2-1 Fahrenheit and Celsius temperature scales

Example 41

Convert 212°F to Celsius.

From the sketch of the two scales in Figure 2-1, you know that the answer should be 100°C. But let's verify it using the three-step process.

The first step is to add 40°:

$$\begin{array}{r} 212° \\ + 40° \\ \hline 252° \end{array}$$

Proceeding to step 2, you must multiply 252° by either 5/9 or 9/5. Since the conversion is to the *Celsius* scale, you will be moving to a number *smaller* than 252. Multiplying by 9/5 is roughly the same as multiplying by 2, which would double 252 rather than make it smaller. In comparison, multiplying by 5/9 is about the same as multiplying by ½, which would cut 252 in half. In this problem, since you wish to move to a smaller number, you should multiply by 5/9:

$$\left(\frac{5}{9}\right)(252°) = \frac{1,260°}{9}$$

$$= 140°$$

The problem can now be completed using step 3 (subtract 40°):

$$\begin{array}{r} 140° \\ -\ 40° \\ \hline 100° \end{array}$$

Therefore, 212°F = 100°C.

Example 42

Convert 0°C to Fahrenheit.

The sketch of the two scales in Figure 2-1 indicates that 0°C = 32°F, which can be verified using the three-step method of conversion.

First, add 40°:

$$\begin{array}{r} 0° \\ +\ 40° \\ \hline 40° \end{array}$$

In this problem, you are going from Celsius to Fahrenheit. Therefore, you will be moving from a smaller number to a larger number, and 9/5 should be used in the multiplication:

$$\left(\frac{9}{5}\right)(40°) = \frac{360°}{5}$$

$$= 72°$$

Subtract 40°:

$$\begin{array}{r} 72° \\ -\ 40° \\ \hline 32° \end{array}$$

Thus, 0°C = 32°F.

Example 43

A thermometer indicates that the water temperature is 15°C. What is this temperature expressed in degrees Fahrenheit?

First, add 40°:

$$\begin{array}{r} 15° \\ + 40° \\ \hline 55° \end{array}$$

Moving from a smaller number (Celsius) to a larger number (Fahrenheit) indicates multiplication by 9/5:

$$\left(\frac{9}{5}\right)(55°) = \frac{495°}{5}$$

$$= 99°$$

Subtract 40°:

$$\begin{array}{r} 99° \\ - 40° \\ \hline 59° \end{array}$$

Therefore, 15°C = 59°F.

Although it is useful to know how to make these temperature conversion calculations, in practical applications you may wish to use a temperature conversion table such as the one in Appendix A. Let's look at a couple of example conversions using the table.

Example 44

Normal room temperature is considered to be 68°F. What is this temperature expressed in degrees Celsius?

Use the Fahrenheit-to-Celsius temperature conversion table. Coming down the Fahrenheit column to 68°, you can see that 68°F = 20°C.

Example 45

Convert 90°C to degrees Fahrenheit.

Use the Celsius-to-Fahrenheit temperature conversion table. Coming down the Celsius column to 90°, you can see that 90°C = 194°F.

Per Capita Water Use

The amount of water a community uses is, of course, of primary importance in water treatment. The water use establishes the amount of water that must be treated. One method of expressing the water use of a community is gallons per capita per day (gpcd). The term *per capita* means "per person." Calculating gallons per capita per day merely determines the average water use *per person*. The mathematical formula used in calculating gallons per capita per day is

$$\text{gpcd} = \frac{\text{water used (gpd)}}{\text{total number of people}}$$

Note that the "total number of people" refers to the total number of people served by the water system; the water system service population is not always the same as the population of the city or town in which the system is located. The following examples illustrate the calculation of gallons per capita per day.

Example 46

During 2008, the average daily water use at a particular water system was 6.3 million gallons per day (mgd). If the water system served a population of 26,900, what was this average daily water use expressed in gallons per capita per day?

$$\text{water use} = \frac{\text{water used (gallons per day)}}{\text{total number of people}}$$

$$= \frac{6{,}300{,}000\text{-gpd flow}}{26{,}900 \text{ people}}$$

$$= 234.2 \text{ gpcd}$$

Note that 6.3 mgd was converted to 6,300,000 gpd in the equation to more easily calculate the numbers.

Example 47

The average daily water use during 2016 for a certain water system serving a population of 14,509 was 3.1 mgd. The percentages of water used in each of four categories are given below. Use this information to determine the 2016 gallons-per-capita-per-day water use for each of the four categories.

- Domestic uses, 46%
- Commercial uses, 16%
- Industrial uses, 25%
- Public uses, 13%

First, calculate the total gallons-per-capita-per-day water use for 2016:

$$\text{water use} = \frac{\text{water used (gallons per day)}}{\text{total number of people}}$$

$$= \frac{3{,}100{,}000\text{-gpd flow}}{14{,}509 \text{ people}}$$

$$= 213.66 \text{ gpcd}$$

Next, use the percentages given to calculate the water use by category:

(213.66 gpcd)(0.46) = 98.28 gpcd for domestic uses

(213.66 gpcd)(0.16) = 34.19 gpcd for commercial uses

(213.66 gpcd)(0.25) = 53.42 gpcd for industrial uses

(213.66 gpcd)(0.13) = 27.78 gpcd for public uses

Example 48

The total water use on a particular day was 4,503,000 L. If the water system served a population of 9,800, what was this water use expressed in liters per day per capita?

$$\text{water use} = \frac{\text{water used (liters per day)}}{\text{total number of people}}$$

$$= \frac{4{,}503{,}000\text{-L/d flow}}{9{,}800 \text{ people}}$$

$$= 495.5 \text{ L/d per capita}$$

Average Daily Flow

The amount of water a community uses every day can be expressed in terms of an *average daily flow (ADF)*. The ADF is the average of the actual daily flows that occur within a period of time, such as a week, a month, or a year. It is expressed mathematically as follows:

$$\text{ADF} = \frac{\text{sum of all daily flows}}{\text{total number of daily flows used}}$$

The ADF can reflect a week's data (weekly ADF), a month's data (monthly ADF), or a year's data (annual ADF, or AADF—the most commonly calculated average).

ADF is important because it is used in several treatment plant calculations. The following examples illustrate the calculation of ADF.

average daily flow (ADF)
A measurement of the amount of water treated by a plant each day. It is the average of the actual daily flows that occur within a period of time, such as a week, a month, or a year. Mathematically, it is the sum of all daily flows divided by the total number of daily flows used.

Example 49

A water treatment plant reported that the total volume of water treated for the calendar year 2005 was 152,655,000 gal. What was the AADF for 2005?

In this problem, the sum of all daily flows has already been determined—a total of 152,655,000 gal was treated during the year. Knowing that there are 365 days in a year, calculate the ADF:

$$\text{ADF} = \frac{\text{sum of all daily flows}}{\text{total number of daily flows used}}$$

$$= \frac{152{,}655{,}000 \text{ gal}}{365 \text{ days}}$$

$$= 418{,}233 \text{ gpd}$$

Example 50

In January 2008, a total of 68,920,000 L of water was treated at a plant. What was the ADF at the treatment plant for this period?

As in the previous example, the sum of all daily flows has already been determined—a *total* of 68,920,000 L was treated for the month. January has 31 days; fill the information into the ADF formula:

$$\text{ADF} = \frac{\text{sum of all daily flows}}{\text{total number of daily flows used}}$$

$$= \frac{68{,}920{,}000 \text{ L}}{31 \text{ days}}$$

$$= 2{,}223{,}226 \text{ L/d}$$

Example 51

The following daily flows (in million gallons per day) were treated during June 2010 at a water treatment plant. What was the ADF for this period?

June	1	6.21	June	11	7.59	June	21	6.43
	2	6.68		12	7.01		22	6.26
	3	7.31		13	6.85		23	6.87
	4	7.80		14	6.43		24	7.27
	5	6.77		15	6.52		25	7.95
	6	6.32		16	6.79		26	7.33
	7	5.96		17	6.91		27	6.72
	8	5.83		18	7.37		28	6.51
	9	6.09		19	7.02		29	5.92
	10	7.22		20	6.88		30	5.90

To calculate the ADF for this time period, first add all daily flows. The total of these flows is 202.72 mil gal. With this information, calculate the ADF for the period:

$$\text{ADF} = \frac{\text{sum of all daily flows}}{\text{total number of daily flows used}}$$

$$= \frac{202.72 \text{ mil gal}}{30 \text{ days}}$$

$$= 6.76 \text{ mgd}$$

In some problems, you will not know actual daily flows for each day during a particular period, but you will know the ADF for that period. ADF information can be used in calculating other ADFs. The following example contrasts the two methods of calculating ADFs.

Example 52

The volume of water (in megaliters [ML]) treated for each day during a 2-week period is listed below. What is the ADF for this 2-week period?

	Week 1		Week 2
Sunday	2.41	Sunday	2.52
Monday	3.37	Monday	3.39
Tuesday	3.44	Tuesday	3.48
Wednesday	3.61	Wednesday	3.88
Thursday	3.23	Thursday	3.19
Friday	2.86	Friday	2.82
Saturday	2.75	Saturday	2.70

Since you are given the actual flows for each day in the 2-week period, you could calculate the ADF in a manner similar to that described in the previous examples. That is,

$$\text{ADF} = \frac{\text{sum of all daily flows}}{\text{total number of daily flows used}}$$

$$= \frac{43.65 \text{ ML}}{14 \text{ days}}$$

$$= 3.12 \text{ ML/d}$$

Suppose, however, that in this problem you do not know the actual flow for each day in the week, but you know the ADF for week 1 (3.10 ML/d) and the ADF for week 2 (3.14 ML/d). The ADF for the 2-week period can still be calculated, but *not* using the ADF equation given above (because in this case you do not know the *sum* of all daily flows). Instead, a similar formula is used:

$$\text{ADF} = \frac{\text{sum of all weekly ADFs}}{\text{total number of weekly ADFs used}}$$

In this case,

$$\text{ADF} = \frac{3.10 \text{ ML/d} + 3.14 \text{ ML/d}}{2}$$

$$= \frac{6.24 \text{ ML/d}}{2}$$

$$= 3.12 \text{ ML/d}$$

Notice that although only ADF information was used, the answer was the same as when all 14 actual daily flows were used in the calculation.

In a similar manner, ADFs for each of 12 months can be used to determine the ADF for a year (i.e., the AADF). Here is the equation:

$$\text{ADF} = \frac{\text{sum of all monthly ADFs}}{\text{total number of monthly ADFs used}}$$

Example 53

The ADF (in million gallons per day) at a treatment plant for each month in the year is given below. Using this information, calculate the annual ADF.

January	10.71	July	11.96
February	9.89	August	12.24
March	10.32	September	11.88
April	10.87	October	11.53
May	11.24	November	11.36
June	11.58	December	10.98

If you knew all 365 flows for the year, the AADF would be calculated using the formula

$$\text{ADF} = \frac{\text{sum of all daily flows}}{\text{total number of daily flows used}}$$

In this problem, however, ADF for each *month* of the year is given. Therefore, the ADF is calculated using the formula

$$\text{ADF} = \frac{\text{sum of all monthly ADFs}}{\text{total number of monthly ADFs used}}$$

Filling in the information given in this problem,

$$\text{ADF} = \frac{134.56 \text{ mgd}}{12}$$

$$= 11.21 \text{ mgd}$$

Basic Pipeline Disinfection Calculations

Flushing Rate

When flushing lines, an adequate rate of flow must be used to ensure proper disinfection. The basic formula used to calculate these flow rates is

$$\text{flow rate (gpm)} = \text{ft/sec} \times 60 \text{ sec/min} \times \text{volume (gal/ft)}$$

The following examples illustrate common flow rate calculations used when flushing lines.

Example 54

What is the flushing flow rate in gpm needed to achieve 2.5 ft/sec velocity in a 6 in. pipeline that is 500 ft long?

You will need to start by calculating the volume of pipeline per foot. Use the following formula:

$$\text{volume of pipeline (gal/ft)} = \pi \times \text{radius}^2 \times \text{length} \times 7.48 \text{ gal/ft}^3$$

The radius of a 6-in. pipe is 3 in., and to convert it to feet, you must divide it by 12 in., resulting in the following calculation:

$$\text{volume (gal/ft)} = 3.14 \times 3/12 \times 3/12 \times 1 \text{ ft} \times 7.48 \text{ gal/ft}^3 = 1.47 \text{ gal/ft}$$

To calculate the flow rate needed to move 2.5 ft/sec, we use the flow rate formula:

$$\text{flow rate (gpm)} = \text{ft/sec} \times 60 \text{ sec/min} \times \text{volume (gal/ft)}$$

$$= 2.5 \times 60 \times 1.47 = 220 \text{ gpm}$$

For practical purposes, when changing water out, you will need to determine how long you need to flush the lines at the given flow rate. The formula for this calculation is

$$\text{time (min)} = \text{volume of pipeline (gal)/flush rate (gpm)}$$

For example, to calculate how long you need to flush the lines at the rate specified in Example 54 to change the water out at least three times, apply the formula as follows:

time (min) = 3 × volume of pipeline (gal)/flush rate (gpm)

= 3 × 1.47 gal/ft × 500 ft/220 gpm = 10 min

Tablet Method

Another important disinfection-related task at which operators must be competent is the calculation of the number of tablets needed in a given length of pipeline. Table 2-2 shows the number of 5-g calcium hypochlorite tablets needed to produce a chlorine residual (if there is no demand) of 25 mg/L in a 20-ft (6-m) pipe length.

Table 2-2 Tablets for 25 mg/L chlorine dosage for each 20-ft (6-m) pipe length

Pipe Diameter		Number of
inches	millimeters	Tablets
4	100	1
6	150	1
8	200	2
10	250	3
12	300	4
16	400	7

Example 55

How many calcium hypochlorite 5-g tables are needed for 100 ft of 8-in. pipeline to achieve a 25-mg/L dosage?

From Table 2-2, it can be seen that two tablets are needed for each 20 ft of pipe. There are five 20-ft lengths of pipe in 100 ft of pipeline (100 ÷ 20 = 5), which gives us 5 × 2 = 10 tablets.

Chlorination Feed Rate

The formula for calculating the rate at which chlorine should be fed is as follows:

chlorine solution feed rate (gpm) × solution strength (mg/L)

= water fill rate (gpm) × chlorine dosage in pipeline (mg/L)

Example 56

A 1 percent chlorine solution is used to disinfect 1,000 ft of 6-in. pipe. Water is being pumped into the pipeline at 20 gpm. A dosage of 25 mg/L is needed. What is the chlorine solution feed rate (gal/h) needed for this procedure? How many gallons of 1 percent chlorine solution are needed to complete the job?

To begin, you need to know the following equivalency:

1% chlorine solution = 10,000 mg/L chlorine

From here, you can complete the feed rate formula as follows:

chlorine solution feed rate (gpm) = 25 mg/L × 20 gpm ÷ 10,000 mg/L

= 0.0025 × 20 gpm

= 0.05 gpm

Converting gallons per minute to gallons per hour, we have

0.05 gpm × 60 min/h = 3 gal/h

To answer the second question, of how many gallons of 1 percent chlorine solution are needed, we must apply the formulas discussed in the Flushing Rate section:

volume of pipeline = π × 3/12 × 3/12 × 1,000 × 7.48 gal/ft3 = 1,468 gal

time to fill = 1,468 gal/20 gpm = 73.4 min

gallons of chlorine 1% needed = solution feed rate × time =
0.05 gpm × 73.4 min = 3.67 gal

Basic Storage Facility Disinfection Calculations

Storage Facility Volume

Calculating the volume of water that can be stored in a given area requires the following formulas:

volume of a cylinder (gal) = π × radius^2 × height × 7.48 gal/ft^3

volume of a rectangle or square (gal) = length × width × height × 7.48 gal/ft^3

Example 57

What is the maximum volume of water that can be stored in a cylindrical tank that is 10 ft in diameter and 20 ft high with an overflow at 18 ft from the base?

volume (gal) = 3.14 × 5 × 5 × 18 × 7.48 = 10,569 gal

Storage Facility Walls Surface Area

The surface area of a storage tank's walls is calculated by multiplying the circumference of the tank by the height of the tank:

wall surface area for a cylindrical tank (ft^2) = π × diameter × height

Example 58

What is the surface area for a cylindrical tank 10 ft in diameter and 20 ft high?

area (ft^2) = 3.14 × 10 × 20 = 648 ft^2

Chlorination Spray Solution

To make a 200-mg/L chlorine spray solution, dilute bleach with a measured volume of water. Usually the strength of liquid sodium hypochlorite bleach is given as a percentage:

$$1\% = 10,000 \text{ mg/L}$$

Example 59

How many liters of 5 percent bleach are needed to make 150 gallons of 200 mg/L chlorine spray solution?

$$5\% \text{ chlorine} = 50,000 \text{ mg/L, so dilute by } 200/50,000 = 0.004$$

Therefore, you need

$$150 \text{ gal} \times 0.004 = 0.6 \text{ gal, or } 0.6 \text{ gal} \times 3.785 \text{ L/gal} = 2.27 \text{ L}$$

Chlorine Amount for Full Facility Method

This method requires a 10-mg/L chlorine dosage for the full volume of the facility. The following equations are used to calculate the amount of chlorination chemical (pure liquid chlorine, liquid sodium hypochlorite bleach, or calcium hypochlorite) needed for this procedure:

$$\text{pure chlorine needed in pounds} = \text{volume (gal)} \times 10 \text{ ppm} \times 8.34 \text{lb}/1,000,000 \text{ gal}$$

$$\text{pure chlorine needed in gallons} = \text{volume (gal)} \times 10 \text{ ppm} \times 1/1,000,000 \text{ gal}$$

$$\text{amount of chlorination chemical in pounds} = \text{lb pure chlorine} \div \% \text{ purity}/100$$

$$\text{amount of chlorination chemical in gallons} = \text{gal pure chlorine} \div \text{purity}/100$$

Example 60

How many gallons of 10 percent sodium hypochlorite bleach are needed to disinfect (full facility method) a 20-ft-diameter cylindrical tank that has a maximum water depth of 30 ft?

$$\text{volume (gal)} = \pi \times \text{radius}^2 \times \text{height} \times 7.48 \text{ gal/ft}^3$$

$$= 3.14 \times 10 \times 10 \times 30 \times 7.48$$

$$= 70,461.6 \text{ gal}$$

$$\text{amount of 10\% sodium hypochlorite (gal)} = \text{volume} \times 10 \text{ ppm} \times 1/1,000,000 \div 10/100$$

$$= 7.1 \text{ gal}$$

Chlorine Amount for Chlorinate-and-Fill Method

This method requires the operator to chlorinate 5 percent of tank volume to 50 mg/L, then fill and hold for at least 24 hours. The following equations can be used to calculate the amount of chlorination chemical needed for this procedure:

$$5\% \text{ volume (gal)} = \text{tank volume (gal)} \times 0.05$$

$$\text{pure chlorine (gal) for 50 mg/L dosage} = 5\% \text{ volume} \times 50/1{,}000{,}000 \text{ ppm}$$

$$\text{pure chlorine (lb) for 50 mg/L dosage} = 5\% \text{ volume} \times 50/1{,}000{,}000 \text{ ppm} \times 8.34 \text{ lb/gal}$$

$$\text{amount of chlorine chemical needed (gal)} = \text{pure chlorine (gal)} \div \% \text{ strength}/100$$

Example 61

How many gallons of 10 percent sodium hypochlorite bleach are needed for the chlorinate-and-fill method for a cylindrical tank 20 ft in diameter and 30 ft of water depth maximum?

$$\text{volume of tank (gal)} = \pi \times \text{radius}^2 \times \text{height} \times 7.48 \text{ gal/ft}^3$$

$$= 3.14 \times 10 \times 10 \times 30 \times 7.48$$

$$= 70{,}461.6 \text{ gal}$$

$$5\% \text{ volume} = 70{,}461.6 \times 0.05 = 3{,}523 \text{ gal}$$

$$\text{pure chlorine (gal) for 50 mg/L dosage} = 3{,}523 \text{ gal} \times 50/1{,}000{,}000$$

$$= 0.18 \text{ gal}$$

$$\text{gal of 10\% bleach} = 0.18 \text{ gal}/0.1$$

$$= 1.8 \text{ gal of 10\% sodium hypochlorite bleach}$$

Calculating Heads

Of the three types of head—pressure, elevation, and velocity—pressure head is the most significant and useful from an operating standpoint. (*Head* is a measure of the energy possessed by water at a given location in the water system, measured in height and expressed in feet; see Chapter 4 for a discussion of head.) The following examples illustrate how pressure heads can be measured and used in operations.

Pressure heads can be measured whether water is standing still or moving. However, dynamic pressure heads will be lower than static heads because of the pressures lost in moving the water.

Example 62

Pressure readings were taken for the system in Figure 2-2A, first when the valve was open and again when it was closed. Determine pressure heads in feet and show the hydraulic grade line for both conditions.

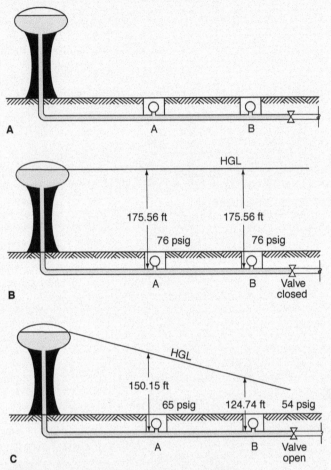

Figure 2-2 Schematic for Example 62

Valve closed: The pressure heads at points A and B are the same, as shown by the gauge readings:

	Valve Open	Valve Closed
Pressure at A	65 psig	76 psig
Pressure at B	54 psig	76 psig

Converting from pounds per square inch gauge (psig) to feet of head, you are moving from the smaller box to the larger box; therefore you should multiply by 2.31:

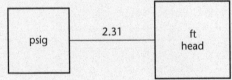

$$\text{pressure head at A and B} = (76 \text{ psig})(2.31 \text{ ft/psig})$$
$$= 175.56 \text{ ft}$$

Valve open: Calculate the pressure heads at points A and B using the same method shown for the closed-valve condition:

$$\text{pressure head at A} = (65 \text{ psig})(2.31 \text{ ft/psig})$$
$$= 150.15 \text{ ft}$$

$$\text{pressure head at B} = (54 \text{ psig})(2.31 \text{ ft/psig})$$
$$= 124.74 \text{ ft}$$

The two pressure heads and the hydraulic grade line (HGL) are shown in Figures 2-2B and 2-2C.

The pressure heads of water that is standing still are higher than those of moving water. The difference represents the amount of energy used by the water in moving from the elevated storage tank to points A and B.

Often in water systems the static pressure head can be found from elevation information that was established by surveyors during construction.

WATCH THE VIDEO
Calculating Head (www.awwa.org/wsovideoclips)

Instantaneous Flow Rate Calculations

The flow rate of water through a channel or pipe at a particular moment depends on the cross-sectional area and the velocity of the water moving through it. This relationship is stated mathematically as follows:

$$Q = AV$$

where

Q = flow rate

A = area

V = velocity

Figure 2-3A illustrates the $Q = AV$ equation as it pertains to flow in an open channel. Because flow rate and velocity must be expressed for the same unit of time, the flow rate in the open channel is expressed as follows:

Q		A		V
(flow rate)	=	(width)	(length)	(velocity)
ft³/time		ft	ft	ft/time

In using the $Q = AV$ equation, the time units given for the velocity must match the time units for cubic-feet flow rate. For example, if the velocity is expressed as feet per second (ft/s), then the resulting flow rate must be expressed as cubic feet per second (ft³/s). If the velocity is expressed as feet per minute (ft/min), then the resulting flow rate must be expressed as cubic feet per minute (ft³/min). And if the velocity is expressed as feet per day (ft/d), then the resulting flow rate must

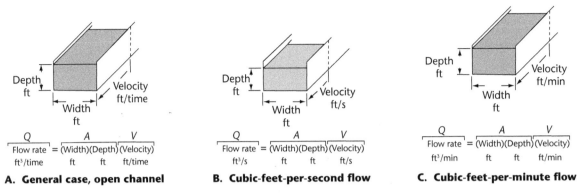
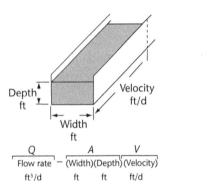
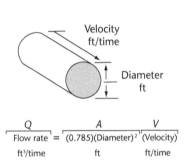

Figure 2-3 The $Q = AV$ equation as it pertains to flow in an open channel

be expressed as cubic feet per day (ft³/d). Figures 2-3B through 2-3D illustrate this concept.

Using the $Q = AV$ equation, if you know the cross-sectional area of the water in the channel (the width and depth of the rectangle), and you know the velocity, then you will be able to calculate the flow rate. This approach to flow rate calculations can be used if the problem involves an open-channel flow with a rectangular cross section, as shown previously, or if the problem involves a pipe flow with a circular cross section, as shown in Figure 2-3E. Only the calculation of the cross-sectional area will differ.

If a circular pipe is flowing full (as it will be in most situations in water supply and treatment), the resulting flow rate is expressed as follows:

Q		A	V
(flow rate) ft³/time	=	(0.785)(diameter)² ft²	(velocity) ft/time

As in the case of the rectangular channel, if the velocity is expressed as feet per second, then the resulting flow rate must be expressed as cubic feet per second. If the velocity is expressed as feet per minute, then the resulting flow rate must be expressed as cubic feet per minute. If the velocity is expressed as feet per day, then the resulting flow rate must be expressed as cubic feet per day.

Example 63

A 15-in.-diameter pipe is flowing full. What is the gallons-per-minute flow rate in the pipe if the velocity is 110 ft/min?

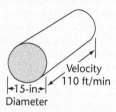

To use the $Q = AV$ equation, the diameter and velocity terms should be expressed using feet. Therefore, convert the 15-in. diameter to feet:

$$\frac{15 \text{ in.}}{12 \text{ in./ft}} = 1.25 \text{ ft}$$

Now use the $Q = AV$ equation to calculate the flow rate. Because the velocity is expressed in feet per minute, first calculate the cubic-feet-per-minute flow rate, then convert to gallons per minute:

$$\underset{\text{(flow rate)}}{Q} = \underset{(0.785)(\text{diameter})^2}{A} \quad \underset{\text{(velocity)}}{V}$$

$$= (0.785)(1.25 \text{ ft})(1.25 \text{ ft})(110 \text{ ft/min})$$

$$= 134.92 \text{ ft}^3/\text{min}$$

Convert cubic feet per minute to gallons per minute:

$$(134.92 \text{ ft}^3/\text{min})(7.48 \text{ gal/ft}^3) = 1,009 \text{ gpm}$$

Example 64

A 305-mm-diameter pipe flowing full is carrying 35 L/s. What is the velocity of the water (in meters per second) through the pipe?

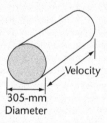

You are asked to determine the velocity V given the flow rate Q and diameter. Note first that the diameter of 305 mm is equivalent to 0.305 m. To use the $Q = AV$ formula to calculate velocity in a pipe, you should use the following mathematical setup:

$$\underset{\text{(flow rate)}}{Q} = \underset{(0.785)(\text{diameter})^2}{A} \quad \underset{\text{(velocity)}}{V}$$

Because you want to know velocity in meters per second, the flow rate must also be expressed in terms of meters and seconds (cubic meters per second). The information given in the problem expresses the flow rate as liters per second. Therefore, the flow rate must be converted to cubic meters per second before you begin the $Q = AV$ calculation:

$$35 \text{ L/s} \times 0.001 \text{ m}^3/\text{L} = 0.035 \text{ m}^3/\text{s}$$

Now determine the velocity using the $Q = AV$ equation:

$$\underset{Q}{0.035 \text{ m}^3/\text{s}} = \underset{A}{(0.785)(0.305 \text{ m})^2} \quad \underset{V}{(x)}$$

And then solve for the unknown value:

$$\frac{0.035 \text{ m}^3/\text{s}}{(0.785)(0.305 \text{ m})^2} = x$$

$$0.48 \text{ m/s} = x$$

Study Questions

1. Determine the detention time in hours for the following water treatment system:
   - Distribution pipe from water plant to storage tank is 549 ft in length and 14 in. in diameter
   - Storage tank averages 2,310,000 gal of water at any given time
   - Flow through system is 6.72 mgd
   a. 7.2 hr
   b. 7.4 hr
   c. 8.0 hr
   d. 8.3 hr

2. If chlorine is being fed at a rate of 260 lb/day for a flow rate of 23 cfs, what should be the adjustment on the chlorinator when the flow rate is decreased to 16 cfs, if all other water parameters remain the same?
   a. 160 lb/day
   b. 180 lb/day
   c. 310 lb/day
   d. 370 lb/day

3. How many gallons of a sodium hypochlorite solution that contains 12.1% available chlorine are needed to disinfect a 1.5-ft diameter pipeline that is 283 ft long, if the dosage required is 50.0 mg/L? Assume the sodium hypochlorite is 9.92 lb/gal.
   a. 0.87 gal sodium hypochlorite
   b. 1.0 gal sodium hypochlorite
   c. 1.3 gal sodium hypochlorite
   d. 1.5 gal sodium hypochlorite

4. A storage tank has a 60.0-ft radius and averages 25.5 ft in water depth. Calculate the average detention time in hours for this storage tank, if flow through the tank averages 2.91 mgd during the month in question.
    a. 17.5 hr
    b. 17.8 hr
    c. 18.6 hr
    d. 19.8 hr

5. A 24.0-in. pipeline, 427 ft long, was disinfected with calcium hypochlorite tablets with 65.0% available chlorine. Determine the chlorine dosage in mg/L, if 7.0 lb of calcium hypochlorite was used. Assume that the hypochlorite is so diluted that it weighs 8.34 lb/gal.
    a. 25 mg/L chlorine
    b. 39 mg/L chlorine
    c. 43 mg/L chlorine
    d. 54 mg/L chlorine

6. A well yields 2,840 gallons in exactly 20 minutes. What is the well yield in gpm?
    a. 140 gpm
    b. 142 gpm
    c. 145 gpm
    d. 150 gpm

7. What is the area of a circular tank pad in ft^2, if it has a diameter of 102 ft?
    a. 6,160 ft^2
    b. 6,167 ft^2
    c. 8,170 ft^2
    d. 8,200 ft^2

8. What is the pressure at 1.85 feet from the bottom of a water storage tank if the water level is 28.7 feet?
    a. 11.6 psi
    b. 12.4 psi
    c. 62.0 psi
    d. 66.3 psi

9. How many gallons are in a pipe that is 18.0 in. in diameter and 1,165 ft long?
    a. 2,060 gal
    b. 10,300 gal
    c. 15,400 gal
    d. 17,200 gal

10. Convert 37.4 degrees Fahrenheit to degrees Celsius.
    a. 3.0°C
    b. 5.3°C
    c. 7.9°C
    d. 9.7°C

11. If 288 is 70.3%, how much is 100%?
    a. 410
    b. 412
    c. 415
    d. 418

12. What resource allows you to convert units, such as feet to inches, simply by following the instructions indicated in the table headings?

13. How many gallons are in a cubic foot?

14. If the water level rises by 15 feet, how many yards has it risen?

15. What is the formula for calculating average daily flow?

16. How do you convert deciliters to milliliters?

Chapter 3
Water Use and System Design

Water Use

Public water supply uses can be grouped into three categories: domestic, industrial, and public.

Domestic Use

The category of domestic water use includes water that is supplied to residential areas, commercial districts, and institutional facilities.

- *Residential uses.* Water provided to residential households serves both interior uses (toilets, showers, clothes washers, faucets, etc.) and exterior uses (car washing, lawn watering, etc.). The amount of water used by households varies according to such factors as the number and ages of occupants in a household, income level, geographic location (which influences climate), and the efficiency of water fixtures and appliances in the home. Typical data for rates of interior water use are presented in Figure 3-1. Outdoor water use is strongly influenced by the annual weather patterns; water use in hot, dry climates is a significantly higher than in colder, wetter climates. Table 3-1 shows a typical breakdown of water use by a typical family of four persons living in a single-family house.
- *Commercial uses.* Water use can vary widely among commercial facilities depending on the type of commercial activity. A small office building would be expected to use less water than a restaurant of the same size, for example. In addition to the amount of water required by employees and customers for drinking and sanitary purposes, commercial users may need water for such diverse uses as cooling, humidifying, washing, ice making, vegetable and produce watering, feeding ornamental fountains, and irrigating landscape.
- *Institutional uses.* Estimates of water demand for facilities such as hospitals and schools are usually based on the size of the facility and the amount of water needed for each user in the facility. Water use by schools is highly dependent on whether the students are housed on campus or are day students.

Industrial Use

In the industrial sector, rates of water use depend on the type of industry, cost of water, wastewater disposal practices, types of processes and equipment, and water conservation and reuse practices. In general, industries that use large

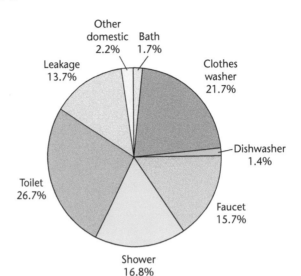

Figure 3-1 Indoor per capita use percentage by fixture

Table 3-1 Typical residential end uses of water by a family of four*

Fixture/End Use	Average Gallons per Capita per Day	Average Liters per Capita per Day	Indoor Use, %	Total Use, %
Toilet	18.5	70	30.9	10.8
Clothes washer	15	56.8	25.1	8.7
Shower	11.6	43.9	19.4	6.8
Faucet	10.9	41.3	18.2	6.3
Other domestic	1.6	6.1	2.7	0.9
Bathtub	1.2	4.5	2.0	0.7
Dishwasher	1	3.8	1.7	0.6
Indoor total	59.8	226.4	100.0	34.8
Leakage	9.5	36	NA	5.5
Unknown	1.7	6.4	NA	1.0
Outdoor	100.8	381.5	NA	58.7
TOTAL	171.8	650.3	NA	100.0

*Data collected from the following cities: Boulder, CO; Denver, CO; Eugene, OR; Seattle, WA; San Diego, CA; Tampa, FL; Phoenix, AZ; Tempe, AZ; Scottsdale, AZ; Cambridge, Ontario, Canada; Waterloo, Ontario, Canada; Las Virgenes, CA; Walnut Valley, CA; and Lompoc, CA.

Adapted from the Water Research Foundation (1999).

quantities of water are located in communities where water quality is good and the cost is reasonable.

Historically, major industrial water users (such as the steel, petroleum products, pulp and paper, and power industries) have provided their own water supply. Smaller industries and industries with low water usage are more inclined to purchase their water from public systems.

Public Use

Municipalities and other public entities provide public services that require varying amounts of water. Water is used for the following:

- Public parks, golf courses, swimming pools, and other recreational areas
- Municipal buildings
- Firefighting
- Public-works uses such as street cleaning, sewer flushing, and water system flushing

Although the total annual volume of water required for firefighting is typically small, the rate of flow required during a fire can be very large. The flow rate needed to fight a large fire in a small community can put a significant strain on the water system and may threaten to reduce pressure in the system substantially. For example, a community that uses an average of 1 mgd (4 ML/d) or 694 gpm (2,600 L/min) might reasonably expect a fire flow demand of 3,000 gpm (11,400 L/min) for up to 3 hours. The rate of flow for firefighting in this case is more than four times the average daily flow, and it will likely be required from a few water mains near the fire.

Variations in Water Use

Water use for a municipality or community can vary due to several factors, including the following:

- Time of day and day of the week
- Climate and season of the year
- Type of community (residential or industrial) and the economy of the area
- Presence or absence of customer meters
- Dependability of supply and dependably high quality of the water
- Condition of the water system (leakage and losses)
- Water conservation/demand management

Time of Day and Day of Week

Water use rates vary considerably according to the time of day and the day of the week. On a typical day in most communities, water use is lowest at night (11 p.m. to 5 a.m.) when most people are asleep. Water use rises rapidly in the morning (5 a.m. to 11 a.m.), and usage is moderate through midday (11 a.m. to 6 p.m.). Usage then increases in the evening (6 p.m. to 10 p.m.) and drops rather quickly around 10:00 p.m. A diurnal curve showing the hourly rate of water use for a community on a typical day is shown in Figure 3-2.

Total water use rates for a community typically vary by the day of the week as well, depending on the habits of the community. Some water systems have significant increases in water use on Monday because in many households Monday is washday. Water systems supplying industries that do not operate on weekends observe significantly reduced usage rates on Saturday and Sunday.

Climate and Season

Water use is typically highest during the summer months when the weather is hot and dry and the need to provide water for exterior activities and landscape irrigation is at its peak. Particularly significant seasonal variations are commonly

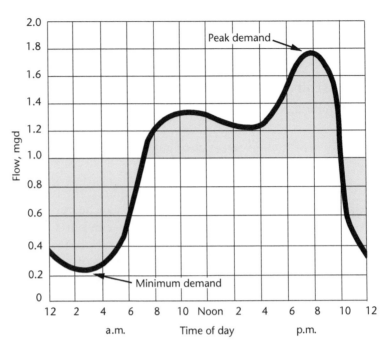

Figure 3-2 Typical daily flow chart showing peak and minimum demands

observed in resort areas, small communities with college campuses, and communities with seasonal commercial or industrial activities.

Most systems have relatively low water use in winter when there is no need for lawn or garden sprinkling and there is less outside activity requiring water. However, in extremely cold weather, consumers may occasionally run water faucets continuously to prevent water pipes from freezing.

Air-conditioning units and cooling towers operate by evaporating water and are used extensively by homes and large commercial and industrial facilities located in hot, dry climates. The demand for water to support air conditioners and cooling towers generally coincides with periods of highest water demand for lawn and garden sprinkling, pushing the annual water use to its peak for the year.

Type and Size of Community

The type of housing in the service area will affect rates of water use. Areas that have individual homes on large lots with gardens and lawns typically have a much higher water use per person than areas with multiple-family dwellings such as townhomes, condominiums, and apartment complexes. Households with automatic washing machines, dishwashers, and garbage disposals tend to use more water than households where these appliances are not installed. Economically depressed areas generally have lower water use per capita.

Community size typically affects the variability of water use rates over time. The ratio of peak use to average use tends to be greater for systems serving small communities than for larger water systems.

Metering

Most water utilities install meters that track all water usage. Other systems do not have customer meters and continue to charge customers a flat rate regardless of the quantity of water they use. The choice of billing scheme significantly influences the amounts of water consumed. Flat-rate customers may use as much as 25 percent more water than is used by customers who have to pay more as they use more. This excess water usage can be caused by unnecessary and unrestricted sprinkling and by personal behavior (e.g., letting a faucet run to get cold water or letting a hose run while washing a car). Another consequence of flat-rate billing is that household leaks are often ignored because wasted water costs the customer nothing.

Dependability and Quality of Water

Customers generally use less water from the public system if the water pressure is poor or if the water has an unpleasant taste, odor, or color. They may also use less if it has a high mineral content. Under such conditions, customers may choose alternative water sources, some of which may be less protective of public health. Bottled water, for example, has been subject to less stringent regulatory standards than tap water. Private wells, likewise, are not as closely regulated as a public utility.

Sewer Connection

The availability of municipal sewer systems also increases water use. Homes with sewer service usually have more water-using appliances and fixtures because there is no concern about overloading an on-site wastewater treatment and disposal system that would be used otherwise. The increase in water use due to sewer availability may be 50–100 percent.

Condition of the Water System

The difference between the average annual water production and the average annual consumption of a system amounts to ongoing unaccounted-for losses. Unaccounted-for losses can result from many factors, including leaks in piping, main breaks, periodic fire hydrant flushing, tank drainage for maintenance purposes, unauthorized use, unmetered services, and inaccurate or nonfunctioning meters. According to the current industry standard, unaccounted-for system leakage and losses should be less than 10 percent of production for new distribution systems (those less than 25 years old) and less than 15 percent for older systems.

Rates of leakage are affected by the age of the system, the materials used in construction, and the level of system maintenance. Water mains are usually tested for watertightness when they are installed. However, as the water mains get older, leakage at pipe joints, valves, and connections increases.

Unless the system is regularly checked for leaks and unless leaks are promptly repaired, the amount of water lost daily can be significant. Such losses not only place an extra strain on the ability of the utility to furnish water to customers, they also represent an important loss of revenue.

Water Rights

Conflicts over who owns a source of water and who gets to use it may arise among individuals, among communities, among states, between states and the federal government, and among countries.

Allocation of Surface Water

In the United States, the legal reasoning used for allocating water from surface lakes and streams follows one of two basic lines of thinking:

- Riparian doctrine is commonly used in the water-rich eastern part of the country.
- Prior appropriation doctrine is used in the more arid western states.

Riparian Doctrine

Riparian doctrine, which is commonly used in the water-rich eastern part of the country, is sometimes called the "rule of reasonable sharing." All states that follow this doctrine subscribe to the basic theory that all property owners who have land abutting a body of water have an equal right to that water. Among the states that follow this doctrine, however, there are substantial differences in how the law is applied.

Under riparian doctrine, each riparian property owner can use as much water as needed for any reasonable purpose, as long as this use does not interfere with the reasonable use by other riparian property owners.

Consequently, when a water shortage occurs, all riparian property owners must accept a reduction in supply. The amount of water allowed for any one property owned is not affected by how long the land has been owned or by how much water the owner has used in the past.

Riparian rights are a common-law doctrine. This means that they are a part of a large body of civil law that, for the most part, was made by judges in court decisions in individual cases. Statutory law, in contrast, is enacted by a legislative body.

Because the possession of rights arises from ownership of land abutting a water body, all of the owners are treated as equals by the courts. Conflicts are then usually resolved by the court, which determines what reasonable use is. Factors considered in the court's decision include the following:

- Purpose of the use
- Suitability of a use with respect to the water body in question
- The economic and social benefits of the use
- The amount of harm caused by the use
- The amount of harm avoided by changing any one party's use
- The protection of existing values

riparian doctrine
A water right that allows the owners of land abutting a stream or other natural body of water to use that water.

prior appropriation doctrine
A water rights doctrine in which the first user has the right to water before subsequent users.

The growth of water use in riparian-doctrine states has led to more frequent conflicts over water sources. Several of these states have tried to remedy weaknesses in the doctrine by enacting statutes requiring that permits be obtained from a state agency. These permits vary widely in the restrictions they apply and the requirements they place on the riparian owner.

The issuance of permits moves a state away from basic riparian doctrine and closer to prior appropriation doctrine. The permit systems are severely criticized by some and considered simply unnecessary by others. They have not undergone serious challenges in the courts, so their eventual success is not guaranteed. The trend toward permit systems, however, is continuing.

Prior Appropriation Doctrine

The appropriation doctrine concept started with the forty-niners in the goldfields of California. When there was too little water for all the mining operations, the principle applied was "first in time, first in right." The principle was given legal recognition by the courts and later made into law by western state legislatures.

The appropriation doctrine rests on two basic principles: priority in time and beneficial use.

Priority in Time When stream flow is less than what is demanded by all water users along the stream, use is prioritized based on who has been using the water for the longest time. Water withdrawals that began more recently must be discontinued so that withdrawals begun earlier can continue. Unlike the riparian doctrine, the appropriation doctrine does not treat riparian owners equally.

Beneficial Use While appropriation doctrine recognizes no water right based on land ownership alone, a user who has been appropriated water is entitled to it only when it can be used beneficially. Waste is therefore theoretically prohibited, and any available water beyond the amount that can be used by one appropriator is available to others. Furthermore, nonuse of the water for a long period may result in loss of the water right by forfeiture or abandonment.

Defining beneficial use remains difficult, and conflicts arise as a result. Beneficial use is determined based on two broad characteristics: the type of use (such as irrigation, mining, or municipal use) and its efficiency. Most types of water uses have been judged to be beneficial, and the majority of court cases have been concerned with whether a particular water use is inefficient or wasteful.

Legal Complications Legal solutions to water conflicts can get quite complex because water, unlike property, is a shared resource, causing the exercise of one user's water rights to affect someone else. Boundaries are not obvious and may be constantly changing.

With the appropriation doctrine, the assignment (or not) of a right at any particular time depends on the observed streamflow, the number of prior existing rights, and the amount of water needed by those who have the prior rights. For many western streams, the exercise of rights might have to be adjusted as often as daily in times of heavy use, such as in summer when irrigation demands are high.

Allocation of Groundwater

There are four systems used in determining rights to groundwater:

1. Absolute ownership
2. Reasonable use
3. Correlative rights
4. Appropriation–permit systems

Absolute Ownership

The absolute ownership system is based on the principle that the owner of the land owns everything beneath that land, all the way down to the center of the earth. The dynamic nature of groundwater movement makes it impossible to apply this ownership literally. For all practical purposes, the system is a rule of capture. That is, the landowner can use all the water that can be captured from beneath his or her land.

There are almost no restrictions involved with this rule. The water can be used for the owner's purposes both on and off the land, and it can be sold to others. There is no liability if pumping reduces a neighbor's supply or even dries up the

---

priority in time
The assigning of water rights based on who has been using the water the longest.

beneficial use
A water rights term indicating that the water is being used for good purposes.

absolute ownership
A water rights term referring to water that is completely owned by one person.

neighbor's well. Some interpretations of absolute ownership allow that water can be wasted as long as it is not done maliciously.

Reasonable Use

As with the rule of absolute ownership, the rule of reasonable use holds that a property owner has the right to pump and use groundwater lying beneath the property. However, if this use interferes with a neighbor's use, it may continue only if it is reasonable. Furthermore, an owner is liable if unreasonable use causes harm to others. This rule amounts to a *qualified right* rather than an *absolute right* to use groundwater.

The determination of what is reasonable is the important aspect of making reasonable use legally workable. Reasonable use for groundwater is much easier to determine than reasonable use for surface water under the riparian doctrine. In general, any use of water that is not wasteful for a purpose associated with the land from which the water is drawn (an overlying use) is reasonable. Conversely, any use off the property may be considered unreasonable if it interferes with the use of the groundwater sources by others.

The overlying use criterion, however, does not resolve disputes if all parties to the dispute are using the water for overlying purposes. In this case, all uses are reasonable, all can continue, and reasonable use becomes the rule of capture. In other words, the owner with the biggest pump gets the water. On the other hand, because a nonoverlying use is considered unreasonable if it interferes with an overlying use, such uses are forbidden no matter how beneficial.

Correlative Rights

The rule of correlative rights holds that the overlying use rule is not absolute but is related to the rights of other overlying users. This rule is used when there is not enough water to satisfy all overlying uses. In this case, the rule requires sharing. This rule has been applied by allocating rights among property owners in proportion to the size of the overlying land parcel.

Appropriation–Permit System

The appropriation–permit system is sometimes called a groundwater appropriation system. The most important aspect of it is the rule of priority: water rights are based on who has used the water for the longest time. Priority is all-important in the appropriation doctrine as applied to surface water because the variation in streamflow requires frequent changes in allowable withdrawals. However, the flow of groundwater in an aquifer fluctuates very slowly under natural conditions, so that frequent adjustment in allocations is not necessary. As such, priority, as applied to groundwater, mainly involves limiting the number of permits to prevent overuse of the aquifer.

The permit system thus amounts to groundwater management and administrative regulation. By placing limitations on pumping rates, well field placement, and well construction standards and by refusing permits when necessary, for example, the administrating agency manages the groundwater in its jurisdiction. The implementation of permit systems varies widely, however. The only generalization that can be made is that permits can be more relaxed if adequate water is available and more stringent if it is not.

---

reasonable use
A water rights term indicating that the water use is acceptable in general terms.

overlying use
The land use that occurs on top of an aquifer.

correlative rights
A rule that contends that the overlying use rule is not absolute but is related to the rights of other overlying users. This rule is used when there is not enough water to satisfy all overlying uses.

appropriation–permit system
A water use system in which permits to use water are regulated so that overdraft cannot occur.

Distribution System Purpose and Planning

Drinking water distribution systems are provided for two primary purposes: (1) consumption and (2) fire protection.

There are many considerations involved in planning for a water distribution system, particularly in terms of which type of system to use, the type of system layout, the sizing of mains, and material selection.

Types of Water Systems

Water supply systems generally can be divided into three categories based on the source of water that they use. The water source, in turn, impacts the design, construction, and operation of the water distribution system. The types of systems, classified by source (Figure 3-3), are as follows:

- Systems with surface water source(s)
- Systems with groundwater source(s)
- Systems with purchased water source(s)

Some of the principal characteristics of each of these systems as they affect the distribution system are described in the following discussion.

Surface Water Systems

In many areas, surface water is readily available for public water system use. The following are some of the special features of surface water systems:

- When a water utility operates its own surface water treatment plant, it may place the distribution system under the same supervision. This arrangement allows personnel and equipment to be shared.

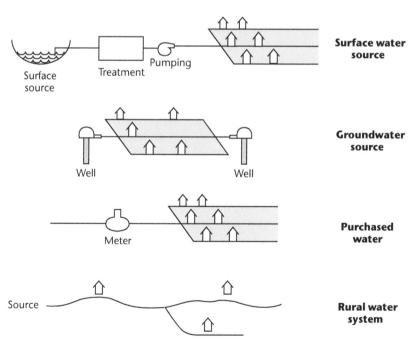

Figure 3-3 Types of water systems classified by source

surface water system
A water system using water from a lake or stream for its supply.

- Surface water must always be treated before it can be distributed to the public, requiring at least "conventional treatment" to remove particles and other contaminants. Surface water must also be disinfected to inactivate harmful microorganisms. Water entering the distribution system therefore has extremely low turbidity and must carry a minimum disinfectant residual.
- The treatment plant is sometimes located on one side of the distribution system, particularly when there are two or more water treatment plants. Large transmission mains are required to carry sufficient quantities of water to the far ends of the system.
- Surface water systems generally attract industries that need large quantities of processed water, such as cooling, cleaning, or incorporation into a product.
- As long as rates are reasonable and use is not restricted, customers tend to use a lot of water. The amount of water used on high-use days may be several times the amount used on an average day.
- By adjusting the treatment process, the water quality can be changed so that it is noncorrosive and nonscale forming.

Groundwater Systems

Although groundwater is generally available in most of North America, the amount available for withdrawal at any particular location is usually limited. Some of the general features of groundwater systems are as follows:

- There are very few large cities that can rely solely on groundwater for their source of supply. Small communities that start out using wells often must change to another source as the population grows and the rate of water use exceeds the aquifer capacity.
- The quality of some groundwater is good enough to use without treatment. However, the water chemistry may be corrosive or scale-forming, or it may have other adverse effects on distribution piping.
- When groundwater has excessive hardness, iron, or other qualities unacceptable to the public, treatment must be provided.
- Some water systems that initially used groundwater have had to change sources or add treatment after discovering that the aquifer they were using was contaminated.
- When groundwater is generally available at any location under a community, wells are usually spaced around the distribution system. This greatly reduces the need for large transmission mains.
- Occasionally, a large quantity of groundwater is available in one area, in which case a "well field" is installed. This situation makes it possible to collect water from all wells and treat it at one location. Usually this approach reduces operating costs and allows closer control over water quality.
- If water enters the distribution system at only one point, the distribution system must be furnished with transmission mains similar to those in a surface water system.

groundwater system
A water system using wells, springs, or infiltration galleries as its source of supply.

Purchased Water Systems

A number of water utilities purchase treated water from another utility. Three of the principal reasons, beyond the economics, these systems have switched to a purchased water supply are (1) their well supplies became inadequate, (2) their water sources were found to be contaminated, or (3) regulatory compliance made the supply and treatment too difficult to continue. Prime examples are the areas surrounding Chicago and Detroit, where a few large treatment plants furnish hundreds of individual water systems with water from the Great Lakes. The following are a few characteristics of purchased water systems:

- The operator's job is limited primarily to operating the distribution system, with little or no treatment required. In some cases, additional disinfection is required and the water must be repumped to boost pressure.
- The quality of water from bulk suppliers is generally good because the water has undergone full treatment.
- Purchasing systems must maintain particularly tight water accountability on their distribution system because all water passing the bulk meter must be paid for, including water from leaks and other wasted water.
- Purchasing systems must frequently provide a greater-than-average water storage capacity because of the possibility that they will temporarily lose their single source. Additional storage may also be required if the water-purchasing agreement involves drawing water at only a limited rate or purchasing water at a cheaper rate during the night.

System Planning Issues

Local conditions have a substantial bearing on system design. They will dictate such details as how deep piping must be buried, how much fire flow is required, and what types of materials are best suited in existing conditions. A few of these considerations are as follows:

- Water availability
- Source reliability
- Water quality
- Location
- State and federal requirements

Policy Considerations

Certain general planning considerations related to new system development or system expansion require local policy decisions. These decisions, which are usually made by the utility or city governing board, are based on an assessment of all relevant details, such as the following:

- Future growth
- Costs
- Financing methods
- Ordinances
- Zoning
- Regulatory issues

Drinking Water Supply and Distribution Systems

The overall conventional water supply system includes all the system components to develop drinking water and distribute it to the customers. Water supply and

purchased water system
A water system that purchases water from another water system and so generally provides only distribution and minimal treatment.

distribution systems are discussed in many publications. Common elements associated with water supply systems in the United States include the following:

- A water source, which may be a surface impoundment such as a lake or reservoir, a river, or groundwater from an aquifer
- Surface supplies, which generally incorporate conventional treatment facilities, including filtration and disinfection
- Transmission systems (including tunnels), reservoirs and/or pumping facilities, and storage facilities
- A distribution system, which carries finished water through a system of water mains and pipes to consumers

Figure 3-4 illustrates the functional components of a conventional water supply system: raw water sources or source development (groundwater and/or surface sources), raw water pumping and transmission, raw water storage, water treatment, high-service pumping, and water distribution.

Water distribution is composed of three major components: distribution piping, distribution storage, and distribution pumping stations. Components represent the largest functional elements in an urban water distribution system, and each of these three major components is composed of one or more subcomponents, as illustrated in Figure 3-5.

Subcomponents represent the basic building blocks of the major components. Pumping stations can be divided into structural, electrical, pumping, and piping subcomponents. Distribution storage can be divided into tank, pipe, and valve subcomponents. Distribution piping can be divided into pipe and valve subcomponents. The sub-subcomponents are composed of one or more subcomponents integrated into a common operational element. One example is the pumping unit subcomponent, which is composed of pipes, valves, pumps, drivers, power transmission, and control sub-subcomponents.

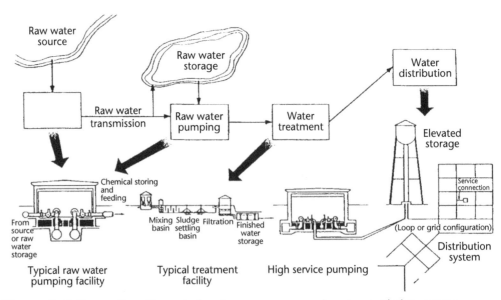

Figure 3-4 Conventional layout of water sources, pumping, transmission, water treatment, water distribution

Source: Cullinane, M. J. Jr. 1989. *Methodologies for the Evaluation of Water Distribution System Reliability/Availability.* PhD Dissertation, University of Texas at Austin.

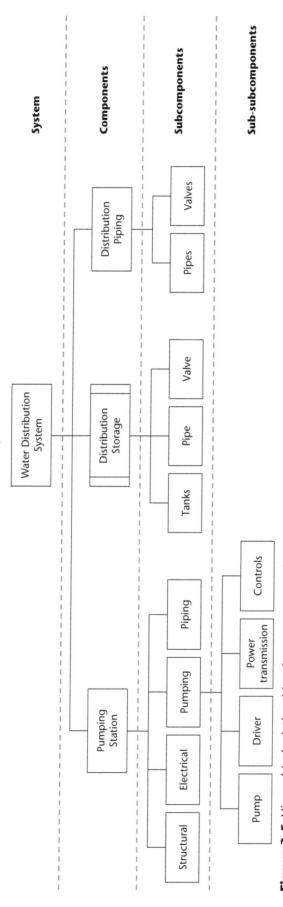

Figure 3-5 Hierarchical relationship of components, subcomponents, and sub-subcomponents for a water distribution system

Source: Cullinane, M. J. Jr. 1989. *Methodologies for the Evaluation of Water Distribution System Reliability/Availability*. PhD Dissertation, University of Texas at Austin.

System Layout

A distribution system layout is usually designed in one of three configurations:

- Arterial-loop system
- Grid system
- Tree system

Most distribution systems are actually a combination of grid and tree systems.

Arterial-Loop System

As illustrated in Figure 3-6, an arterial-loop system attempts to surround the distribution area with larger-diameter mains. The large mains then contribute water supply within the grid from several different directions.

Grid System

As illustrated in Figure 3-7, a grid system depends primarily on the fact that all mains are interconnected, so water drawn at any point can actually flow from several different directions. The distribution mains in the general grid system are usually 6 or 8 in. (150 or 200 mm) in diameter. They are then reinforced with larger arterial mains, and the general area is fed by still larger transmission mains.

Tree System

As illustrated in Figure 3-8, a tree system brings water into an area with a transmission main, which then branches off into smaller mains. The smaller mains generally end up as dead ends. This is not considered a good distribution system design and is generally not recommended. However, in many cases, site-specific conditions result in the selection of this type of system.

Dead Ends

In general, dead-end water mains should be avoided if at all possible. Problems caused by dead ends include the following:

- A consumer located on a dead-end main can draw only through the single line, so flow is restricted by the single main length and pipe size. This limited flow is a problem if a large volume of water is needed, such as in the event of a fire. The advantage of a grid system is that water can flow to the point of demand from several different directions when a high flow rate is required.
- Mains are sized to provide adequate fire flow. Because the domestic use of a few houses requires much less water than fire flow, the water in the mains moves very slowly. Therefore, water often becomes stagnant or degrades in dead ends. Many water systems find it necessary to set up special programs to flush these dead-end mains periodically to maintain acceptable water quality. Mains that are looped generally flow continually in one direction or another, so they do not experience this problem.
- If repair work must be performed on a dead-end main, all customers beyond the repair lose service until the main is repaired, disinfected, and placed back in service. With a looped system, only the customers between the two closest valves would lose service.

---

arterial-loop system
A distribution system layout involving a complete loop of arterial mains (sometimes called trunk mains or feeders) around the area being served, with branch mains projecting inward. Such a system minimizes dead ends.

grid system
A distribution system layout in which all ends of the mains are connected to eliminate dead ends.

tree system
A distribution system layout that centers around a single arterial main, which decreases in size with length. Branches are taken off at right angles, with subbranches from each branch.

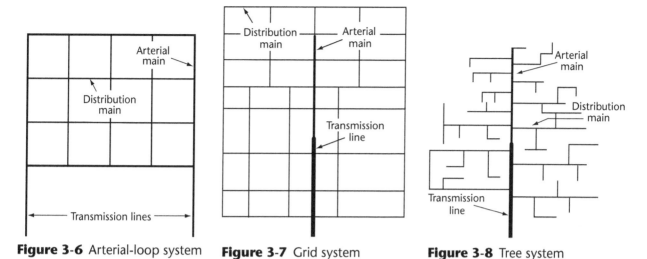

Figure 3-6 Arterial-loop system   **Figure 3-7** Grid system   **Figure 3-8** Tree system

Mapping

As an initial step in system planning, a basic map should be developed to show both the existing system and the areas that may have to be served in the future. Details essential to the mapping process include the following:

- Information about the existing system (including main size, water pressure, valves, and hydrants)
- Existing and planned streets
- Areas outside the system designated for future expansion
- Ground-level elevations, contour lines, and topographic features
- Existing underground utility services, including sanitary sewers; storm sewers; water, gas, and steam lines; and underground electric, telephone, and television cables
- Population densities (present and projected)
- Normal water consumption (present and projected)
- Proposed additions or changes to the system

Figure 3-9 illustrates a section of a typical water system map. Water system operators should realize that information about buried mains, valves, and services must be immediately recorded after installation—not simply committed to memory. It can be extremely expensive to recreate records of "lost" underground facilities. Computerized geographical information systems (GIS) are commonly used to create maps and record location information. These systems can provide needed information to system operators.

Valving

Shutoff valves should be provided so that areas within the system can be isolated for repair or maintenance. To minimize service interruptions, valves should be located at regular intervals and at all branches from the arterial mains. Where mains intersect in a grid, at least two (preferably three) of the branch lines should be provided with valves. The distribution system should be planned so that most of the flow is maintained even if any section of the system is taken out of service.

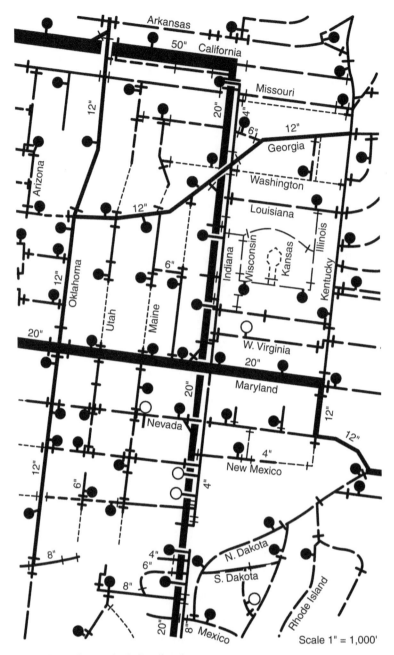

Figure 3-9 Portion of a typical distribution system map

Air-and-vacuum relief valves are required at high points, and blowoff valves are required at low points. Cross-connections between the domestic supply and other water sources should not be used. Backflow prevention devices are required by applicable regulations to prevent contamination from nonpotable sources. Chapter 9 provides additional information on valves.

Sizing Mains

The size of a water main determines its carrying capacity. Main sizes must be selected to provide the flow (capacity) to meet peak domestic, commercial, and industrial demands in the area to be served. They must also provide for fire flow at the necessary pressure.

Quantity Requirements

The quantity of water that must be carried by a water main usually depends on two things: consumption (domestic, industrial, commercial, institutional, etc.) and fire flow requirements.

Domestic Use

Requirements for domestic use can be determined either from past records or from general usage figures for the area. If rates are reasonable and water use is not restricted, the use of water for irrigation may cause domestic use to be much higher in summer months (depending on climate) than it is in winter. The determination of use must also consider projected growth factors to ensure that the system's design capacity will meet future demand.

Fire Flow Requirements

The determining factor in sizing mains, storage facilities, and pumping facilities for communities with a population of less than 50,000 is usually the need for fire protection. Fire flow requirements for each community are set by the Insurance Services Office (ISO), which is an organization representing fire insurance underwriters. This group determines the minimum flow that the system must be able to maintain for a specified period of time in order for the community to achieve a specified fire protection rating. Fire insurance rates are then based, in part, on this classification.

Many rural water systems are designed to serve only domestic water needs. Fire protection is not provided, and fire flow requirements are not considered in the design of these systems.

Fire insurance underwriters recommend that no main in the distribution system be less than 6 in. (150 mm) in diameter. They also suggest that minimum pipe sizes be governed, in part, by the type of area to be served. For example:

- High-value districts (such as sports stadiums, shopping centers, and libraries) should have minimum pipe sizes of 8 and 12 in. (200 and 300 mm).
- Residential areas should have minimum sizes of 6 and 8 in. (150 and 200 mm).
- Mains smaller than 6 in. (150 mm) should be used only when they complete a desirable grid.

These recommendations may be difficult to follow in practice, but they should be a goal when system improvements are being made. The nearest ISO representative should be consulted if more specific information on fire flow requirements or a fire insurance rating is needed.

Pressure Requirements

In areas requiring high fire flow capacity, the minimum static pressure required at all fire hydrants is generally 35 psi (240 kPa) or higher. The required pressure during fire flow conditions should not drop below 20 psi (140 kPa). Some insurance companies require that both a minimum water pressure and a minimum flow requirement be maintained to the buildings they insure.

Water systems located in an area with widely varying elevations usually find it necessary to divide the distribution system into two or more pressure zones. Residential water pressure of between 50 and 75 psi (345 and 517 kPa) is usually considered most desirable. Customers do not generally like higher pressure because water comes out of a quickly opened faucet with too much force.

In addition, higher pressure contributes to more main and service leaks and also hastens the failure of water heaters. Excessive pressure may also damage automatic dishwashers and washing machines. Homes having very high water pressure may need to install household pressure-reducing valves.

Velocity Requirements

Flow velocity is another factor to consider when determining pipe capacity and required pipe size. Velocities should normally be limited to about 5 ft/sec (1.5 m/sec) in order to minimize friction loss as water flows through the pipe. For example, water passing through 100 ft (30.5 m) of cast-iron pipe in relatively good condition will have a head loss of about 12 ft (4 m) when flowing at a velocity of 5 ft/sec (1.5 m/sec). If the velocity is increased to 9 ft/sec (2.7 m/sec), the head loss will be tripled to about 36 ft (11 m).

It should be a goal to keep velocities low under normal operating conditions, but velocities may exceed the guidelines under fire flow conditions. Another goal concerning velocities is that water quantity and quality goals need to be balanced. Oversized pipes may cause water quality problems.

Network Analysis

The carrying capacity of pipe depends on a combination of factors, including pipe size, pressure, flow velocity, and head loss resulting from friction. The amount of friction loss depends on the pipe roughness, flow velocity, and pipe diameter.

The required pipe size can be calculated when the other requirements and characteristics are known. A common method for calculating pipe size is the Hazen–Williams formula. The Hazen–Williams formula for the velocity of flow in a pipe is expressed as

$$V = 1.318 C R^{0.63} S_f^{0.54}$$

where V is the flow velocity in feet per second, C is the Hazen–Williams coefficient, R is the hydraulic radius in ft, and S_f is the friction slope in feet per feet. Charts and computer programs that use this formula (or are based on it) have been developed for various sizes and types of pipe to help in selecting proper pipe size.

When distribution system expansion is to be extensive, it is usually prudent, if not necessary, to analyze the entire system. Consideration needs to be given to both existing and projected water demands. This analysis requires a plot of pressures and flows at points throughout the system. Most comprehensive system evaluations are performed using computerized modeling techniques.

Quality Requirements

Maintaining water quality throughout the distribution system is a primary goal of system operators. The design of the piping network greatly impacts the operating and maintenance procedures that must be employed to achieve this goal. The type of materials, the design of storage facilities, the size and location of mains, and the quality of water entering the distribution system are critical factors that influence water quality. Chapter 8 describes the practices that operators can employ to maintain acceptable water quality. The distribution system design should include equal emphasis on quantity and quality factors.

Hazen–Williams formula
A formula for the velocity of flow in a pipe.

Study Questions

1. Head is measured in
   a. absolute pressure.
   b. gauge pressure.
   c. feet.
   d. foot-pounds.

2. A plat is
   a. a map.
   b. a corrosion point on a pipe.
   c. an organelle found in some protozoans.
   d. a highly corrosive soil type.

3. At which time of day is the age of the water stored in the distribution system the highest?
   a. Early morning
   b. Late morning
   c. Early afternoon
   d. Late evening

4. Water mains should primarily be sized based on
   a. earthquake size potential.
   b. peak domestic and commercial demands.
   c. peak commercial and industrial demands.
   d. adequate fire flow at an appropriate pressure.

5. The most desirable residential pressure ranges from
   a. 20 to 35 psi (138 to 241 kPa).
   b. 35 to 50 psi (241 to 345 kPa).
   c. 50 to 75 psi (345 to 517 kPa).
   d. 75 to 90 psi (517 to 621 kPa).

6. Domestic water use includes water that is supplied to residential areas, _____, and institutional facilities.
   a. public parks
   b. commercial districts
   c. industrial facilities
   d. municipal buildings

7. Riparian doctrine is sometimes called the "rule of
   a. first come, first serve."
   b. reasonable sharing."
   c. equal use."
   d. eminent domain."

8. During which season is water use typically the highest?

9. Of the common types of distribution systems, which type is considered least desirable?

10. What type of valves should be provided so that areas within the system can be isolated for repair or maintenance?

Chapter 4
Hydraulics

Fluids at Rest and in Motion

Hydraulics is the study of fluids in motion or under pressure. An understanding of hydraulics is necessary for the proper operation of a water distribution system. In this chapter, the subject will be confined to the behavior of water in water distribution systems.

Static Pressure

Water flows in a water system when it is under a force that makes it move. The force on a unit area of water is termed pressure. The pressure in a water system is a measure of the height to which water theoretically will rise in an imaginary standpipe open to atmospheric pressure. The pressure can be static; that is, it exists although the water does not flow. Pressure can also be dynamic, existing as "moving energy."

All objects have weight because they are acted on by gravity. When a 1-lb brick is placed on a table with an area of 1 square inch (1 in.²), it exerts a force of 1 pound per square inch (psi) on the table. Two stacked bricks on the 1-in.² table would exert a force of 2 psi. But if the size of the table is doubled, the pressure is halved. And if the table size is tripled, the pressure in pounds per square inch is reduced by one-third.

Likewise, a column of water 10 ft high exerts a total force of 4.33 psi. If you connect a pressure gauge at the bottom of a water tube with 10 ft of water in it, the gauge will read 4.33 psi. If you connect the pressure gauge to the bottom of a larger-diameter column with 10 ft of water in it, the gauge will still read 4.33 psi (Figure 4-1). Water pressure is dependent only on the height of the column. However, the total weight exerted on the floor by the water in the large column will obviously be much more.

Dynamic Pressure

If the water in the column is permitted to empty horizontally from the bottom of a column, the water will begin to flow under the hydrostatic pressure applied by the height of the column. The flowing water will have little hydrostatic pressure, but it will have gained moving, dynamic pressure, or kinetic energy. The hydrostatic pressure is static potential energy converted into moving energy.

One can add energy to a water system and thereby increase hydrostatic and dynamic pressure. A pump does this when it pumps water into elevated storage. The hydrostatic pressure (height) to which the water can be pumped is equivalent to pressure (less losses) at the pump discharge.

---

hydraulics
The study of fluids in motion or under pressure.

pressure
The force on a unit area of water.

static pressure
Pressure that exists in water although the water does not flow.

dynamic pressure
Pressure that exists in water as moving energy.

pounds per square inch (psi)
A measure of pressure.

hydrostatic pressure
The pressure exerted by water at rest (for example, in a nonflowing pipeline).

89

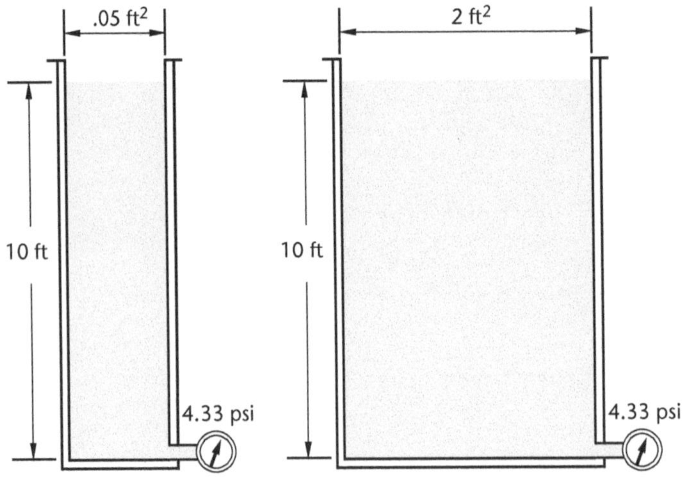

Figure 4-1 Hydraulic head depends only on column height

Pressure is usually measured in either pounds per square inch or feet of head in US units, or as kilopascals (kPa) of pressure or meters (m) of head in metric units. A pressure of 1 psi is equal to approximately 6.895 kPa.

Velocity

The speed at which water moves is called **velocity**, usually abbreviated V. The velocity of water is usually measured in feet per second (ft/sec) in US units and meters per second (m/sec) in metric terms. For comparison, a rapidly moving river might move at about 7 ft/sec (2.13 m/sec).

The quantity of water (Q) that flows through a pipe depends on the velocity (V) and the cross-sectional area (A) of the pipe. This relationship is stated mathematically as the formula $Q = A \times V$. Or, in terms of velocity,

$$V = \frac{Q}{A}$$

For example, a flume is 2 ft wide and 2 ft deep, so the cross-sectional area of the flume is 4 ft². The flume is flowing full of water and the quantity is measured at 12 ft³ in 1 second (12 ft³/sec). The velocity of the water would therefore be calculated as follows:

$$V = \frac{Q}{A}$$

$$= \frac{12 \text{ ft}^3/\text{sec}}{4 \text{ ft}^2} = 3 \text{ ft/sec}$$

Measuring Pressure

As mentioned, water pressures are directly related to the height (depth) of water. The discussion that follows will illustrate the role of water height in calculating water pressure.

Suppose, for example, you have a container 1 ft by 1 ft by 1 ft (a cubic-foot container) that is filled with water. What is the pressure on the square-foot bottom of the container?

velocity
The speed at which water moves; measured in ft/sec or m/sec.

Pressure in this case is expressed in pounds per unit area. In this case, because the density of water is 62.4 lb/ft², the force of the water pushing down on the square foot surface area is 62.4 lb (Figure 4-2).

From this information, the pressure in pounds per square inch can also be determined. Convert pounds per square foot to pounds per square inch:

$$\frac{62.4 \text{ lb}}{\text{ft}^2} = \frac{62.4 \text{ lb}}{(1 \text{ ft})(1 \text{ ft})}$$

$$= \frac{62.4 \text{ lb}}{(12 \text{ in.})(12 \text{ in.})}$$

$$= \frac{62.4 \text{ lb}}{144 \text{ in.}^2}$$

$$= 0.433 \frac{\text{lb}}{\text{in.}^2}$$

$$= 0.433 \text{ psi}$$

This means that a foot-high column of water over a square-inch surface area weighs 0.433 lb, resulting in a pressure of 0.433 psi (Figure 4-3). The factor 0.433 allows you to convert from pressure measured in feet of water to pressure measured in pounds per square inch. A conversion factor can also be developed for converting from pounds per square inch to feet.

Because 1 ft is equivalent to 0.433 psi, set up a ratio to determine how many feet of water are equivalent to 1 psi (that is, how many feet high a water column must be to create a pressure of 1 psi):

$$\frac{1 \text{ ft}}{0.433 \text{ psi}} = \frac{x \text{ ft}}{1 \text{ psi}}$$

Then solve for the unknown value:

$$\frac{(1)(1)}{0.433} = x$$

$$2.31 \text{ ft} = x$$

Therefore, 1 psi is equivalent to the pressure created by a column of water 2.31 ft high (Figure 4-4).

Because the density of water is assumed to be a constant 62.4 lb/ft³ in most hydraulic calculations, the height of the water is the most important factor in determining pressures. It is the height of the water that determines the pressure over

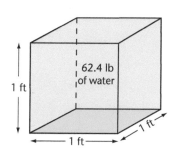

Figure 4-2 A cubic foot of water

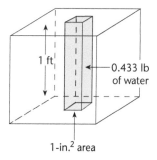

Figure 4-3 A 1-foot column of water over a 1-in.² area

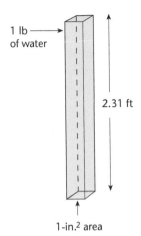

Figure 4-4 A 2.31-ft water column (which creates a pressure of 1 psi at the bottom surface)

head
(1) A measure of the energy possessed by water at a given location in the water system, expressed in feet. (2) A measure of the pressure or force exerted by water, expressed in feet.

gauge pressure
The water pressure as measured by a gauge. Gauge pressure is not the total pressure. Total water pressure (absolute pressure) also includes the atmospheric pressure (about 14.7 psi at sea level) exerted on the water. However, because atmospheric pressure is exerted everywhere (against the outside of the main as well as the inside, for example), it is generally not written into water system calculations. Gauge pressure in pounds per square inch is expressed as "psig."

pounds per square inch gauge (psig)
Pressure measured by a gauge and expressed in terms of pounds per square inch.

the square-inch area. Pressure measured in terms of the height of water (in meters or feet) is referred to as head. As long as the height of the water stays the same, changing the shape of the container does not change the pressure at the bottom or any other level in the container. For example, see Figure 4-5. Each of the containers is filled with water to the same height. Therefore, the pressures against the bottoms of the containers are the same. And the pressure at any depth in one container is the same as the pressure at the same depth in either of the other containers.

In water system operation, the shape of the container can help to maintain a usable volume of water at higher pressures. For example, suppose you have an elevated storage tank and a standpipe that contain equal amounts of water. When the water levels are the same, the pressures at the bottom of the tanks are the same (Figure 4-6A). However, if half of the water is withdrawn from each tank, the pressure at the bottom of the elevated tank will be greater than the pressure at the bottom of the standpipe (Figure 4-6B).

Because of the direct relationship between the pressure in *pounds per square inch* at any point in water and the *height in feet* of water above that point, pressure can be measured either in pounds per square inch or in feet (of water), called head.

When the water pressure in a main or in a container is measured by a gauge, the pressure is referred to as gauge pressure. If measured in pounds per square inch, the gauge pressure is expressed as pounds per square inch gauge (psig). The gauge pressure is not the total pressure within the main. Gauge pressure does not show the pressure of the atmosphere, which is equal to approximately 14.7 psi at sea level. Because atmospheric pressure is exerted everywhere (against the outside of the main as well as the inside, for instance), it can generally be neglected in water system calculations. However, for certain calculations, the

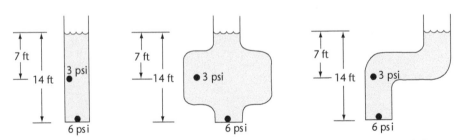

Figure 4-5 Pressure depends on the height of the water, not the shape of the container

Hydraulics **93**

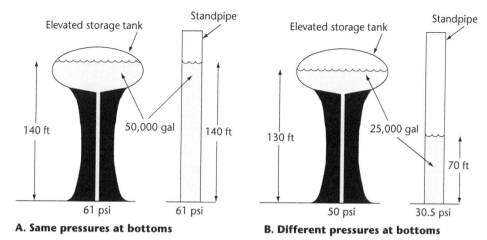

A. Same pressures at bottoms   **B. Different pressures at bottoms**

Figure 4-6 A 2.31-ft water column (which creates a pressure of 1 psi at the bottom surface)

total (or absolute) pressure must be known. The **absolute pressure** (expressed as pounds per square inch absolute [psia]) is obtained by adding the gauge and atmospheric pressures. For example, in a main under 50-psi gauge pressure, the absolute pressure would be 50 psig + 14.7 psi = 64.7 psia. A line under a partial vacuum, with a gauge pressure of –2, would have an absolute pressure of (14.7 psi) + (–2 psig) = 12.7 psia (Figure 4-7).

Pressure gauges can also be calibrated in feet of head. A pressure gauge reading of 14 ft of head, for example, means that the pressure is equivalent to the pressure exerted by a column of water 14 ft high. The equations that relate gauge pressure in pounds per square inch to pressure in feet of head are given below. In this discussion, conversions from one term to another will use the first equation only.

> 1 psig = 2.31 ft head
> 1 ft head = 0.433 psig

The following examples illustrate conversions from feet of head to pounds per square inch gauge and from pounds per square inch gauge to feet of head. The method used is similar to the conversion approach (box method) discussed in Chapter 2.

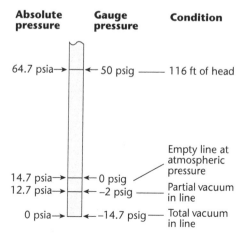

Figure 4-7 Gauge versus absolute pressure

absolute pressure
The total pressure in a system, including both the pressure of water and the pressure of the atmosphere (about 14.7 psi, at sea level).

Example 1

Convert a gauge pressure of 14 ft to pounds per square inch gauge.

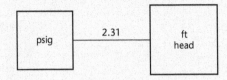

1 psig = 2.31 ft head

Using the diagram, in moving from feet of head to pounds per square inch gauge, you are moving from a larger box to a smaller box. Therefore, you should divide by 2.31:

$$\frac{14 \text{ ft}}{2.31 \text{ ft/psig}} = 6.06 \text{ psig}$$

Example 2

What would the pounds-per-square-inch gauge pressure readings be at points A and B in the diagram in Figure 4-8?

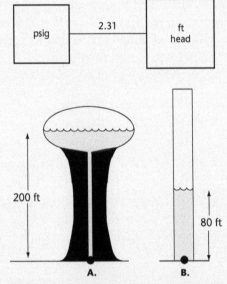

Figure 4-8 Schematic for Example 2

In each case, the conversion is from feet of head to pressure in pounds per square inch gauge. Therefore, when you move from a larger box to a smaller box, division by 2.31 is indicated:

$$\text{pressure at A} = \frac{200 \text{ ft}}{2.31 \text{ ft/psig}}$$

$$= 86.58 \text{ psig}$$

$$\text{pressure at B} = \frac{80 \text{ ft}}{2.31 \text{ ft/psig}}$$

$$= 34.63 \text{ psig}$$

Head

Head is one of the most important measurements in hydraulics. It is used to calculate the hydraulic forces acting in a pipeline and to determine a pump's capacity to overcome or pump against these forces.

Head is a measurement of the energy possessed by the water at any particular location in the water system. In hydraulics, energy (and therefore head) is expressed in units of foot-pounds per pound, written

$$\text{head} = \frac{\text{ft-lb}}{\text{lb}}$$

These somewhat cumbersome units of measure cancel out so that head can be expressed in feet:

$$\text{head} = \frac{\text{ft-lb}}{\text{lb}}$$
$$= \text{ft}$$

When head is expressed in feet, as it normally is, the measurement can always be considered to represent the height of water above some reference elevation. The height of water in feet can also be expressed as pressure in pounds per square inch gauge by dividing the feet-of-water height by 2.31.

Types of Head

There are three types of head:

- Pressure head
- Elevation head
- Velocity head

Pressure Head

Pressure head is a measurement of the amount of energy in water *due to water pressure*. As shown in Figure 4-9, it is the height above the pipeline to which water will rise in a piezometer. The water pressure is easily measured by a pressure gauge, and the resulting pressure reading in pounds per square inch gauge may then be converted to pressure head in feet.

Pressure head describes the vertical distance from the point of pressure measurement to the hydraulic grade line (HGL). So, if you were interested in locating the HGL for a particular pipeline, you would take gauge pressure readings at

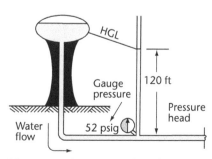

Figure 4-9 Pressure head

> **pressure head**
> A measurement of the amount of energy in water due to water pressure.
>
> **hydraulic grade line (HGL)**
> A line (hydraulic profile) indicating the piezometric level of water at all points along a conduit, open channel, or stream. In an open channel, the HGL is the free water surface.

several critical points along the pipeline, convert those pressure readings to feet of pressure head, and then plot the HGL.

The normal range of pressure heads in water transmission and distribution systems can vary from as little as 50 ft (about 20 psig) to over 1,000 ft (about 450 psig). At certain locations within the treatment plant, pressure heads can be very small, perhaps 2 ft (about 1 psig) or less.

Elevation Head

Elevation head is a measurement of the amount of energy that water possesses *because of its elevation*. It is measured as the height in feet from some horizontal reference line or benchmark elevation (such as sea level) to the point of interest in the water system. For example, a reservoir located 500 ft above sea level is said to have an elevation head of 500 ft relative to sea level. The concept is illustrated in Figure 4-10. Elevation head is quite useful in design but has little day-to-day operating significance.

Velocity Head

Velocity head is a measurement of the amount of energy in water *due to its velocity, or motion*. The greater the velocity, the greater the energy and, therefore, the greater the velocity head. Anything in motion has energy because of that motion.

You might think of the energy of motion in terms of the effort you would need to apply to stop the motion. For example, stopping a car traveling at 55 mph requires that the car's brakes do much more work than if the car were traveling 5 mph, because the faster-moving car has more energy of motion.

Like all other heads, velocity head is expressed in feet. It is determined by multiplying velocity (in feet per second) by itself and then dividing by 64.4 feet per second squared (ft/s^2), as follows:

$$\text{velocity head} = \frac{V^2}{64.4 \text{ ft/s}^2}$$

For example, if the velocity of the water is 30 ft/s, the velocity head is calculated as follows:

$$\text{velocity head} = \frac{(30)(30)}{64.4}$$

$$= 13.98 \text{ ft}$$

elevation head
The energy possessed per unit weight of a fluid because of its elevation above some reference point (called the "reference datum"). Elevation head is also called position head or potential head.

velocity head
A measurement of the amount of energy in water due to its velocity, or motion.

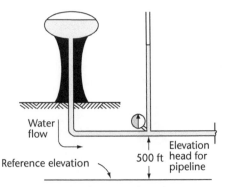

Figure 4-10 Elevation head

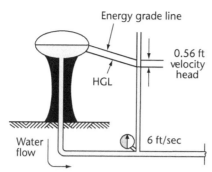

Figure 4-11 Velocity head

This means that if you pour water from a bucket held 13.98 ft in the air, then the water will be moving 30 ft/s when it hits the ground.

Velocity head is also a measurement of the vertical distance from the HGL to the energy grade line (EGL), as shown in Figure 4-11. EGLs are an advanced concept and will not be discussed further in this basic text.

The velocity, or speed, at which water travels in a water system is usually at least 2 ft/s and generally not more than 10 ft/s. These speeds would produce velocity heads of

$$\text{velocity head at 2 ft/s velocity} = \frac{(2)(2)}{64.4}$$

$$= 0.06 \text{ ft}$$

$$\text{velocity head at 10 ft/s velocity} = \frac{(10)(10)}{64.4}$$

$$= 1.55 \text{ ft}$$

Because velocity heads are so small relative to pressure heads (which usually range from 50 to 1,000 ft), they can usually be ignored in operations without causing a significant error. However, although velocity head can be ignored in most cases, it is important that you recognize that velocity head should be considered whenever it is greater than 1 or 2 percent of the pressure head—that is, whenever it is large enough to be a significant part of the total head. This scenario will usually occur either in pipelines or pump systems when the pressure head is very small (less than 70 ft) and velocities are in the range of 7–10 ft/s.

 WATCH THE VIDEO
Operator Chemistry—Hydraulics (www.awwa.org/wsovideoclips)

Study Questions

1. Head is measured in
    a. absolute pressure.
    b. gauge pressure.
    c. feet.
    d. foot-pounds.

> **energy grade line (EGL)**
> A line joining the elevations of the energy heads; a line drawn above the hydraulic grade line by a distance equivalent to the velocity head of the flowing water at each section along a stream, channel, or conduit.

2. If the pressure head on a fire hydrant is 134 ft, what is the pressure in psi?
   a. 50 psi
   b. 52 psi
   c. 54 psi
   d. 58 psi

3. A meter indicates the water flow from a fire hydrant is 5.5 ft³/min. How many gallons will flow from the hydrant in 20 minutes?
   a. 820 gal
   b. 850 gal
   c. 880 gal
   d. 920 gal

4. Records for a pump show that on June 1 at exactly 9:00 a.m. the number of pumped gallons was 71,576,344 and on July 1 at exactly 9:00 a.m. it was 72,487,008 gallons. Determine the average gallons pumped per day (gal/day) for this month to the nearest gallon.
   a. 18,605 gal/day
   b. 25,875 gal/day
   c. 30,355 gal/day
   d. 34,325 gal/day

5. The velocity of water is calculated as the quantity of water that flows through a pipe
   a. divided by the cross-sectional area of the pipe.
   b. divided by the time the water takes to reach its destination.
   c. divided by the water's weight in cubic feet.
   d. multiplied by resistance to flow.

6. What is the term for water pressure in a main or in a container that is measured by a gauge?

7. In hydraulics, how is head expressed?

8. What is the term for pressure measured in terms of the height of water (in meters or feet)?

9. What are the three types of head?

Chapter 5
Pipe

Pipe Material Selection

In developing or maintaining a water system, it is essential to understand the factors that affect the choice of pipe materials. Incorrect selections may result in premature failure or a deterioration of water quality.

For example, if components constructed of dissimilar metals are directly connected, a corrosion cell may be formed if the components are immersed in a conducting fluid such as water or damp soil. An insulating spacer (nonconducting material between the two metal surfaces) placed between the connection can eliminate this problem.

Selecting unprotected iron or steel pipe for use in a corrosive soil may result in pipe failure. Properly protecting the ferrous pipe or choosing a resistant material may reduce this problem. These are only two of the many possible consequences of improper material selection. Fortunately, a number of standards apply to many distribution system components. All components and materials must conform to the appropriate standard. There are also certification programs designed to verify that a product or material meets the requirements of the standard. Utilities that require certified products or materials provide added assurance that the system components perform as expected and the drinking water quality is protected.

The following general factors need to be considered in the selection of pipe (Ysusi, 2000):

- Service conditions
  - Pressure (including surges and transients)
  - Soil loads, bearing capacity of soil, potential settlement
  - Corrosion potential of soil
  - Potential corrosive nature of some waters
- Availability
  - Local availability and experienced installation personnel
  - Sizes and thicknesses (pressure ratings and classes)
  - Compatibility with available fittings
- Properties of the pipe
  - Strength (static and fatigue, especially for water hammer)
  - Ductility
  - Corrosion resistance
  - Fluid friction resistance (more important in transmission pipelines)

- Economics
  - Cost (installed cost, including freight to jobsite and installation)
  - Required life
  - Cost of maintenance and repairs

ANSI/AWWA Standards

AWWA has developed consensus standards since 1908. Currently, AWWA has more than 170 standards for products and procedures used in the water industry. The standards are developed under procedures accredited by the American National Standards Institute (ANSI) and are thus registered by this international standards body. AWWA standards define the minimum requirements for many aspects of drinking water systems. Compliance with these standards is not mandated by AWWA, but the standards have been made mandatory by utilities and regulatory agencies. AWWA does not test any product in developing or approving a standard, and no products are "AWWA approved." AWWA standards are available in these major categories:

- Wells
- Treatment
  - Filtration
  - Softening
  - Disinfection chemicals
  - Coagulation
  - Scale and corrosion control
  - Taste and odor control
  - Fluorides
- Pipe and accessories
  - Ductile-iron pipe and fittings
  - Steel pipe
  - Concrete pipe
  - Valves and hydrants
- Pipe installation
- Disinfection of facilities
- Meters
- Service lines
- Plastic pipe
- Water storage
- Pumps
- Plant equipment
- Utility management

ANSI/NSF Standard 61 and Certification

The ANSI/NSF Standard 61 covers materials that come in contact with potable water and focuses on the issue of whether contaminants leach or migrate from the product/material into the drinking water, resulting in unacceptable levels of contaminants. Certain materials that may have the potential for microbiological growth (e.g., some coatings, gaskets, lubricants) are also evaluated. The standard covers valves, pipes, protective materials, joining and sealing materials, mechanical devices, process media, and related products. Most materials or components that may come in contact with drinking water are covered in this standard.

There are several certification programs, conducted by various organizations (e.g., NSF International, Underwriters Laboratories [UL], Canadian Standards Association), based on this standard that include product testing, evaluation of materials, and factory inspections. Testing is conducted in accordance with the requirements of the standard. Contaminant concentrations in laboratory testing are converted to tap levels and assessed against the maximum allowable level in the standard. If the material or product satisfies the requirements, it is certified by the testing company as acceptable for use in potable water systems. To remain certified, production facilities may be inspected at any time and periodic retesting is required (usually on an annual basis).

Many utilities and most state and provincial regulatory agencies require certification to ANSI/NSF Standard 61 for all system components that come in contact with drinking water. Utilities should be aware of these requirements and check products to make sure they comply.

Selecting Pipeline Materials

Some questions that should be considered when selecting pipe material include the following:

- What pipe materials are available?
- Does one material perform better than another?
- Are all materials acceptable?
- What pipeline material does the distribution system have now?
- Why was the present pipe material originally selected?
- Are the materials for the existing pipe and the proposed new pipe compatible?

The following sections discuss general requirements for distribution pipe.

Pipe Characteristics

Characteristics of a pipe that need to be considered include the following:

- Strength
- Pressure rating
- Durability
- Corrosion resistance
- Smoothness of the inner surface
- Ease of tapping and repair
- Water quality maintenance

Strength Water distribution pipe must have adequate strength to handle external load and internal pressure.

External load is the pressure exerted on a pipe after it has been buried in a trench. This pressure is a result of the backfill (the material that is filled back into the trench after the pipe has been laid), the traffic load (the weight or impact of the traffic passing over the pipe), and longitudinal thrust loads.

The pipe must be able to resist a reasonable amount of external damage from impact during installation. It must also be capable of resisting crushing or undue deflection due to external load. External load is expressed in pounds per linear foot (lb/lin ft) or pounds per square inch (psi). (In metric units, external load is expressed in kilograms per linear meter [kg/m (linear)] or kilopascals [kPa].)

> **external load**
> Any load placed on the outside of the pipe from backfill, traffic, or other sources.

Internal pressure is the hydrostatic pressure within the pipe. Normal water pressure depends on local conditions and requirements but is usually in the 40–100 psi (280–690 kPa) range.

Surge, also known as **water hammer**, is a sudden repeated increase and decrease in pressure that continues until dissipated by friction losses. It occurs from a sudden change in water velocity often caused by the opening or closing of valves or hydrants too quickly, or the sudden starting or stopping of pumps due to loss of power. Water hammer is a transient pressure wave that travels very rapidly through the pipe and may cause pressure that is several times greater than normal pressure. It can cause extensive damage, such as ruptured pipe and damaged fittings.

Pipe strength is expressed in terms of tensile strength and flexural strength. **Tensile strength** is a measure of the resistance a material has to longitudinal, or lengthwise, pull before that material fails. **Flexural strength** is a measure of the ability of a material to bend or flex without breaking.

Pipe shear breakage or beam breakage may occur when a force exerted on a pipe causes stresses that exceed the material's tensile or flexural strength. A shear break occurs when the earth shifts. Beam breakage, which resembles a shear break, may occur when a pipe is unevenly supported along its length. A pipe that is resting on a rock, for instance, may break if it is weak in tensile and flexural strength. Figure 5-1 shows these effects.

Pressure Rating Pipe should be carefully chosen or designed to ensure that its pressure rating is adequate for handling the pressures in a specific system. Pressure ratings can be calculated using various formulas and tables found in current AWWA standards. Distribution system pipe should have a pressure rating of 2.5 to 4 times the normal operating pressure. When a section of pipe is being replaced, the new piece must have a pressure rating equal to or greater than that of the piece being replaced.

Specific minimum requirements or standards for all types of pipe have been established and published by AWWA to ensure adequate and consistent quality of water mains. Other agencies that have established standards for pipe include federal and state governments, UL, NSF International, ASTM International, and the manufacturers themselves. These standards, which cover the method of design, manufacture, and installation in detail, should be used for selecting pipe.

Durability Durability is the degree to which a pipe will provide satisfactory and economical service under the conditions of use. It implies long life, toughness, and the ability to maintain tight joints with little or no maintenance.

Corrosion Resistance Consideration must be given to a pipe's resistance to both internal and external corrosion. If a pipe is made of material that might be corroded by the water being carried, the pipe is usually lined. Some pipe materials will be vulnerable to corrosive soil unless special coatings or wrappings are applied to the pipe exterior and/or cathodic protection is provided.

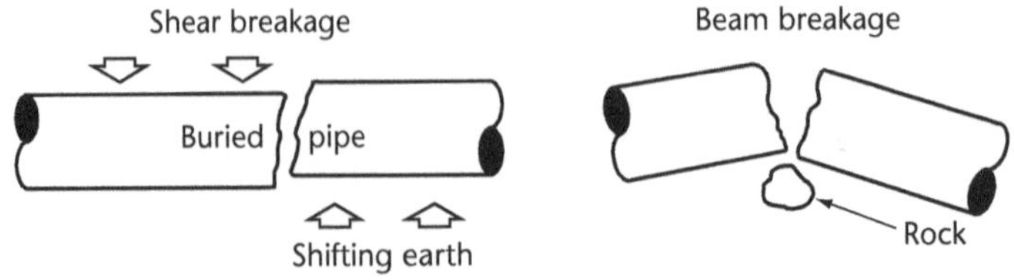

Figure 5-1 Shear and beam breakage

---

internal pressure
The hydrostatic pressure within a pipe.

water hammer
The potentially damaging slam, bang, or shudder that occurs in a pipe when a sudden change in water velocity (usually as a result of someone too-rapidly starting a pump or operating a valve) creates a great increase in water pressure.

tensile strength
A measure of the ability of pipe or other material to resist breakage when it is pulled lengthwise.

flexural strength
The ability of a material to bend (flex) without breaking.

Smooth pipe
Courtesy of J-M Manufacturing Co., Inc.

Tuberculated pipe
Courtesy of Girard Industrie

Figure 5-2 Effect of tuberculation

Smoothness of the Inner Surface Smooth pipe walls ensure maximum flow capacity for water pipe. The C value of a pipe is a measure of the pipe wall's roughness, which determines the extent to which a pipe retards flow because of friction. Higher C values correspond to smoother pipe. Figure 5-2 illustrates the difference between smooth pipe and pipe with an interior surface roughened by tuberculation (the buildup of corrosion products on the pipe walls).

Ease of Tapping and Repair The pipe selected should be easy to repair and tap for service connections. It should hold the service connection firmly without cracking, breaking, or leaking. The tapping connection should be easily replaceable or at least repairable.

Water Quality Maintenance The pipe must be able to maintain the quality of the water distributed by the system. It should not add taste, odor, chemicals, or other undesirable qualities to the water.

Economics

Several considerations in addition to price must be made in determining the most economical choice of pipe for an installation. Some questions that should be considered include the following:

- Are the sizes and types of pipe being considered readily available?
- Are the necessary tees, elbows, and other accessories readily available?
- Is there a difference in the cost of installing different types of pipe?

Installation cost must be carefully assessed. Installation is a major part of the cost of a project, whether the utility does the work with its own forces or contracts the work to outside companies. The projected cost, as well as the additional unplanned costs or savings, will vary significantly for the same job based on factors such as amount and quality of advance planning and engineering, reliability of delivery times, and weather conditions. Installation considerations include the following:

- Variability of pressure requirements over the pipeline route
- Pipe section length
- Weight of the pipe for handling
- Coatings and linings that have to be added during installation
- Pipe strength
- Ease of assembling joints
- Maximum deflection allowed by the joints

tuberculation
The growth of nodules (tubercles) on the pipe interior, which reduces the inside diameter and increases the pipe roughness.

Pipe installation conditions that might have to be considered include the following:

- Unusual soil conditions
- Uneven terrain
- High groundwater and dewatering requirements
- High bedrock
- River or highway crossings
- Proximity to sewer lines
- Proximity to other utility services
- Cathodic protection
- Thrust design
- Air-release valves
- Soil types and need for imported materials for pipe base, pipe zone, and backfill
- Site restoration and backfill requirements
- Tolerance for trench settlement

Piping Systems

Four general types of piping systems are used by water utilities (Figure 5-3). Each use has certain characteristics that dictate, to some degree, the system that will be most economical and best suited for the installation. For example:

- Transmission lines carry large quantities of water from a source of supply to a treatment plant, or from a treatment plant or pumping station to a distribution system. They generally run in a rather straight line from point to point, have few side connections, and are not tapped for customer services.
- In-plant piping systems are the pipes located in pump stations and treatment plants. The piping is generally exposed and has many valves, outlets, and bends that must be secured against movement.
- Distribution mains are the pipelines that carry water from transmission lines and distribute it throughout a community. They have many side connections and are frequently tapped for customer connections.
- Service lines, or "services," are small-diameter pipes that run from the distribution mains to the customer's premises. They are discussed in Chapter 8.

transmission line
The pipeline or aqueduct used for water transmission, i.e., movement of water from the source to the treatment plant and from the plant to the distribution system.

in-plant piping system
The network of pipes in a particular facility, such as a water treatment plant, that carry the water or wastes for that facility.

distribution main
Any pipe in the distribution system other than a service line.

service line
The pipe (and all appurtenances) that runs between the utility's water main and the customer's place of use, including fire lines.

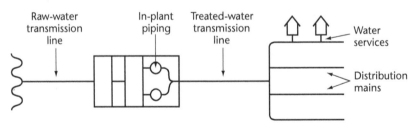

Figure 5-3 Types of piping systems

Types of Pipe Materials

Very early water systems often had water mains made of wood. These mains were usually logs with a hole burned or bored down the center, and the logs were joined with metal sleeves. Older city water systems still occasionally dig up pieces of log pipe during distribution system repairs. The logs were not usually reinforced, so the pipe did not withstand much pressure. Log mains were therefore primarily used to furnish water to local fountains or reservoirs where residents could draw water, rather than to distribute water to homes as we see today. A section of log pipe is shown in Figure 5-4.

From the late 1800s until about World War II, cast-iron, wrought steel, riveted steel, and wood stave pipe were common. The principal types of pipe in use today for water system transmission, distribution, and plant piping fall into the following general categories and the appropriate AWWA standards:

- Ductile iron (DIP)   AWWA C151
- Steel   AWWA C200
  - Cement–mortar-lined steel pipes   AWWA C200, C205, C207, and C208
- Polyvinyl chloride (PVC)   AWWA C900 and C905
- Reinforced concrete pressure pipe (RCPP)
  - Steel cylinder   AWWA C300
  - Prestressed, steel cylinder   AWWA C301
  - Noncylinder   AWWA C302
  - Pretensioned, steel cylinder   AWWA C303
    (also called concrete cylinder pipe [CCP])
- High-density polyethylene (HDPE)   AWWA C906

Asbestos–cement pipe (ACP) was available in the United States starting around 1930 but is no longer used in the water industry because of concerns regarding the hazards of asbestos. ACP is made by mixing portland cement and asbestos fiber under pressure and heating it to produce a hard, strong, machinable product. It is estimated that over 300,000 miles of ACP is in service in the United States (Ysusi, 2000). The US Environmental Protection Agency (USEPA) banned most uses of asbestos in 1989. This ban is not so much due to the

Figure 5-4 Section of an old log pipe

danger of specific products as it is to the overall exposure of people involved in the mining, production, installation, and ultimate removal and disposal of asbestos products.

Gray cast-iron pipe (CIP) is another type of pipe that is no longer used in the United States. The oldest installation of CIP on record is in Versailles, France, dating to 1664. Gray CIP has been used for water distribution systems in the United States and Canada for many years. There are more miles of this pipe in use today than of any other type. Many water systems have cast-iron mains that are more than 100 years old and still function well in daily use. In the 1920s, manufacturers began to make gray CIP using the centrifugal process (Figure 5-5). In this method, molten iron is distributed into a rotating horizontal mold, which forms the outside of the pipe. Lead joints were commonly used to connect gray CIP until the 1920s.

Standards for the design, manufacture, and installation of pressure pipe have been developed by AWWA, ANSI, and ASTM International. These specifications have been established by experimentation, testing, and experience in practice. Pipe manufacturers also publish product literature that is useful in pipe selection and installation. Further details on standards and approval of materials are given in Appendix B. The names and addresses of associations representing manufacturers of the various types of pipes are listed in Appendix C.

Table 5-1 summarizes the advantages and disadvantages of pipeline materials, and Table 5-2 summarizes the types of joints used and their applications. Other considerations that may have a bearing on the type of pipe selected for use in a system include the following:

- State and local regulations
- Local soil conditions
- Local weather conditions
- Likelihood of earthquake activity
- Whether the system serves a community with fire protection or rural customers without fire protection
- The type of pipe already in use in the system
- Whether the pipe is to be exposed to the weather or sunlight
- The possibility of the pipe being exposed to fire

Figure 5-5 Pipe being centrifugally cast by the de Lavaud process
Courtesy of American Cast Iron Pipe Company.

Table 5-1 Comparison of transmission and distribution pipeline materials

Material	Common Sizes—Diameter in.	(mm)	Normal Maximum Working Pressure psi	(kPa)	Advantages	Disadvantages
Ductile iron	3–64	(76–1,625)	350	(2,413)	Durable, strong, high flexural strength, good corrosion resistance, lighter weight than cast iron, greater carrying capacity for same external diameter, easily tapped	Subject to general corrosion if installed unprotected in a corrosive environment
Concrete (reinforced)	12–168	(305–4,267)	250	(1,724)	Durable with low maintenance, good corrosion resistance, good flow characteristics, O-ring joints are easy to install, high external load capacity, minimal bedding and backfill requirements	Requires heavy lifting equipment for installation, may require special external protection in high-chloride soils
Concrete (prestressed)	16–144	(406–3,658)	350	(2,413)	Same as for reinforced concrete	Same as for reinforced concrete
Steel	4–120	(102–3,048)	High		Lightweight, easy to install, high tensile strength, low cost, good hydraulically when lined, adapted to locations where some movement may occur	Subject to general corrosion if installed unprotected in a corrosive environment; poor corrosion resistance unless properly lined, coated, and wrapped
Polyvinyl chloride	4–36	(102–914)	200	(1,379)	Lightweight, easy to install, excellent resistance to corrosion, good flow characteristics, high tensile strength and impact strength	Difficult to locate underground so tracer tape can be used, requires special care during tapping, susceptible to damage during handling, requires special care in bedding
High-density polyethylene	4–63	(102–6000)	250	(1,724)	Lightweight, very durable, very smooth, liners and wrapping not required, can use ductile-iron fittings	Relatively new product, thermal butt-fusion joints, requires higher laborer skill

Table 5-2 Pipe joints and their applications

Type of Material	Type of Joint	Application
Ductile iron	Push-on or mechanical	General use where flexibility is required
	Flanged	Where valves or fittings are to be attached in vaults or above grade
	Flexible ball	River crossings or in very rugged terrain
	Restrained	To resist thrust forces and in unstable soils
Concrete	Galvanized steel ring, bell-and-spigot types, or their variations with elastomeric gaskets	All locations
Plastic, polyvinyl chloride (PVC)	Bell and spigot type	Most commonly used for typical municipal uses (ASTM F 477)
	Solvent weld	Only for small lines
Plastic, high-density polyethylene (HDPE)	Thermal butt-fusion, flange assemblies	ASTMD 2657
	Mechanical methods recommended by manufacturer	Joining HDPE pipe to valves and ductile-iron fittings
Steel	Mechanical sleeve coupling	All diameters, but especially on pipe too small for a person to enter
	Rubber gasket joints	Low-pressure applications
	Welded joints	High-pressure applications, 24-in. (610-mm) and larger pipes
	Flanged joints	Where valves or fittings are to be attached
	Expansion joints	Allows movement so that expansion or contraction is not cumulative over several lengths

Ductile-Iron Pipe

Ductile-iron pipe (DIP) resembles CIP and has many of the same characteristics. It is produced in the same manner, but it differs in that the graphite is distributed in the metal in spheroidal or nodular form rather than in flake form (Figure 5-6). Adding an inoculant, usually magnesium, to the molten iron allows this distribution to be achieved. DIP is much stronger and tougher than CIP.

Although unlined ductile iron has a certain resistance to corrosion, aggressive waters can cause the pipe to lose carrying capacity through corrosion and tuberculation. The development of a process for lining pipe with a thin coating of cement mortar has virtually eliminated these problems. The cement–mortar lining adheres closely to the pipe wall, as illustrated in Figure 5-7. The lined pipe may be cut or tapped without damage to the lining. Bituminous external coatings and polyethylene wraps are commonly used to reduce external corrosion.

DIP is available in standard pressure classes ranging from 150 to 350 psi (1,034–2,413 kPa) and in diameters of 3 to 64 in. (76–1,625 mm). The standard lengths of DIP are 18 and 20 ft (5.5 and 6.1 m).

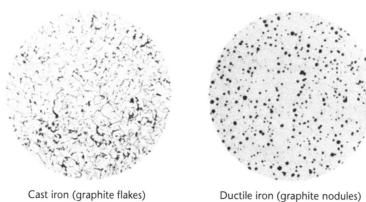
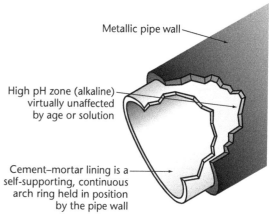

Figure 5-6 Microphotographs of cast iron and ductile-iron
Courtesy of the Ductile Iron Pipe Research Association.

Figure 5-7 Placement of cement–mortar lining on pipe
Courtesy of Mainlining Services, Inc.

Advantages of DIP include the following:

- Good durability and flexural strength
- Smooth interior (C value of 140)
- Fracture resistance
- Good exterior corrosion resistance in most soils

Disadvantages of DIP include the potential for external corrosion in aggressive environments if the pipe is not protected (see AWWA Standard C105/A21.5, *Polyethylene Encasement for Ductile-Iron Pipe Systems* [most recent edition]).

 WATCH THE VIDEO
Ductile-Iron Pipe (www.awwa.org/wsovideoclips)

Ductile-Iron Pipe Joints

The following types of joints are generally used today for connecting DIP and fittings:

- Flanged joints
- Mechanical joints
- Ball-and-socket or submarine joints
- Push-on joints
- Restrained joints
- Grooved and shouldered joints

Details of a mechanical joint are shown in Figure 5-8.

Flanged Joints (AWWA C115) Flanged joints consist of two machined surfaces that are tightly bolted together with a gasket between them. They are primarily used in exposed locations where rigidity, self-restraint, and tightness are required, such as inside treatment plants and pump stations. They are also used where valves or other flanged appurtenances need to be installed. Flanged joints should not normally be used for buried pipe because of their lack of flexibility to compensate for ground movement.

> **flanged joint**
> A pipe joint that consists of two machined surfaces that are tightly bolted together with a gasket between them.

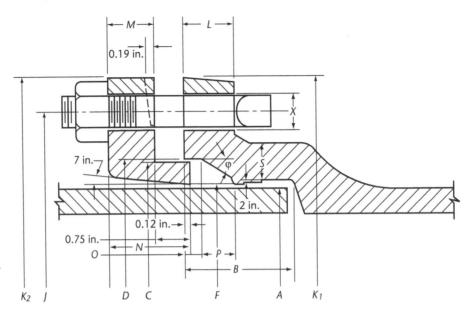

Figure 5-8 Mechanical-joint example for sizes 3–48 in.

Mechanical Joints A mechanical joint is made by bolting a movable follower ring on the spigot to a flange on the bell. The follower ring compresses a rubber gasket to form the seal. Mechanical joints are more expensive than some other joints, but they make a very positive seal and require little technical expertise by installers. The joints also allow some deflection of the pipe at installation. They provide considerable flexibility in the event there is ground settlement after the pipe is installed.

Ball-and-Socket Joints Ball-and-socket joints are special-purpose joints, most commonly used for intakes and river crossings. Their great advantages are that they provide for a large deflection (up to 15 degrees) and are positively connected so they won't come apart. The large amount of allowable deflection also makes this type of joint useful for pipelines laid across mountainous terrain. Both bolted and boltless flexible pipe joints designed on the ball-and-socket principle are available.

Push-on Joints The most recently developed (and now the most popular) joints used in water distribution systems are push-on joints. The joint consists of a bell with a specially designed recess to accept a rubber ring gasket. The spigot end must have a beveled edge so it will slip into the gasket without catching or tearing it.

Workers assemble the joint by lubricating the gasket and spigot end with special nontoxic lubricant and pushing the spigot into the bell. Push-on joints are both less expensive to manufacture and quicker to assemble than mechanical joints because there are no bolts to install. When the pipe joint is assembled, the rubber ring gasket is compressed to produce a watertight seal.

Push-on joints are available in several designs and permit considerable flexibility in pipe alignment. Small diameters may be assembled by hand. Larger sizes usually require mechanical assistance, such as pulling the spigot end of the pipe into the bell with a come-along or pushing it home with a backhoe.

Restrained Joints Various types of restrained joints are available from different manufacturers. They are used where it is necessary to ensure that joints do not separate. Some manufacturers offer special versions of the push-on joint with a restraining feature, and others have separate devices for this purpose.

---

mechanical joint
A pipe joint for ductile-iron pipe that uses bolts, flanges, and a special gasket.

ball-and-socket joint
A special-purpose pipe joint that provides for a large deflection (up to 15 degrees) and is positively connected so it won't come apart.

push-on joint
A pipe joint consisting of a bell with a specially designed recess to accept a rubber ring gasket. The spigot end must have a beveled edge so it will slip into the gasket without catching or tearing it.

Grooved and Shouldered Joints The grooved joint utilizes a bolted, segmental, clamp-type, mechanical coupling with a housing that encloses a U-shaped rubber gasket. The housing locks the pipe ends together and compresses the gasket against the outside of the pipe ends. The ends of the pipe are machine-grooved to accept the housing (Figure 5-9). The shouldered joint is similar except that the pipe ends are shouldered instead of grooved. These joints are covered in AWWA Standard C606, *Grooved and Shouldered Joints* (most recent edition).

An explanation of AWWA standards is provided in Appendix B. Information on DIP is available from the Ductile Iron Pipe Research Association at the address listed in Appendix C.

Fittings for Iron Pipe

A wide variety of ductile-iron and cast-iron fittings are available for use with iron pipe, as illustrated in Figure 5-10.

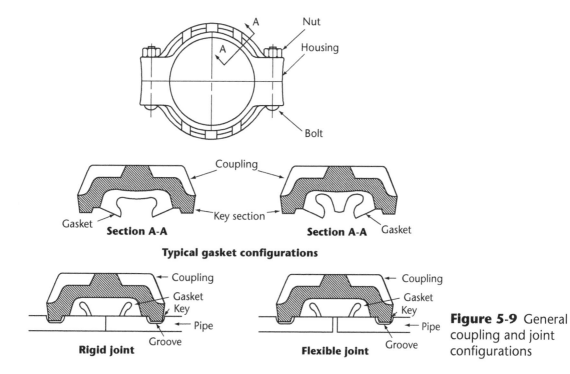

Figure 5-9 General coupling and joint configurations

Figure 5-10 Typical iron pipe fittings

Courtesy of US Pipe and Foundry Company.

Steel Pipe

Steel pipe has been used in US water systems since 1852. It is frequently used where there is particularly high pressure or where very large-diameter pipe is required. Figure 5-11 shows a picture of riveted steel pipe being installed in 1907.

Advantages and Disadvantages

Advantages of steel pipe include the following:

- Relatively light weight
- Competitive price, particularly in larger diameters
- Ease of fabricating special configurations
- Designs that will withstand high internal pressures and loads if necessary
- Relative ease of transporting and installing
- Resistance to shock loads and ability to bend to some degree without buckling

Disadvantages of steel pipe include the following:

- Potential for both internal or external corrosion if not properly protected
- The need to carefully consider external loads in the installation design
- Potential for a partial vacuum (caused by rapidly emptying, relatively thin-walled pipe) to cause pipe distortion or complete collapse, as shown in Figure 5-12 (The design for proper vacuum relief should be checked by methods detailed in AWWA Manual M11, *Steel Pipe—A Guide for Design and Installation*.)

Because steel pipe is generally most competitive in sizes larger than 16 in. (400 mm), it is primarily used for feeder mains in water distribution systems and for long-distance transmission mains. In special cases, steel pipe has been fabricated in diameters of 30 ft (9 m). The length of fabricated pipe varies depending on the diameter and shipping restrictions, but 45-ft (14-m) lengths are usually considered optimal for shipping and handling. The thickness of steel plate used to fabricate the pipe varies depending on both the internal water pressure of the pipeline and the external loads that will be exerted on the pipe.

Figure 5-11 Riveted steel pipe being installed in Philadelphia in 1907

Courtesy of American Iron and Steel Institute.

Figure 5-12 Collapsed steel pipe

Courtesy of the Los Angeles Department of Water and Power.

The interior of steel pipe is usually protected with either cement mortar or epoxy, as specified in AWWA Standard C205, *Cement–Mortar Protective Lining and Coating for Steel Water Pipe—4 In. (100 mm) and Larger—Shop Applied*, and AWWA Standard C210, *Liquid-Epoxy Coating Systems for the Interior and Exterior of Steel Water Pipelines* (most recent editions). The exterior of water main pipe must be protected from corrosion and also against abrasion if the pipe is to be buried. AWWA standards provide for a variety of plastic coatings, bituminous materials, and polyethylene tapes depending on the degree of protection required. For the highest degree of protection, a coated pipeline is commonly also provided with cathodic protection.

WATCH THE VIDEO
Steel Pipe (www.awwa.org/wsovideoclips)

Steel Pipe Joints and Fittings

Pipe lengths are commonly joined by welding for pipe diameters of 600 mm (24 in.) or larger using butt-welded or lap-welded joints. Steel pipe can also be joined by various types of mechanical joints similar to the joints used for other types of pipe, as illustrated in Figures 5-13 and 5-14. In installations where steel pipe is expected to experience a high degree of expansion and contraction, expansion joints can be installed, as illustrated in Figure 5-15.

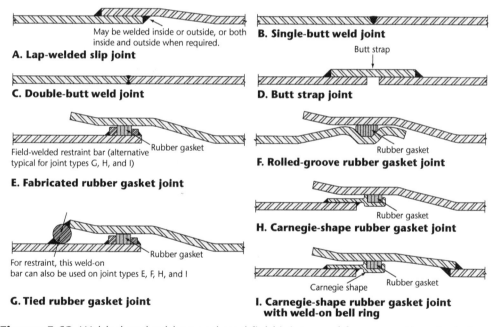

Figure 5-13 Welded and rubber-gasketed field joints used for connecting steel pipe

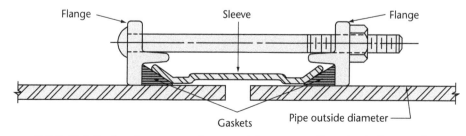

Figure 5-14 Detail of a sleeve coupling used for connecting steel pipe sections

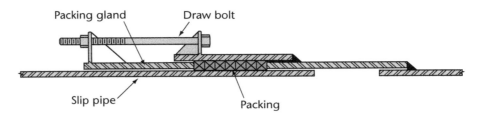

Figure 5-15 Detail of one type of expansion joint for steel pipe

Cast-iron, ductile-iron, fabricated steel, and stainless-steel bends and fittings are usually used for standard changes in either direction or size for smaller-diameter steel pipe. Long-radius bends can also be made in smaller-size pipe by the use of "wrinkle bending," which is done by heating the pipe with a welding flame at several points on the inside of the bend. Pipe fabricators can also furnish any form of special fittings desired.

Steel Plant Piping

Steel piping is frequently used for plant piping because it is inexpensive and relatively light for handling during installation. In addition, special sizes, shapes, and outlets can be fabricated either offsite or onsite by welding (Figure 5-16). External corrosion of piping in a plant can be minimized if the pipe is protected with a properly applied coating and also if the plant is dehumidified.

Information on steel pipe is available from the Steel Tube Institute of North America at the address listed in Appendix C.

Plastic Pipe

Plastic pipe is widely used by water utilities. Plastic pipe was first introduced in the United States around 1940. There are many different types of thermoplastic and thermoset plastic materials that can be manufactured to have various properties. Two properties are of particular importance for water supply piping. First, the plastic piping must have a long-term hydrostatic stress rating to withstand both internal and external pressures. Second, there must not be any harmful substances in the plastic that will leach into the water to cause tastes, odors, or adverse health effects.

Figure 5-16 Complex piping fabricated from steel pipe
Courtesy of L. B. Foster Co.

Plastic Materials

All plastic pipe used to convey potable water must be certified for conformance with the NSF International Standard 61 for potable water use. It must also meet applicable AWWA or other industry standards. More details on standards are included in Appendix B.

Plastic materials used for fabricating water main pipe include PVC, polyethylene (PE), and polybutylene (PB). Composite plastic pipe is also available in the form of plastic pipe reinforced with fiberglass. Plastic materials will not usually react with water on the inside or with soil on the outside, so no extra corrosion protection is necessary.

Permeation

Research and actual occurrences have documented that organic compounds can pass through the walls of plastic pipe, even though the pipe is carrying water under pressure. The process by which the molecules pass through the plastic is called **permeation**. If gasoline, fuel oil, or other organic compounds have saturated the soil around plastic pipe, a disagreeable taste or odor will often be created in the water. More importantly, the contamination can pose a significant health threat to customers using the water. The organic compounds will also soften the plastic pipe, leading to its eventual failure.

Plastic pipe should therefore not be installed in areas where there is known soil contamination by organic compounds or where future contamination is possible, such as close to old petroleum storage tanks.

PVC Pipe

PVC pipe is manufactured by an extrusion process and is available in various pressure ratings. It is by far the most widely used type of plastic pipe material for small-diameter water mains. The AWWA C900 standard for PVC pipe in sizes 4–12 in. (100–300 mm) is based on the same outside diameter as for DIP. In this way, standard DIP bends, tees, and other fittings can be used with PVC pipe. However, AWWA C905 for PVC transmission pipe in sizes 14–48 in. (350–1,200 mm) is available in outside-diameter sizes based on either iron pipe size or cast-iron pipe size. Information on PVC water main pipe is available in AWWA Manual M23, *PVC Pipe—Design and Installation*.

Advantages of PVC pipe include the following:

- Exceptionally smooth interior (C value of at least 150)
- Chemical inertness
- Generally lower cost and greater ease in shipping
- Greater ease of installation because of lighter weight
- Moderate flexibility, so it will adapt to ground settlement
- Ease of handling and cutting

Disadvantages of PVC pipe include the following:

- Susceptibility to damage (gouges deeper than 10 percent of the wall thickness can seriously affect pipe strength)
- Difficulty in locating underground because it is nonconductive
- Inability to be thawed electrically
- Need for rigid adherence to the use of proper tools and procedures when service taps are made

permeation
The process by which organic compounds pass through plastic pipe.

- Susceptibility to damage from the ultraviolet radiation in sunlight
- Susceptibility to permeation
- Need for careful bedding during installation to maintain pipe shape
- Susceptibility to buckling under a vacuum

WATCH THE VIDEO
PVC Pipe (www.awwa.org/wsovideoclips)

PVC Joints and Fittings PVC pipe may be joined by a bell-and-spigot push-on joint in large diameters. In the push-on method, either (1) a rubber gasket set in a groove in the bell end or (2) a double bell coupling provides the necessary seal.

Injection-molded PVC fittings are available in sizes up to 8 in. (200 mm), and fabricated fittings are available in larger sizes. Cast-iron, ductile-iron, and cast-steel fittings are also often used. Figure 5-17 illustrates the use of a restrained fitting on PVC pipe.

PE and PB Pipe

Extruded PE and PB are primarily used for the manufacture of water service pipe in small sizes (see Chapter 8). The use of PB has decreased remarkably because of structural difficulties causing premature pipe failures.

Details of large-diameter PE pipe are found in AWWA Standard C906, *Polyethylene (PE) Pressure Pipe and Fittings, 4 In. (100 mm) Through 63 In. (1,600 mm), for Water Distribution and Transmission* (most recent edition). Information on PVC, PE, PB, and other types of plastic pipe is available from some of the organizations listed in Appendix C.

WATCH THE VIDEO
PE Pipe (www.awwa.org/wsovideoclips)

Fiberglass Pressure Pipe

Fiberglass pipe is available for potable water use in diameters ranging from 1 to 144 in. (25–3,600 mm), in five pressure classes ranging from 50 psi to 250 psi (345–1,724 kPa). There are also several different stiffness classes available that are incorporated into the design, depending on the exterior loading that will be applied to the pipe.

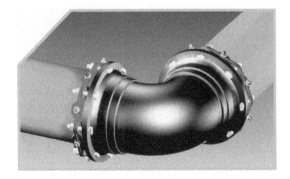
Figure 5-17 Restrained fitting
Courtesy of EBAA Iron, Inc.

Advantages of fiberglass pipe include the following:

- Corrosion resistance
- Light weight
- Low installation cost
- Ease of repair
- Hydraulic smoothness

Disadvantages of fiberglass pipe include the following:

- Susceptibility to mechanical damage
- Low modulus of elasticity
- Lack of a standard jointing system

One method of manufacturing the pipe is called filament winding. A continuous glass-fiber roving saturated with resin is wound around a mandrel in a carefully controlled pattern and under controlled tension. The inside diameter of the pipe is fixed by the mandrel diameter, and the wall thickness is governed by the pressure and stiffness class desired.

The other manufacturing method is centrifugal casting. The resin and fiberglass reinforcement are applied to the inside of a mold that is rotated and heated. The outside diameter of the pipe is determined by the mold and the inside diameter varies depending on the wall thickness. Fittings are made by filament winding, by spraying chopped fiberglass and resin on a mold, or by joining cut pieces of pipe to make mitered fittings.

Several different methods are used by various manufacturers to join pipe sections and fittings. One method is to butt the sections together and wrap the joint with fiberglass material and resin. The other methods include a variety of tapered bell-and-spigot joints that are bonded with adhesives. Pipe, fitting, and adhesive for joining usually are not interchangeable between pipe from different manufacturers. Fiberglass pipe is covered in AWWA Standard C950, *Fiberglass Pressure Pipe* (most recent edition).

Concrete Pipe

The use of concrete pressure pipe has grown rapidly since 1950. The pipe provides a combination of the high tensile strength of steel and the high compressive strength and corrosion resistance of concrete. The pipe is available in diameters ranging from 10 to 252 in. (250–6,400 mm) and in standard lengths from 12 to 40 ft (3.7–12.2 m).

Concrete pipe is available with various types of liners and reinforcement. Four types are commonly used in the United States and Canada:

1. Prestressed concrete cylinder pipe
2. Bar-wrapped concrete cylinder pipe
3. Reinforced concrete cylinder pipe
4. Reinforced concrete noncylinder pipe

Prestressed Concrete Cylinder Pipe

Prestressed concrete cylinder pipe has been made in the United States since 1942. Two general types are manufactured: lined-cylinder pipe (Figure 5-18) and embedded-cylinder pipe (Figure 5-19).

> **prestressed concrete**
> Reinforced concrete placed in compression by highly stressed, closely spaced, helically wound wire. The prestressing permits the concrete to withstand tension forces.

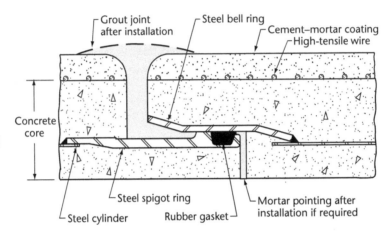

Figure 5-18 Prestressed concrete lined-cylinder pipe
Drawing furnished by American Concrete Pressure Pipe Association.

Figure 5-19 Prestressed concrete embedded-cylinder pipe
Drawing furnished by American Concrete Pressure Pipe Association.

For either type, manufacturing starts with a full length of welded steel cylinder. After joint rings are attached to each end, the pipe is hydrostatically tested to verify watertightness. For embedded-cylinder pipe, the steel cylinder is cast within a concrete core.

After the pipe core has cured on embedded-cylinder pipe, hard-drawn steel wire is helically wrapped around the exterior of the cylinder. The wire spacing is accurately controlled to produce a predetermined compression of the concrete and steel core. The core is then covered with a cement slurry and a dense coating of mortar to provide corrosion protection for the steel cylinder and wire. The pipe is ready for use after the mortar has cured.

For lined-cylinder pipe, the concrete core is cast within the steel cylinder, which is then wrapped with high-strength steel wire and coated with portland-cement mortar.

Lined-cylinder pipe is generally available in diameters from 16 to 60 in. (406–1,524 mm). Embedded-cylinder pipe is commonly available in diameters from 24 to 144 in. (610–3,658 mm). By variations of design, pipe can be fabricated to withstand more than 400 psi (2,758 kPa) of pressure and more than 100 ft (30 m) of cover. Prestressed concrete cylinder pipe is covered in AWWA Standard C301, *Prestressed Concrete Pressure Pipe, Steel-Cylinder Type*, for manufacture, and by AWWA Standard C304, *Design of Prestressed Concrete Cylinder Pipe* (most recent editions).

 WATCH THE VIDEO
Concrete Pressure Pipe (www.awwa.org/wsovideoclips)

Bar-Wrapped Concrete Cylinder Pipe

Bar-wrapped concrete cylinder pipe is manufactured using a hydrostatically tested steel cylinder with welded attached joint rings. The steel cylinder is lined with concrete and wrapped with mild steel bar. The entire assembly is coated with mortar for corrosion protection (Figure 5-20). The core is then protected with a cement-mortar coating.

Bar-wrapped concrete cylinder pipe is manufactured mainly in Canada and the western and southwestern United States. It is normally available in diameters of 10–72 in. (250–1,800 mm). This pipe design is covered by AWWA Manual M9, *Concrete Pressure Pipe*. Manufacture is covered by AWWA Standard C303, *Concrete Pressure Pipe, Bar-Wrapped, Steel-Cylinder Type* (most recent edition).

Reinforced Concrete Cylinder Pipe

Reinforced concrete cylinder pipe is manufactured by casting mild steel reinforcing cages and a steel cylinder with welded joint rings within a thick concrete core (Figure 5-21). Pipe of this design is generally available in diameters of 24–144 in. (610–3,658 mm). The design is covered by AWWA Manual M9, *Concrete Pressure Pipe*. Manufacture is covered by AWWA Standard C300, *Reinforced Concrete Pressure Pipe, Steel-Cylinder Type* (most recent edition).

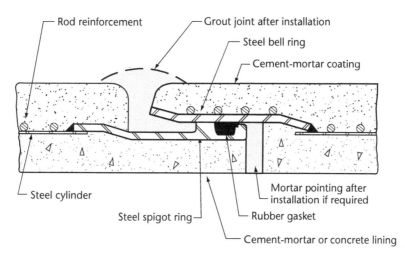

Figure 5-20 B concrete cylinder pipe

Drawing furnished by American Concrete Pressure Pipe Association.

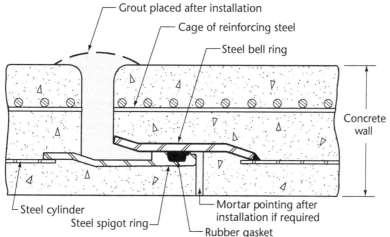

Figure 5-21 Reinforced concrete cylinder pipe

Drawing furnished by American Concrete Pressure Pipe Association.

Reinforced Concrete Noncylinder Pipe

Reinforced concrete noncylinder pipe does not contain an internal watertight membrane (steel cylinder). Therefore, its use is limited to internal pressures of less than 55 psi (379 kPa).

The reinforcement consists of one or more steel cages of welded wire fabric or helically wrapped rods welded to longitudinal rods. The pipe is made by placing concrete around the reinforcement by vertical casting. Noncylinder pipe is covered by AWWA Standard C302, *Reinforced Concrete Pressure Pipe, Noncylinder Type* (most recent edition) for manufacture, and AWWA Manual M9 for design.

The general advantages of concrete pipe include the following:

- Can be manufactured relatively inexpensively in larger sizes
- Can be manufactured to withstand relatively high internal pressure or external load
- Resistant to both internal and external corrosion
- Properly manufactured and installed, has a very long, trouble-free life
- Subject to minimal bedding requirements during installation

Disadvantages of concrete pipe include the following:

- Heavy weight makes shipping long distances expensive.
- Special handling equipment is required due to weight.
- Exact pipe lengths and fittings are required for installation and must be carefully laid out in advance.
- It is more difficult to tap and repair.

Joints and Bends

As shown in Figures 5-18 through 5-21, pipe joints are sealed with a rubber gasket that is compressed as the steel joint rings are pushed together. The metal parts are then protected from corrosion with a covering of mortar.

Installers can achieve horizontal and vertical deflections by ordering pipe lengths with standard beveled ends. Bends are made with fabricated concrete and steel fittings made up to standard angles or to meet special conditions.

Additional information on concrete pipe is available from the American Concrete Pressure Pipe Association at the address listed in Appendix C.

Study Questions

1. The highest degree of protection for the exterior of a coated steel pipe is
   a. cathodic protection.
   b. bituminous materials.
   c. plastic coatings.
   d. polyethylene tapes.

2. The C value is a measure of a pipe wall's
   a. smoothness.
   b. smoothness, which allows even flow.
   c. smoothness, which retards turbulent flow.
   d. roughness, which retards flow due to friction.

3. Which of the following is a type of joint for ductile iron piping?
   a. Expansion joint
   b. Push-on joint
   c. Bell and spigot with rubber O-ring
   d. Rubber gasket joint

4. What is the term for a sudden repeated increase and decrease in pressure that continues until dissipated by friction losses?
   a. Pipe knocking
   b. Water hammer
   c. Pipe shear
   d. Beam breakage

5. Which of the following is *not* a type of joint generally used today for connecting DIP and fittings?
   a. Push-on joint
   b. Flanged joint
   c. Riveted joint

6. What measurement of a pipe's strength refers to its ability to withstand the pressure exerted on it after it has been buried in a trench?

7. What term refers to the process by which organic compounds pass through plastic pipe?

8. What type of cylinder pipe is manufactured by casting mild steel reinforcing cages and a steel cylinder with welded joint rings within a thick concrete core?

Chapter 6
Water Main Installation and Rehabilitation

Water mains must be properly installed to maximize their service life and minimize future maintenance problems. Improper installation of water mains can result in frequent temporary loss of service to customers, compromised fire protection, and unnecessary costs for water system maintenance and repair. Construction operations for installing distribution system mains and accessories can vary depending on the pipe material used. The various operations are discussed in general in this chapter without reference to specific materials. Additional details on procedures for specific piping materials are available in the form of published recommendations from the manufacturer and in the references and sources listed in Appendixes B and C.

Pipe Shipment

Pipe may be shipped from the factory to the jobsite using various types of transportation. Determining the most economical method depends on several factors, including the distance that the pipe must be transported, the size and weight of the pipe, the amount of pipe to be used on the job, and the availability of various types of transportation. Figures 6-1, 6-2, and 6-3 show pipe being transported by truck, railroad, and barge.

Figure 6-1 Pipe being transported by truck
Courtesy of US Pipe and Foundry Company.

Figure 6-2 Large-diameter pipe on railroad flatcars

Photograph furnished by American Concrete Pressure Pipe Association.

Figure 6-3 Pipe being delivered by barge

Photograph furnished by American Concrete Pressure Pipe Association.

From the user's standpoint, direct truck delivery is the most satisfactory because the pipe can usually be unloaded directly at the jobsite. Many utilities require pipe to be delivered with the ends covered (with plastic wrap or some other material). This procedure keeps the pipe protected from foreign material that could be picked up during transportation or storage. It also prevents drying and cracking of the cement-mortar lining of metallic pipes. When pipe is shipped by rail or barge, it is usually necessary to stockpile the pipe at an unloading point and then reload it onto trucks to deliver it to the construction site.

Pipe Handling

There are four general steps in handling pipe:

1. Inspection
2. Unloading
3. Stacking
4. Stringing

Pipe and Fitting Inspection

Pipe normally receives a final inspection before leaving the factory. It can, however, be damaged in transit, when the carrier still has responsibility for it. Therefore, pipe should be inspected as it is unloaded. All pipe, fittings, gaskets, and accessories should be checked against the shipping list for proper size, class, number, and condition before delivery is accepted. Any missing, damaged, or improper materials should be acknowledged in writing by the driver and reported to the shipping company. Damaged material should be saved, and a claim should be made according to the shipping company's instructions.

Unloading

Pipe handling is extremely important. Any type of pipe can be damaged by rough handling. It is generally best to use appropriate mechanical equipment to unload pipe to protect it from damage. Care should be taken to prevent abuse and

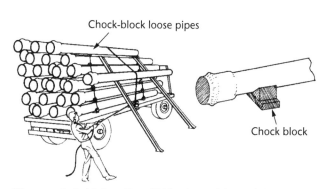

Figure 6-4 Unloading PVC pipe with snubbing ropes

Figure 6-5 Pipe being unloaded with power equipment
Courtesy of US Pipe and Foundry Company.

damage to the pipe, no matter what method of unloading is used. Pipe should not be dropped or allowed to strike other pipe. Lined and coated pipe must be handled particularly carefully to avoid damaging the lining and coating.

Small-diameter pipe may be unloaded with skids and snubbing ropes, as shown in Figure 6-4. Using a derrick or other power equipment makes unloading both safer and quicker (Figure 6-5). Large-diameter pipe must be unloaded with heavy-duty equipment.

Use of a forklift for unloading pipe is not generally recommended unless the forks are padded to prevent pipe damage. Rubber-covered hooks can be used for handling pipe, but fabric slings are even better. Plastic pipe is particularly vulnerable to damage from rough handling. Scratches on the exterior can grow into cracks over time. Such scratches can eventually be the cause of pipe failure. Polyvinyl chloride (PVC) pipe is susceptible to impact damage and is more easily damaged in cold weather.

Fittings should be kept as clean as possible while in storage. It is best to store them in a secure location to protect them from vandalism. They may then be brought to the jobsite as needed. Gaskets should be protected from dirt, oil, excessive heat, and excessive exposure to sunlight. Gaskets are also subject to damage from ozone, so they should not be stored near electric motors or other operating electrical equipment. Always store gaskets in a clean, secure location until they are needed.

 WATCH THE VIDEO
Pipe Handling Safety for Field Crews (www.awwa.org/wsovideoclips)

Stacking

If pipe is to be stockpiled, stacking should be done in accordance with the manufacturer's published instructions. Following are some general recommendations:

- Build stockpiles on a flat base and off the ground to minimize contamination (Figure 6-6).
- Support the bottom layer uniformly on timbers so that the bells do not touch the ground. Use secured blocks at each end to prevent rolling

Figure 6-6 Pipe stockpiles kept off ground surface
Courtesy of the Ductile Iron Pipe Research Association.

- For ductile-iron and concrete pipe, place boards with blocks at each end between each layer of pipe.
- When bell-end pipe is being stacked, project the bell ends over the end of the barrels in alternate layers.
- Group pipe of the same sizes and classes together.
- Group short lengths, fittings, and adapters separately.
- Leave PVC pipe in shipping units until needed. Do not stack pipe more than three bundles high. If it is loose, do not stack more than 3 ft (0.9 m) high. Protect the pipe from sunlight with a canvas tarpaulin or other opaque material. Do not use clear plastic sheets, and allow air to circulate beneath the cover.

Stringing

When pipes are distributed (strung) along the trench in preparation for installation, observe the following procedures:

- Lay or place pipe as near to the trench as possible to avoid excess handling (Figure 6-7).
- If the trench is already open, string pipe on the side opposite the spoil bank (i.e., the side without the pile of dirt). This way, the pipe can easily be moved to the edge of the trench for lowering into position.

Figure 6-7 Pipe strung out along a jobsite
Courtesy of CertainTeed Corporation.

- If the trench is not open, determine which side excavated earth will be thrown to. Then string the pipe on the opposite side, leaving room for the excavating equipment.
- Place pipe where it will be protected from traffic and heavy equipment.
- Place the bells in the direction of installation progress.
- Place and secure each pipe so there is no chance of it rolling into the trench.
- If there is a danger of vandalism or other damage, string out only enough pipe for one day's laying.
- If pipe must be strung in advance in residential areas, there is a particular danger of children playing around the pipe and injuring themselves. The pipe might also roll onto the street. In these cases, take special care to provide secure blocking.
- Place pipe where dirt will not get into the ends. If this is not possible, cover the pipe ends to prevent contamination.

Excavation

This section discusses the many issues that must be taken into consideration before and during the excavation (trench-digging) process.

Preparations for Excavation

Project plans should include details of the alignment, grade, and depth of mains; the locations of valves, hydrants, and fittings; and details of all known obstructions. Usually, the design engineer submits the plans to the state drinking water control agency for approval before work is started. Additional approval may also be needed from the county, city, state highway department, railroad companies, or other authorities. If it will be necessary to enter onto private property to do the construction work, easements or other property access and usage rights must be obtained from the property owner in advance.

Before any digging is done, all other utilities must be notified to locate and mark their underground pipes and cables. Some utilities may take several days to do this if they are particularly busy. They should be provided with ample notice before construction is to begin. Most states now have an underground excavation alert system through which a single phone call will notify all utilities about the water company's need to know pipe and cable locations.

For the purposes of traffic control and community outreach, it is best to inform the public. It is a good idea to mark the work area in advance with signs to inform the public of what will be done. It is also good public relations to write a brief letter to all adjoining property owners to let them know what will be done, why it is being done, when work is expected to begin, and when it will be completed. An apology can also be made for any inconvenience they will suffer as a result of the construction work. The excavation site should be properly marked with barricades, flashing lights, warning or detour signs, and a flagger if necessary to protect the work crew and the public. Work-area protection is discussed in Chapter 7. Regulatory agencies in some locations may require erosion and sedimentation control.

The selection and use of appropriate excavation equipment are important considerations. Machines that are too large will probably cause unnecessary damage. Machines that are too small will move the job too slowly, which not only is uneconomical but may cause problems such as subsequent settlement due to frozen backfill, side-slope instability, and prolonged public inconvenience.

Most trenching for water main installation is done by hydraulic backhoes because they are easy to operate and provide excellent control. The best bucket width for the job should be determined and provided. The width of excavation should be minimized, yet it must be wide enough to maintain side support and allow workers to work properly in the trench. Most buckets can be changed very quickly, and an appropriate size can usually be rented if necessary. Trenching machines may be used for large installations requiring high productivity. Specialized equipment may be required in rocky terrain.

Asphalt roads and driveways should be scored in advance with an air hammer and a flat spade cutter to give a relatively smooth cut. Concrete can also be scored with an air hammer before it is broken out, but this usually leaves a rather jagged edge for patching. A smooth cut can be made with a diamond-edged power saw, which minimizes the amount of pavement replacement. It is usually most economical to employ a firm specializing in concrete cutting to do this work.

Large pieces of concrete and asphalt debris should generally not be used for backfill. They must be hauled away for disposal elsewhere before the excavation starts. The debris from asphalt removed with a rotomill will be small enough that it can be replaced in the upper part of the trench backfill. If traffic cannot be diverted around the area where the hard surface was removed, the hole can be filled with gravel or crushed stone or covered with steel plates to make a temporary driving surface.

Trenching

Several factors control the depth and width of trench excavation. Some of the principal considerations include the following:

- Ground frost conditions
- Groundwater conditions
- Traffic load that will be over the pipe
- Soil type
- Size of pipe to be installed
- Economics
- Surface restoration requirements
- Depth of other utility lines that must be crossed or paralleled by the water main

The most expensive part of pipe installation is the excavation. The trench should therefore be as shallow and as narrow as safety and soil conditions permit.

Trench Depth

In colder climates, the maximum depth of frost penetration that can be expected governs the depth at which mains are buried. During an extremely severe winter, frost may penetrate to twice the average winter frost depth. A section of water main can generally be surrounded by frost for a short time without freezing as long as water is moving in the pipe. The greatest danger of installing mains too shallowly is that frost heaving (expansion of the frozen soil) will later increase the frequency of broken water mains. When frost penetrates to the depth of the mains, it is also likely that water services connected to the main will freeze.

Consideration must also be given to the fact that frost usually penetrates much deeper under pavements and driveways because they are kept free of snow. Surfaces that have continual snow cover often have less frost penetration, though this is not always the case. Most areas occasionally have a winter with little snow cover during extremely cold periods. Frost can penetrate quite deeply within a few days.

If adequate depth is not possible, a main may be insulated. Closed-cell Styrofoam insulation board, 2 in. (50 mm) thick and 2–4 ft (0.6–1.2 m) wide, works well. The insulation should be placed on stable fill, a few inches above the pipe, for as long of a distance as necessary. Bead-board-type sheets should not be used. They are less expensive, but they have low insulating value and will absorb water, thereby further reducing their effectiveness.

In warmer climates where there is no frost, pipe must still have a minimum amount of cover to protect it from damage. This cover will also protect the pipe from highway wheel loads and other impact loads at the surface. The minimum required cover is usually 2.5 ft (0.8 m) for mains and 18 in. (0.5 m) for services.

When deciding on trench depth, the project engineer should also consider whether there is any possibility of future changes in ground surface elevation. For instance, if there is any chance of changes in grade due to future road construction or erosion of soil over the installed main, it is far less expensive to bury a main a little deeper initially than to have to re-lay it in the future.

Trench Width

The width of trench necessary for pipe installation is generally governed by the following considerations:

- Minimum width to allow for proper joint assembly and compaction of soil around the pipe
- Safety considerations
- Economics
- The need to minimize the external loading on the pipe

Trench width below the top of the pipe should generally be no more than 1–2 ft (0.3–0.6 m) greater than the outside diameter of the pipe. This width provides workers with enough room to make up the joints and tamp the backfill under and around the pipe.

The pressure from the tire load of vehicles passing over a trench spreads out through the soil. Below a certain depth (a few feet below the surface) the loading is generally not a danger to buried pipe. Maintaining trench width as narrow as possible in paved areas will reduce the possibility of pipe damage due to vehicle loads.

Wide trenches should be avoided for small-diameter pipe if at all possible, particularly in hard clay soils. Tables 6-1 and 6-2 list recommended trench widths for smaller-diameter pipe. If a wider trench width is necessary or if exceptionally heavy loads may be imposed over the pipeline, a design engineer or pipe manufacturer should be consulted for recommendations.

When pipe must be laid on a curve, the maximum permissible deflection of the joints limits the degree of curvature. Trench widths for curved pipelines must, of course, be wider than usual.

Trenching Operations

As the trenching process proceeds, excavated soil should be piled on one side of the trench, preferably between the trench and the traffic. It must be placed far enough away from the trench so that it will not fall back into the excavation. In addition, it must be placed far enough away that it will not significantly increase the weight on the trench wall. Soil that is too close will increase the likelihood of cave-in. There must also be enough room for workers to walk alongside the trench.

Table 6-1 Trench widths for ductile-iron mains

Normal Pipe Size		Recommended Trench Widths	
in.	(mm)	in.	(m)
3	(80)	27	(0.69)
4	(100)	28	(0.70)
6	(150)	30	(0.76)
8	(200)	32	(0.81)
10	(250)	34	(0.86)
12	(300)	36	(0.92)
14	(350)	38	(0.96)
16	(400)	40	(1.0)
18	(450)	42	(1.07)
20	(500)	44	(1.12)
24	(600)	48	(1.22)
30	(750)	54	(1.37)
36	(900)	60	(1.52)
42	(1,050)	66	(1.68)
48	(1,200)	72	(1.83)
54	(1,350)	78	(1.98)
60	(1,500)	84	(2.13)

Table 6-2 Trench widths for PVC pipe

Pipe Diameter		Trench Width			
		Minimum		Maximum	
in.	(mm)	in.	(m)	in.	(m)
4	(100)	18	(0.46)	29	(0.74)
6	(150)	18	(0.46)	31	(0.79)
8	(150)	21	(0.53)	33	(0.84)
10	(250) and greater	24	(0.31) greater than outside diameter of pipe	36	(0.61) greater than outside diameter of pipe

    The bottom of the trench must be dug to the specified depth while maintaining the specified grade (the elevation of the bottom of the pipe as specified in the plans). Both depth and grade should be double-checked frequently. Installers usually do this by lining a "story pole" with a string line stretched out along the side of the ditch. This line is set an even number of feet above the grade and provides a reference line for installers. The bottom of the trench should form a continuous, even support for the pipe. A proper trench bottom will usually require some handwork and possibly some special fill material to provide proper bedding.

    Excavation of the trench should not extend too far ahead of pipe laying. Keeping the excavation just ahead of the pipe installation will minimize the possibility of cave-ins or trench flooding during rainy weather. An open trench presents a danger not only to the workers but also to traffic, pedestrians, and children. The dangers of an open trench are often greater after working hours than during construction hours. These dangers can be minimized if open sections of trench are kept as short as possible and if written warnings, proper barricades, signals, and flaggers are used.

Local regulations often require that the trench be filled or protected in a specific way overnight. Long sections of open trench may also cause disruptions and inconvenience to local residents, certain municipal services, and emergency vehicles.

Special Excavation Problems

Some possible problems encountered during excavation include rocky ground, poor soil, and groundwater.

Rock Excavation

The term *rock* generally applies to solid rock, ledge rock such as hardpan or shale, and loose boulders more than 8 in. (200 mm) in diameter. In any type of rock formation, the rock must be excavated to a level 6–9 in. (150–230 mm) below the grade line of the pipe bottom. Proper bedding material must then be added for pipe support. Excavated rock should be hauled away and not used for backfill.

Some rock may require blasting. If blasting is necessary, it is best to employ a professional firm that is experienced and specially insured to do this type of work. A detailed record should be kept of dates when blasting is done, the condition of surrounding property prior to blasting, and details of any damages or injuries incurred as a result of the blasting.

Poor Soil

Where poor soil conditions exist (e.g., coal mine debris, cinders, sulfide clays, mine tailings, factory waste, or garbage), the soil should be excavated well below the grade line, hauled away, and properly disposed of. The excavation should then be filled with more suitable material for a foundation under the pipe bedding.

Groundwater

Groundwater will enter the trench when the trench bottom is below the water table. Pipe should not be laid or joined in groundwater. In many locations, the groundwater level fluctuates during the year, so the elevation at which pipe is to be installed may be above the groundwater table only at certain times of the year. If this is the case, the pipe installation should be timed, if possible, for the driest time of the year.

When pipe must be installed below the groundwater table, disposal of the water greatly increases costs. In addition, working in saturated ground can be hazardous because of the danger of trench cave-in. To dewater the ground in advance of excavation, a system of well points can be placed at intervals along the trench, as shown in Figure 6-8. The points have a screen at the bottom and are installed below the trench bottom on one side (and sometimes both sides) of the trench line before excavation. The points are then connected to a pipe manifold and pumping system. It is important to check with local regulatory agencies to ensure proper disposal of the pumped water.

Avoiding Trench-Wall Failure

This section discusses the reasons trench walls fail and ways of minimizing the risks.

Types of Soil

Variations in soil types, water content, and slope stability require modifications in trenching methods. When the digging takes place in firm soil, there is usually

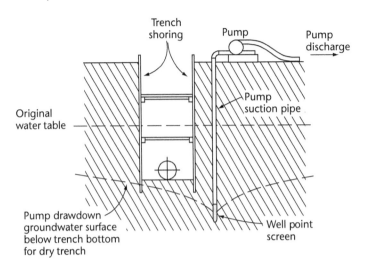

Figure 6-8 Well point pumping to keep trench dry

a tendency to make trench walls too steep and trenches too narrow in order to minimize the amount of excavation. When the digging occurs in wet silts or free-running soils, workers may dig too deep instead of taking steps to restrain the soil in the excavation.

Soils are generally characterized as clays, tills, sands, and silts. Firm clay and tills with low moisture content can usually be excavated easily and safely, but they require careful control during backfill. Dry silt, which is fairly uncommon, behaves in the same manner. Operating in dry sand requires special care during excavation because sand can slip or run easily. Wet silts, which are common, often require special treatment because of their unpredictability and their potential for caving in.

Causes of Trench Failure

The possibility of trench-wall failure and cave-in is undoubtedly the greatest danger for workers on a main-installation job. The primary causes of failure include the following:

- The hydrostatic pore pressure
- External loads caused by construction equipment operating near the edge of the trench
- The weight (load) of excavated soil that is piled too close to the trench
- Trench walls that are too steep for the type of soil being excavated
- Cleavage planes (fissures or cuts) in the soil caused by previous excavations

Failures occur more often in winter and early spring when ground moisture is higher. These times should be considered particularly dangerous for excavation. Failures usually give little warning and occur almost instantaneously.

Danger Signs

Workers should be on the lookout for the following dangers signs:

- Tension cracks in the ground surface parallel to the trench, often found a distance of one-half to three-quarters of the trench depth away from the edge
- Material crumbling off the walls
- Settling or slumping of the ground surrounding the trench
- Sudden changes in soil color, indicating that a previous excavation was made in the area

Water Main Installation and Rehabilitation   133

Trench-wall failures can be prevented if problem areas and conditions are recognized before they occur.

Methods of Preventing Cave-in

Four basic means of cave-in prevention and protection are sloping, shielding, shoring, and sheeting.

Sloping Sloping involves excavating the walls of the trench at an angle. This approach means that the downward forces on the soil are never allowed to exceed the soil's cohesive strength. For any section of an excavation, there will be a certain angle, called the angle of repose, for which the surrounding earth will not slide or cave back into the trench. The angle varies with the type of soil, the amount of moisture the soil contains, and the surrounding conditions, especially vibration from machinery. Figure 6-9 shows the approximate angles of repose for different common soil conditions. The angle ratios shown in the figure indicate the horizontal distance compared with the vertical. For instance, an excavation at a 2:1 ratio indicates that the cut is 2 ft (0.6 m) out from the edge of the ditch for each foot of excavation depth. Figure 6-10 shows an excavation with sloped trench walls.

Shielding Shielding involves the use of a steel box—open at the top, bottom, and ends—that is placed into the ditch so workers can work inside it, as illustrated in Figure 6-11. As the work progresses, the protective box is pushed or towed along the trench to provide a constant shield against any caving of the trench walls. This box is also called a trench shield, portable trench box, sand box, or drag shield. The shield is constructed of steel plates and bracing, welded or bolted together. It is important that the shield extend above ground level or that the trench walls above the top of the shield be properly sloped.

Shielding does not prevent a cave-in. The shield cannot fit tightly enough in the trench to hold up the trench walls; if it did, it would be impossible to move. However, if a cave-in does occur, workers within the shield are protected.

A major disadvantage of a shield is that workers have a tendency to leave its protection in order to check completed work, to help adjust pipe placement, or

> **sloping**
> A method of preventing cave-ins that involves excavating the sides of the trench at an angle (the angle of repose) so that the sides will be stable.
>
> **shielding**
> A method to protect workers against cave-ins through the use of a steel box open at the top, bottom, and ends. Allows the workers to work inside the box while installing water mains.

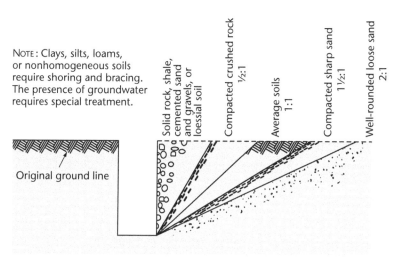

Figure 6-9 Approximate angles of repose for different soil types being excavated

Figure 6-10 Pipe excavation with sloped trench walls

Photograph furnished by American Concrete Pressure Pipe Association.

Figure 6-11 Pipe being installed using a shield

Photograph furnished by American Concrete Pressure Pipe Association.

just to get out of the way of the job in progress. The shield offers protection in only a limited area. As soon as work in an area is completed, the trench should be progressively excavated, the shield moved forward, and the trench backfilled.

Shoring Shoring, if properly installed, will actually prevent the caving in of trench walls. It is basically a framework support system of wood, metal, or both that maintains pressure against both trench walls. It is important that the type and condition of the soil be checked before excavation begins. This allows the correct shoring to be selected. The type and condition of the soil also need to be checked for changes as construction progresses. Soil conditions can vary greatly within a few feet, and conditions may also be changed by weather or vibration.

A shoring assembly has three main parts, as illustrated in Figure 6-12.

1. *Uprights* are the vertically placed boards that are in direct contact with the faces of the trench. The spacing required between them will vary depending on soil stability.
2. *Stringers* are the horizontal members of the shoring system to which the braces are attached. The stringers are also known as whalers, because they are similar to the reinforcements in the hulls of wooden ships.
3. *Trench braces* are the horizontal members of the system that run across the excavation to keep the uprights separated. They are usually either timbers or adjustable trench jacks.

shoring
A framework of wood and/or metal constructed against the walls of a trench to prevent cave-in of the earth walls.

To prevent movement and failure of the shoring, there should not be any space between the shoring and the sides of the excavation. After the shoring has been installed, all open spaces should be filled in and the material compacted. A basic principle of shoring operations is that shoring and bracing should always be installed from the top down, then removed from the bottom up after the operation has been completed. This approach will give maximum protection to

Water Main Installation and Rehabilitation   135

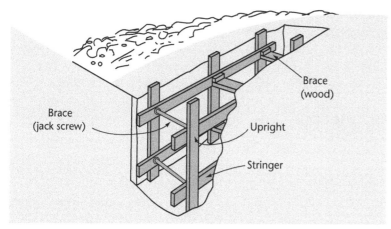

Figure 6-12 Principal parts of a shoring assembly

Figure 6-13 Typical shoring installation

workers constructing or dismantling the shoring. A shoring installation is shown in Figure 6-13.

When a pipe installation within a shored excavation is finished, it may be necessary to leave the shoring in place while the trench is backfilled and compacted. The recommended procedure for removing shoring is to raise the shoring a few feet, then place and compact a layer of backfill, then raise the shoring a few feet more and place more backfill. In some installations, uprights that are above the top of the pipe can be left in place until all or most of the backfill has been placed. Then they can be removed with power equipment.

When no compacting of backfill is needed and the shoring is to be removed, ropes should be used to pull jacks, braces, and other shoring parts out of the trench. This way, workers do not have to enter the trench after the protection has been removed.

Sheeting Sheeting is the process of installing tightly spaced upright planks against each other to form a solid barrier against the faces of the excavation, as illustrated in Figure 6-14. Under normal soil conditions, the sheeting can be installed as the excavation progresses. When soil conditions are very poor or for very deep excavations, steel sheet piling may have to be driven into the ground

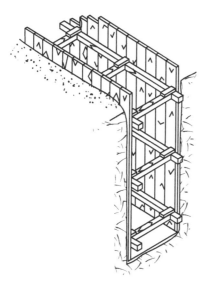

Figure 6-14 Tight sheeting in soft ground

> **sheeting**
> A method to protect workers against cave-ins by installing tightly spaced upright planks against each other to form a solid barrier against the faces of the excavation.

before excavation begins. Sheeting is generally removed as backfilling progresses, much the same as with shoring.

Local regulations and the Occupational Safety and Health Administration regulations require shoring or shielding in certain excavation situations. Regulatory requirements must be satisfied for all pipeline installation procedures.

 WATCH THE VIDEO
Trenching and Shoring (www.awwa.org/wsovideoclips)

Avoiding Other Utilities

To avoid the pipes and cables of other utilities, a water utility must either adjust the line and grade of a new water main or arrange for the other utility to relocate its lines. The grade of a sewer line is usually fixed, so the water main grade must be adjusted. If gas, electric, telephone, or existing water services are in a location where a new water main unavoidably must go, those services can usually be relocated. However, there will be delays in the main installation unless the conflicts can be anticipated and the adjustment made in advance. The other utility may charge a fee for making the changes. A conflict in grade with a pipe that cannot be relocated can also be avoided if several bends are installed in the water main (Figure 6-15).

Working around other utilities requires careful work with machines and by hand. The pipe must not touch or rest on other pipes or be used to support another structure. Accidentally cutting another utility line can be costly and dangerous. If gas from a punctured gas main is ignited, a serious fire can result. A buried electric cable often looks much like a tree root, and a worker in a damp ditch can be electrocuted by breaking the insulation with a shovel. Cutting a fiber-optic communication cable can require repairs costing thousands of dollars. In addition, each time another utility is damaged, it is often necessary to stop the water main construction job until the repair is completed.

Potentially serious problems can result when sanitary sewers and water mains are buried close together. Sewer lines may leak, and a water main could be surrounded by sewage-contaminated soil. If there is also a leak in the water main at the same location, it is possible for sewage to be drawn into the main if a main break or fire creates a vacuum in the main.

In general, potable water pipes and sanitary sewers should not be laid in the same trench. Wherever a water main crosses a sewer line, the water main should be at least 18 in. (0.45 m) above or below the sewer line. For 10 ft (3 m) on either side of the crossing, the sewer line should be made of either (1) ductile iron with mechanical or push-on joints or (2) PVC with mechanical-joint couplings. If water and sewer lines are parallel to each other, the distance between the two should be at least 10 ft (3 m). Refer to local or regulatory agency rules for

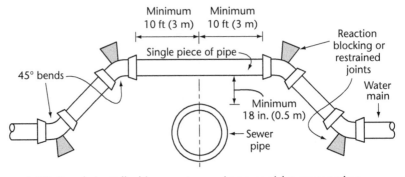

Figure 6-15 Bends installed in a water main to avoid a sewer pipe

minimum separation distances and specific requirements such as pipe materials, joints, and connections.

Bedding

The bottom of the trench excavated for water main installation must be properly leveled and compacted so that the barrel of the pipe will have continuous, firm support along its full length. A leveling board should be used to ensure that there are no voids or high spots and that the grade is correct. Any high spots should be shaved off, and voids should be filled with well-tamped soil. The practice of laying pipe on blocks or earth pads to allow room and position for joining pipe is not recommended.

There are some situations in which special pipe bedding may be specified by the design engineer. This usually occurs where the local soil contains many large rocks or is soft or unstable. If special bedding material is required, it should be a clean, well-graded, granular material up to 1 in. (25 mm) in size. It should contain no lumps or frozen ground and have no more than 12 percent clay or silt that can be sensitive to water. The bedding material should be spread over the trench bottom to the full width of the trench.

When natural trench bedding is unable to structurally support the pipe, such as in very soft soils, it may be necessary to dig an extra 12–24 in. (300–600 mm) deeper. Then coarse granular material in well-mixed sizes up to 3 in. (75 mm) can be used as backfill on top of geotextile fabric. This material should be compacted, and more added if necessary, until the trench bottom has been brought up to the proper grade. This approach is especially important in areas where there is an upward flow of groundwater or in muddy material. The trench bottom must be stabilized before the pipe can be laid. In extreme cases, pilings or timber foundations may be required.

"Haunching" (compacting backfill beneath the pipe curvature) is necessary to increase the load-bearing capacity of the pipe (Figure 6-16). If the pipe is supported only over a narrow width, as with a round pipe on a flat-bottom trench, there could be a very intense load at the bottom of the pipe. Failure is then

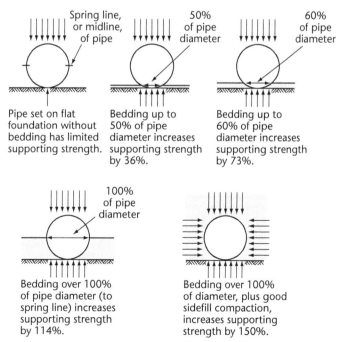

Figure 6-16 Effect of bedding on a pipe's load-bearing strength (for rigid pipe)

Figure 6-17 Good and bad pipe bedding
Courtesy of J-M Manufacturing Co., Inc.

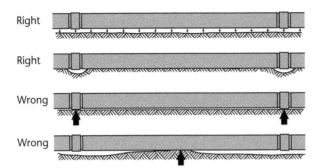

possible. Distributing the load over a wider area will reduce the load intensity beneath the pipe. Proper haunching tamped up to the pipe spring line (middle of the pipe) greatly increases the load-bearing strength of the pipe.

As illustrated in Figure 6-17, it is also important that there is not undue weight placed on the pipe bell. Normal practice is to excavate a "bell hole" at the proper place so that the bell is completely free for assembly of the joint.

Laying Pipe

This section gives guidelines for laying pipe once the length of trench is ready. It discusses inspection and placement, jointing, connecting to existing mains, tunneling, thrust restraints, and air vents.

Inspection and Placement

After the trench bottom has been prepared, the pipe may be set in place. The proper procedure varies somewhat with the type of pipe, but the following general directions for laying pipe apply to all types.

Inspection

Before the pipe is lowered into the trench, it should be inspected for damage. Any unsatisfactory sections should be rejected.

Cleaning

The inside of each pipe length should also be inspected for dirt, oil, grease, animals, and other foreign matter that must be removed before installation. If the pipe has been tapped before being placed in the trench, the holes or corporation stops in the pipe should be covered to keep out dirt during pipe placement.

If mud and surface water have been permitted to stand or flow through strung-out pipe, the inside should be swabbed with a strong hypochlorite solution. This will save time and expenses later when the pipe is disinfected. All gaskets should be kept clean and dry until they are ready for use.

Placement

The pipe should be lowered into the trench by hand or mechanical equipment. It should never be rolled into the trench from the top. Smaller-diameter pipe may be lowered into the trench by two people using ropes, similar to the method illustrated in Figure 6-4 for unloading pipe from a truck. Larger pipe sizes must be handled with power equipment.

When pipe is lowered by machinery, it is usually supported by a sling in the middle of the pipe. The sling must be removed once the pipe is down, which

Figure 6-18 Pipe being lifted with a pipe clamp
Courtesy of the Ductile Iron Pipe Research Association.

usually requires some hand excavation to free the sling. The space created under the pipe should be backfilled and tamped. The use of a lifting clamp greatly facilitates handling pipe (Figure 6-18).

Pipe that is joined by couplings may be laid in either direction. Bell-end pipe is normally laid with the bells facing the direction in which the work progresses; however, this approach is not mandatory. When a main is being laid on a slope steeper than 6 percent, contractors often find it is easier to lay the pipe with the bells facing uphill.

Installers should make a conscious effort to keep the inside of the pipe clean during laying. When laying is not in progress, the open ends of installed pipe should be plugged to prevent animals, dirt, and trench water from entering. A piece of plywood placed over the end of the pipe at the end of the day will not always be enough to keep out animals, dirty water, or children who wish to throw rocks into the pipe. The most satisfactory method of plugging is with a standard pipe plug made for the type of pipe joint being used.

Jointing

Jointing pipe is an important, sensitive part of pipe installation. It must be done correctly to minimize leakage and ensure long and satisfactory service. The method for creating a joint depends on the type of pipe material and the type of joint. In all cases, sand, gravel, dust, tar, and other foreign material should be carefully wiped from the gasket recesses in the bells. Otherwise the joint may leak. The spigots must be smooth and free of rough edges.

Bell holes must be dug by hand where pipes will be joined. These spaces will allow room for the joint to be installed while the remainder of the pipe rests on the bed. Bells or couplings should not be allowed to support the pipe. The size of the bell hole will vary with the type of pipe and joint used. Bell holes that are too large result in undue stress on the pipe.

Published instructions from the pipe manufacturer for making up joints should be followed. General directions for the more common joint types are given in this section.

Push-on Joints

The four general steps in assembling ductile-iron or PVC pipe push-on joints are illustrated in Figure 6-19. The most common cause of joint failure is not having the joint completely clean. For instance, a small leaf caught between the spigot

1. Thoroughly clean the groove and the bell socket of the pipe or fitting; also clean the plain end of the mating pipe. Using a gasket of the proper design for the joint to be assembled, make a small loop in the gasket and insert it in the socket, making sure the gasket faces the correct direction and that it is properly seated. NOTE: In cold weather, it is necessary to warm the gasket to facilitate insertion.

2. Apply lubricant to the gasket and plain end of the pipe in accordance with the pipe manufacturer's recommendations. Lubricant is furnished in sterile containers, and every effort should be made to protect against contamination of the container's contents.

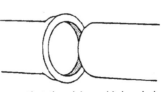
3. Be sure that the plain end is beveled; square or sharp edges may damage or dislodge the gasket and cause a leak. When pipe is cut in the field, bevel the plain end with a heavy file or grinder to remove all sharp edges. Push the plain end into the bell of the pipe. Keep the joint straight while pushing. Make deflection after the joint is assembled.

4. Small pipe can be pushed into the bell socket with a long bar. Large pipe requires additional power, such as a jack, level puller, or backhoe. The supplier may provide a jack or level puller on a rental basis. A timber header should be used between the pipe and jack or backhoe bucket to avoid damage to the pipe.

Figure 6-19 Push-on joint assembly

end and gasket will usually leak enough that the installation will not pass the pressure test; yet, such a leak may be so small that it is difficult to locate. Finding a small leak and correcting it can be quite time-consuming. It is well worth the time to make up joints carefully.

Another potential problem is not pushing the spigot end the full distance into the bell. The spigot ends of all pipe sections are provided with a painted line to indicate when the pipe is fully "home." If a short piece of pipe is cut on the job, a similar mark should be made on the new spigot end with a piece of chalk or felt marker to ensure that the joint is properly made up.

The spigots of all push-on joint pipe are provided with a beveled end to allow for easy slipping into the joint gasket. If special pipe lengths are cut from full lengths in the field, the new spigots must be provided with a similar bevel. Although the bevel can be made with a rasp file or hand grinder, a special tool for beveling will save a lot of time if there are many cut lengths used on a job.

If a push-on joint is to be deflected, it is best to first push it home in line with the previous pipe before deflecting it. If installers try to push the pipe home on an angle, the pipe may resist going in or may not make a good joint. Table 6-3 lists the maximum deflections that can be obtained for full lengths of standard ductile-iron push-joint pipe.

Small-diameter pipe may be pushed home by hand or using a pry bar, as illustrated in Figure 6-20. A block of wood should always be used between the bar and the pipe bell to avoid damaging the bell. Larger joints should be pulled together using a come-along or chain hoists, as illustrated in Figure 6-21.

Table 6-3 Maximum joint deflection,* full-length pipe—push-on type joint pipe

Nominal Pipe Size		Deflection Angle*	Maximum Offset—S† in. (m)		Approx. Radius of Curve—R† Produced by Succession of Joints ft (m)	
in.	(mm)	degrees	L† = 18 ft (5.5 m)	L† = 20 ft (6 m)	L† = 18 ft (5.5 m)	L† = 20 ft (5 m)
3	(76)	5	19 (0.48)	21 (0.53)	205 (62)	230 (70)
4	(102)	5	19 (0.48)	21 (0.53)	205 (62)	230 (70)
6	(152)	5	19 (0.48)	21 (0.53)	205 (62)	230 (70)
8	(203)	5	19 (0.48)	21 (0.53)	205 (62)	230 (70)
10	(254)	5	19 (0.48)	21 (0.53)	205 (62)	230 (70)
12	(305)	5	19 (0.48)	21 (0.53)	205 (62)	230 (70)
14	(356)	3	11 (0.28)	12 (0.30)	340 (104)	380 (116)
16	(406)	3	11 (0.28)	12 (0.30)	340 (104)	380 (116)
18	(457)	3	11 (0.28)	12 (0.30)	340 (104)	380 (116)
20	(508)	3	11 (0.28)	12 (0.30)	340 (104)	380 (116)
24	(610)	3	11 (0.28)	12 (0.30)	340 (104)	380 (116)
30	(762)	3	11 (0.28)	12 (0.30)	340 (104)	380 (116)
36	(914)	3	11 (0.28)	12 (0.30)	340 (104)	380 (116)
42	(1,067)	3	11 (0.28)	12 (0.30)	340 (104)	380 (116)
48	(1,219)	3		12 (0.30)		380 (116)
54	(1,400)	3		12 (0.30)		380 (116)
60	(1,500)	3		12 (0.30)		380 (116)
64	(1,600)	3		12 (0.30)		380 (116)

*For 14-in. (356-mm) and larger push-on joints, maximum deflection angle may be larger than shown here. Consult the manufacturer.

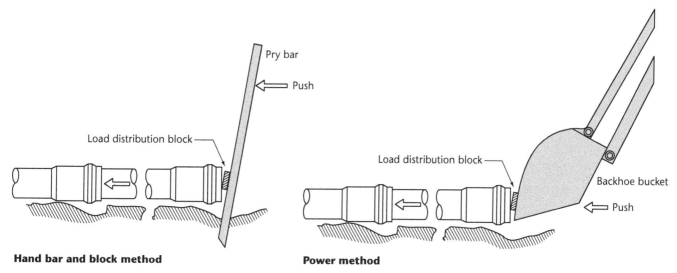

Figure 6-20 Assembling small-diameter push-joint pipe

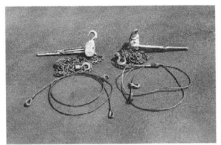

1. For pipe up to 24 in. (610 mm) in diameter, use two 1-ton (900-kg) chain hoists with 25 ft (7.6 m) of chain.

   For larger size pipe, use 2.5-ton (2,300-kg) hoists.

2. Wrap two choker slings around the pipe just behind the bell, with one on each horizontal center line.

3. Double-wrap the chains of the two chain hoists around the pipe barrel about 6 ft (2 m) from the spigot end.

4. Attach the hook of each chain hoist into the eye of each bell choker sling.

5. Operate the chain hoists evenly to pull the spigot into the bell.

Figure 6-21 Using chain hoists to assemble push-on joints

Courtesy of US Pipe and Foundry Company.

Mechanical Joints

The assembly of a mechanical joint is illustrated in Figure 6-22. Although these joints are more expensive and take longer to assemble, they make a very positive seal and allow some deflection. Table 6-4 indicates the proper torque that should be used in tightening mechanical joint bolts of various sizes.

Connecting to Existing Mains

Installers can connect new mains to existing mains either by inserting a new tee or by making a pressure tap.

Tee Connections

To insert a tee fitting in an existing main, the main must be out of service for several hours. The valves on either side of the location should be tested in advance to be sure they will work, and customers with water services connected within the section must be notified. The type of pipe and outside diameter must also be determined in advance; the correct fittings and gasket sizes are sure to be available once the main is shut down. Two methods are commonly used for inserting a tee.

The first method is illustrated in Figure 6-23. After a section of main is cut out, a cutting-in sleeve is slipped all the way to one side over the existing pipe. The tee is then installed on the other side of the cut, and the sleeve is slid back so that the spigot end engages the joint of the tee. The length of existing main that must be cut out is quite specific. Instructions on how to compute this length are included with the inserting sleeve. The normal procedure for cutting out a piece of main is to make a cut or snap at the outside edges of the section to be removed. Then a third cut is made midway between them. When the pipe is hit with a sledge in the center, it will easily break out.

Water Main Installation and Rehabilitation **143**

1. Clean the socket and the plain end. Lubrication and additional cleaning should be provided by brushing both the gasket and the plain end with soapy water or an approved pipe lubricant meeting the requirements of ANSI/AWWA C111/A21.11 just prior to slipping the gasket onto the plain end for joint assembly. Place the gland on the plain end with the lip extension toward the plain end, followed by the gasket with the narrow edge of the gasket toward the plain end.

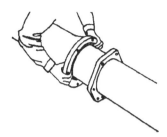

2. Insert the pipe into the socket and press the gasket firmly and evenly into the gasket recess. Keep the joint straight during assembly.

3. Push the gland toward the socket and center it around the pipe with the gland lip against the gasket. Insert bolts and hand-tighten nuts. Make deflection after joint assembly but before tightening bolts.

4. Tighten the bolts to the normal range of bolt torque while at all times maintaining approximately the same distance between the gland and the face of the flange at all points around the socket. This can be accomplished by partially tightening the bottom bolt first, then the top bolt, next the bolts at either side, finally the remaining bolts. Repeat the process until all bolts are within the approximate range of torque. In large sizes (30–48 in. [762–1,219 mm]), five or more repetitions may be required. The use of a torque-indicating wrench will facilitate this procedure.

Figure 6-22
Mechanical-joint assembly

Table 6-4 Mechanical-joint bolt torque

Joint Size		Bolt Size		Range of Torque	
in.	(mm)	in.	(mm)	ft•lb	(N•m)
3	(76)	⅝	(16)	45–60	(61–81)
4–24	(102–610)	¾	(19)	75–90	(102–122)
30–36	(762–914)	1	(25)	100–120	(136–163)
42–48	(1,067–1,219)	1¼	(32)	120–150	(163–203)

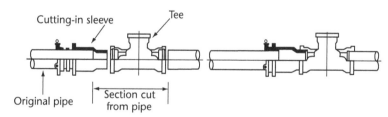

1. Cut a section from the existing pipe equal to the length of the tee plus the length specified by the sleeve manufacturer.

   Place the sleeve gland and gasket on the pipe. Slide the sleeve as far as it will go on the main.

   Install the tee with one bell mounted on the opposition side of the cut-out section.

2. Pull the sleeve spigot end toward the valve, and place it all the way into the bell of the tee.

   Assemble the gaskets and glands, and tighten all bolts.

Figure 6-23 Steps in using a cutting-in sleeve to install a tee

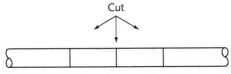
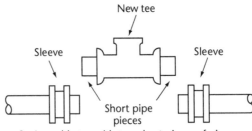

Figure 6-24 Steps in using short pieces of new pipe and rubber-joint sleeves to install a tee

The second method, illustrated in Figure 6-24, uses two short pieces of new pipe (one on either side of the tee) and two rubber-joint sleeves. The tee and new pieces of pipe can be made up in advance. The section of old main is then cut out to match the same laying length. It is essential to verify that proper-size sleeve gaskets are available before the existing main is broken out. For example, if the old main is sand-cast pipe and the short pieces on either side of the tee are ductile-iron pipe, each sleeve will need two different gasket sizes.

Pressure Taps

Today most connections of new mains to existing mains are made with a pressure tap, which offers the following advantages:

- Customers are not inconvenienced by having their water turned off.
- There is a much lower probability of contaminating the water system, particularly if working conditions in the excavation are poor.
- Fire protection in the area remains in service.
- There are likely to be fewer complaints of discolored water.
- There is no water loss.
- Taking into account the time required to close valves and notify customers before inserting a tee, making a pressure tap is much faster.

Once the appropriate excavation has been prepared, a pressure tap is completed in the following steps:

1. The surface of the existing main is cleaned. The tapping sleeve is installed on the main, with the face of the flange vertical.
2. The special tapping valve is attached to the sleeve's projecting flange.

3. The tapping machine is installed on the valve, with the shell cutter attached. The machine is supported on temporary blocks (Figure 6-25).
4. With the tapping valve open, the cutter is advanced to cut a hole in the main.
5. The cutter is retracted to beyond the valve, and the valve is closed.
6. The tapping machine is removed, and the new main is connected (Figure 6-26).

As shown in Figure 6-27, the shell cutter cuts a round "slug" or "coupon" from the existing main and retains it on the cutter. These coupons should be tagged with the date and location of the tap because they provide a good record of the condition of the interior of water mains in the system. Figure 6-28 shows a pressure tap being made on large-diameter concrete pressure pipe.

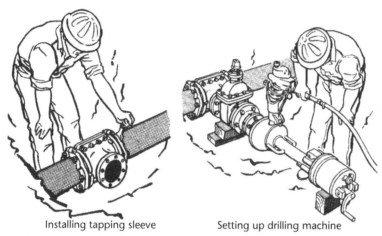

Installing tapping sleeve   Setting up drilling machine

Figure 6-25 Preparing to tap a large connection
Courtesy of Mueller Company, Decatur, Illinois.

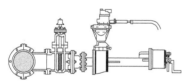 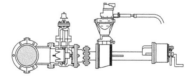

1. The tapping sleeve and valve are first attached to the main. Then the drilling machine, with a shell cutter fastened to its boring bar, is attached to the tapping sleeve and valve by an adapter. The assembly should be pressure-tested before the cut is made.

2. With the tapping valve open, the shell cutter and boring bar advance to cut the main.

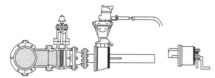 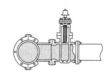

3. The boring bar is retracted and the tapping valve closed to control the water pressure.

4. With the machine removed, the lateral is connected and the tapping valve opened to pressurize the lateral and place it in service.

Figure 6-26 Using a tapping sleeve to install a large lateral main
Courtesy of Mueller Company, Decatur, Illinois.

> **coupon**
> In tapping, the section of the main cut out by the drilling machine.

Figure 6-27 Coupon cut from main
Courtesy of T. D. Williamson, Inc.

Figure 6-28 Tap being made on a large-diameter main
Photograph furnished by American Concrete Pressure Pipe Association.

When workers are making direct pressure taps on PVC mains, they should strictly follow the pipe manufacturer's published recommendations.

Tunneling

When pipelines are installed under railroad tracks, major highways, and other obstructions, the following requirements are usually imposed on the installation:

- If the pipe should break or leak, the water must be carried to one side of the roadway. If water from a broken main were to come up in the middle of a railroad track or major highway, it could cause a serious accident.
- If the water main should require repair or replacement, such action must be possible without the need for excavating in the roadway.
- The pipe must be protected from excessive dead and impact loads, which might otherwise cause the pipe to fail.

The most common method of installing a water main under these conditions is to install a casing pipe and then place the water main inside it. When a casing pipe is used for highway or railroad crossings, the project must meet applicable federal, state, and local regulations, as well as any requirements of the highway department or railroad company.

The casing pipe is usually made of steel or concrete pipe. For casing diameters up to 36 in. (920 mm), boring is done to create the casing hole, with a maximum length of about 175 ft (55 m). Jacking is used for diameters of 30–60 in. (760–1,500 mm) and for lengths of about 200 ft (60 m). Tunneling is done for pipes 48 in. (1,200 mm) and larger for longer lengths. The casing pipe diameter should be 2–8 in. (50–200 mm) larger than the outside diameter of the water main bells.

After the casing is in place, runners or skids are attached at each pipe joint with steel bands. The chocks must be of sufficient size that the bells will clear the casing. The water main is then pushed or pulled through the casing.

Water Main Installation and Rehabilitation

Thrust Restraints

Water under pressure and water in motion can exert tremendous forces inside a pipeline. The resulting thrust pushes against fittings, valves, and hydrants and can cause joints to leak or pull apart. All tees, bends, reducers, caps, plugs, valves, hydrants, and other fittings that stop flow or change the direction of flow should be restrained or blocked.

Thrust Locations

Thrust almost always acts perpendicular to the inside surface it pushes against. As illustrated in Figure 6-29, the thrust acts horizontally outward, tending to push the fitting away from the pipeline. If uncontrolled, this thrust can cause movement in the fitting or pipeline that will result in leakage or complete separation of a joint. Although upward or downward bends are more unusual than horizontal bends in water pipelines, the same type of thrust should be anticipated and controlled to prevent the pipe joints from separating under a surge situation.

Plastic pipe has a particularly smooth surface. It is especially prone to sliding out of a push-on joint if not firmly restrained. In addition, when polyethylene bags are placed around ductile-iron pipe for corrosion protection, the surrounding soil provides much less restraint for the pipe. In other words, if there is water hammer in the line, the pipe may slide within the plastic if it is not well restrained. Installers must be particularly careful to provide good blocking of joints in plastic pipelines or plastic-encased pipelines.

Thrust Control

Thrust can be controlled by restraining the outward movement of a pipeline by four general methods: installation of thrust blocks, installation of thrust anchors,

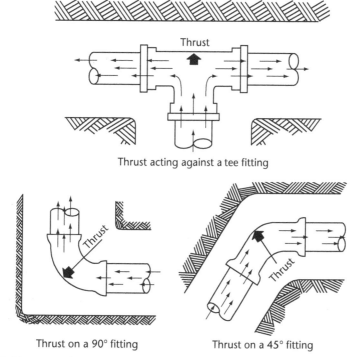

Figure 6-29 Thrust actions

> **thrust**
> (1) A force resulting from water under pressure and in motion. Thrust pushes against fittings, valves, and hydrants; it can cause couplings to leak or to pull apart entirely. (2) In general, any pushing force.

use of restraining fittings or joints, or use of batter piles (which is not discussed below, but refers to piles driven at an angle to prevent lateral movement).

Thrust blocks are masses of concrete that are cast in place between a fitting being restrained and the undisturbed soil at the side or bottom of the pipe trench. Figure 6-30 illustrates thrust blocks installed for various pipeline bend situations. It is important that the block be centered on the thrust force. The block should also partially cradle the fitting to distribute the force, but it should not cover the joint fittings.

It is also extremely important to cast the block against undisturbed soil. Disturbed soils cannot generally be expected to provide adequate support. For this reason, it is important to not over-excavate a trench at bends in the pipeline.

Thrust anchors may be used when a thrust block cannot be used because there is an obstruction or there is no undisturbed soil to block against. They are also used on vertical bends. As illustrated in the typical designs illustrated in Figure 6-31, steel rods connected to relatively massive blocks of concrete restrain the piping bends.

Tie rods are frequently used to restrain mechanical-joint fittings that are located close together. As shown in Figure 6-32, threaded rods are usually used. Nuts on either side of each joint connection take the place of the mechanical joint bolt that they replace. If nonalloyed carbon steel is used for tie rods, the rods should be properly coated or covered for protection against corrosion. Restraining fittings, which make use of clamps and anchor screws, are coming into more general use for restraining joints because they are so easy to use. They are also particularly useful in locations where other existing utilities or structures are so numerous that thrust blocks are precluded.

thrust block
A mass of concrete cast in place between a fitting to be anchored against thrust and the undisturbed soil at the side or bottom of the pipe trench.

thrust anchor
A block of concrete, often a roughly shaped cube, cast in place below a fitting to be anchored against vertical thrust, and tied to the fitting with anchor rods.

tie rod
A device frequently used to restrain mechanical-joint fittings that are located close together when thrust blocks cannot be used.

restraining fitting
A device for restraining joints that are particularly useful in locations where other existing utilities or structures are so numerous that thrust blocks are precluded.

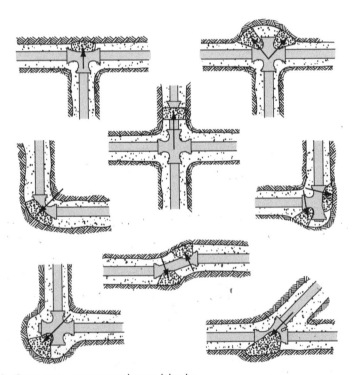

Figure 6-30 Common concrete thrust blocks
Courtesy of J-M Manufacturing Co., Inc.

Water Main Installation and Rehabilitation

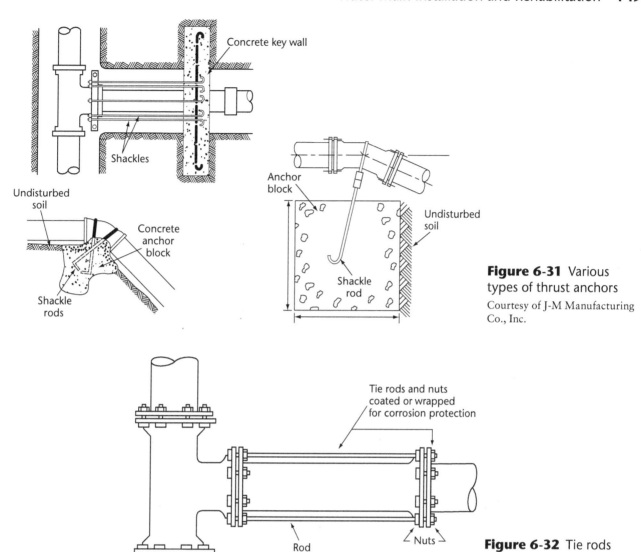

Figure 6-31 Various types of thrust anchors
Courtesy of J-M Manufacturing Co., Inc.

Figure 6-32 Tie rods used to secure piping where blocking cannot be used

Air Vents

Whenever a water main is laid on uneven ground, there will be air trapped at each high point as the main is filled. If the "bubbles" of air are allowed to remain at these locations, they will constrict the flow much the same as if there were a throttled valve in the line. For small-diameter mains, opening one or more fire hydrants will ordinarily produce enough velocity in the water to blow out the air, but with large mains, this is not possible. One method of providing for a release of air is to install fire hydrants at the high points. As illustrated in Figure 6-33, a special tangent tee or connection should be used so that all air will be released from the top of the pipe. The only problem with this method is that there is some entrained air in most water that will gradually refill the air pockets. A regular schedule should be set up for manually operating the hydrants to release accumulated air. Another method of releasing air is to provide automatic air-relief valves at each high point. Figure 6-34 illustrates one design. Although the valves operate automatically, they must periodically be inspected to make sure they are operating properly.

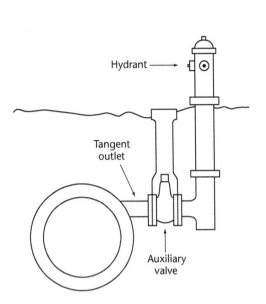

Figure 6-33 Fire hydrant installed on a tangent outlet located on a high point of a large-diameter water main

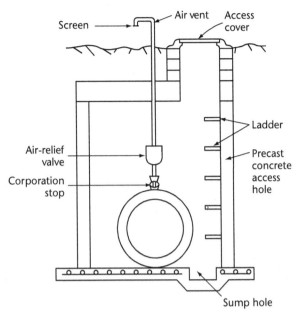

Figure 6-34 Installation of an air-relief valve at a high point on a large-diameter water main

Study Questions

1. When PVC pipe is stacked loose, it should not be stacked more than how high?
    a. 2.0 feet
    b. 3.0 feet
    c. 5.0 feet
    d. 7.5 feet

2. What is the most common cause for pipe joint failure (leaking) in newly laid pipe?
    a. The use of a cracked gasket
    b. Not pushing the spigot end the full distance into the bell
    c. Not having the joint completely clean
    d. An incorrect trench bedding angle

3. The backfill material for a pipe installation should contain enough _____ to allow for thorough compaction.
    a. moisture
    b. sand
    c. gravel
    d. mixed sizes

4. What is the approximate angle of repose for average soils when using the sloping method for the prevention of cave-ins? (Note: horizontal to vertical distance, respectively)
   a. 0.5:1.0
   b. 1.0:1.0
   c. 1.5:1.0
   d. 2.0:1.0

5. Which of the following statements is true regarding unloading gaskets?
   a. Always have gaskets as near the trench as possible for easy installation.
   b. Always store gaskets in a clean, secure location until they are needed.
   c. Store gaskets near electric motors or other operating electrical equipment.
   d. Don't worry about getting gaskets dirty while unloading, as long as they're cleaned before installation.

6. Wherever a water main crosses a sewer line, the water main should be at least _____ above or below the sewer line.
   a. 12 in. (0.3 m)
   b. 18 in. (0.45 m)
   c. 24 in. (0.61 m)
   d. 30 in. (0.76 m)

7. What form of pipe shipment is preferable from the user's standpoint because the pipe can usually be unloaded directly at the jobsite?

8. What is the greatest expense relating to pipe installation?

9. What is the term for the process of installing tightly spaced upright planks against each other to form a solid barrier against the faces of the excavation?

Chapter 7
Backfilling, Main Testing, and Installation Safety

After a section of new water main has been installed, it must be carefully backfilled to protect and support the pipe and fittings. Finally, the main must be tested for leaks, flushed, disinfected, and tested for bacteriological quality before the water can be used by the public. Safety for the workers as well as the public is also a major concern for a construction project; it must be kept in mind from the time the project is planned until the final work is completed.

Backfilling

Backfill material is placed directly around and over the pipe to accomplish the following:

- Provide support for the pipe
- Provide lateral stability between the pipe and the trench walls
- Carry and transfer surface loads to the side walls

Placing Backfill

Backfill is usually placed using mechanical equipment to meet the requirements of the appropriate authority. Granular material or selected soil should be used for the first layer of backfill. If soil is used, it must be either carefully selected excavated soil or imported material. Native soil can be used if it does not contain large rocks, roots, or organic material. Backfill material should contain enough moisture to permit thorough compaction. The backfill should have no large rocks, roots, construction debris, or frozen material in it.

Compacting

The pipe embedment and backfill placed in a trench are generally compacted by a process of tamping, vibrating, or saturating the soil with water, depending on the type of soil or material used.

Tamping

The first layer of backfill should be placed equally on both sides of the pipe, joints, valves, and fittings up to the center line of the pipe. Then it should be compacted. This process is sometimes referred to as haunching. A hand tamper, which is often used for this purpose, is illustrated in Figure 7-1. Pneumatic tampers are also used.

Above the spring line (the pipe midline), backfill practices vary considerably depending on local conditions and regulatory requirements. In general, an initial backfill should be placed around the upper half of the pipe. This will prevent

> **backfill**
> (1) The operation of refilling an excavation, such as a trench, after the pipeline or other structure has been placed into the excavation. (2) The material used to fill the excavation in the process of backfilling.

153

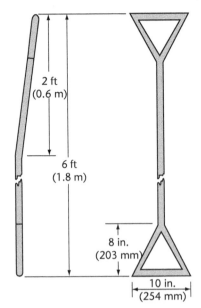

Figure 7-1 Hand tamper

damage to the pipe and prevent the pipe from moving or becoming buoyant until the remaining backfill is placed. The compacted covering layer should be 6–12 in. (150–300 mm) deep for pipe smaller than 8 in. (200 mm) in diameter and 12–24 in. (300–610 mm) for larger pipe.

If the final layer of backfill is to be mechanically compacted, workers should backfill the remainder of the trench by placing the material in layers and compacting it thoroughly. This backfill does not need to be selected, placed, or compacted quite as carefully as the material placed adjacent to the pipe—unless the specifications or regulatory agency says otherwise. However, the fill should be uniformly dense, and unfilled spaces should be avoided.

Saturating With Water

Where water is available at a reasonable cost and the soil drains relatively freely, water settling can be used to compact the backfill. However, simply flooding the backfilled or partially backfilled trench will result in good compaction only when granular material with few fines is being used. If the native soil is relatively dense, only the upper backfill may be compacted by the initial saturation. The backfill may then continue to settle over a period of years.

A process known as jetting involves repeatedly pushing a pressurized water pipe vertically to near the bottom of the loose fill at intervals along the excavation, as shown in Figure 7-2. This approach provides good compaction if done thoroughly. The use of water will generally compact the backfill to within 5 percent of the maximum density.

In areas where the surface doesn't need to be restored immediately to its original condition, it may be adequate to simply mound excess fill over the ditch and let the fill be settled by rain and gravity. It will then be necessary for someone to return periodically to provide additional fill material to compensate for the material that has settled.

Granular Backfill

When trenches are located in areas that must be repaved—such as near roads, sidewalks, or driveways—excavated soil must be very well compacted to minimize

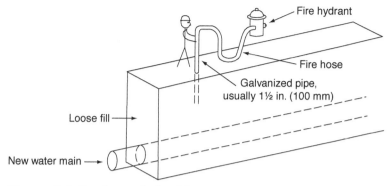

Figure 7-2 Settling a ditch with water. A worker plunges the pipe into the loose fill repeatedly to completely saturate the new fill.

future settlement. A common practice in many areas is to backfill these sites with imported clean granular material or processed material, instead of native soil. If cost is the primary consideration, the expense of purchasing the new backfill and disposing of the excess excavated material must be considered. Some engineers or local authorities specify that granular or processed material must be used under surfaces that are to be repaved. This allows the pavement to be restored to use more quickly. It also avoids any future pavement settlement resulting from inadequately compacted soil.

Shoring Removal

When soil must be compacted in shored trenches, the shoring timbers and braces should be withdrawn in stages to match the layers of earth being placed. When shoring extends below the spring line of the pipe, workers may not be able to withdraw it without disturbing the pipe bedding. In these cases, the only option is to cut the shoring off near the top of the pipe.

Where poor ground conditions exist from the bottom of the trench up to the surface, it may be necessary to leave shoring in place for the entire depth to maintain the installation's stability during backfilling. In this case, the top of the shoring is usually cut off about 2 ft (0.6 m) below the surface. Regardless of which compaction method is used, all voids caused by removing the shoring must be filled. In some locations, any shoring that will be left in place must be made of treated lumber.

Backfilling and Tamping Equipment

The goal in backfilling and tamping is to have a good balance in the sizes and types of equipment on the job. A small bulldozer or a loader works well in combination with a self-propelled vibratory roller for clay-like or silty soils. Vibratory compactors work better for granular materials. Trying to use the same machine both for backfilling and as a crane for lowering pipe is not usually a good idea if the job needs to move at a good pace.

The following are some types of mechanical compacting equipment and their appropriate applications:

- Irregular drum tampers are used for clays, tills, and silts.
- Hand-controlled plate tampers (Figure 7-3) are used for sand in shallow lifts.
- Boom-mounted plate tampers are used for clays or sands in deep, narrow trenches that prevent workers from going into the trench.

Figure 7-3 Mechanical plate tamper
Courtesy of the WACKER Group.

Pressure and Leak Testing

After the trench has been partially backfilled, water mains should be tested for their ability to hold pressure and to determine if there is any leakage. Testing can be done between valves after individual pipe sections are completed, or it can be done after the entire pipeline has been laid. In either case, these tests should be done before the trench is completely closed so that any leakage can be observed and easily repaired.

Years ago, when only lead joints were used, joints were rarely perfect. It was assumed that there would be a little leakage from many of them. Accordingly, testing standards were written to specify just how much leakage should be acceptable. However, experience has shown that if the joints of a pipeline are strictly mechanical, push-on, or other rubber-gasket joints that have been properly made up, there should be very little joint leakage. If there is some leakage, it is most likely from another source.

Testing Procedure

Leakage is defined as the volume of water that must be added to the full pipeline to maintain a specified test pressure within a 5-psi (34-kPa) range. The water added to the pipeline will be equal to the amount of water that leaks out. This test process assumes, of course, that all air has been expelled from the line. The specified test pressure and the allowable leakage for a given length and type of pipe and joint are given in the AWWA standards and manuals.

Although pressure and leakage tests can be performed separately, the usual practice is to run one test combining both. The following procedure is used:

1. Allow at least 5 days for the concrete used for thrust blocks to cure, unless high-early-strength cement was used.
2. Install a pressure pump equipped with a makeup reservoir (a container of additional water to be pumped into the pipeline during the test), a pressure gauge, and a method for measuring the amount of water pumped.
3. Close all appropriate valves.
4. Slowly fill the test section with water while expelling air through valves, hydrants, and taps at all high points. If chlorine tablets are to be used for disinfection, take particular care to fill the pipe slowly; otherwise, the tablets will be dislodged.

5. Start applying partial pressure with the positive-displacement pump. Before bringing the pressure to the full test value, bleed all air out of the mains by venting it through service connections and air-relief valves. Corporation stops may have to be installed at high points in the pipe to release all air properly.
6. Once the lines are full and all air has been bled, leave partial pressure on and allow the pipe to stand for at least 24 hours to stabilize.
7. For pressure testing, subject the test line to the hydrostatic pressure specified in the applicable AWWA standard. A pressure of either 1.5 times the operating pressure or 150 psi (1,030 kPa) for a period of 30 minutes is usually the minimum. Sometimes a pressure chart recorder is connected to the main to keep a record of pressure changes during the test period.
8. Examine the installed pipe and fittings for visible leaks or pipe movement. Any joints, valves, or fittings that show leakage should be checked, adjusted, or repaired as needed. Repeat the test after any adjustments or repairs.
9. After the test pressure has been maintained for at least 2 hours, conduct a leakage test by using the makeup reservoir and measuring the amount of water that has to be pumped into the line in order to maintain the specified test pressure.
10. Compare the amount of leakage to the suggested maximum allowable leakage given in the appropriate AWWA standards and manuals. If a swift loss of pressure occurs, the leak is likely due to a break in the line or a major valve being open. If a slow loss of pressure occurs, the leak may be due to a leaking valve or leaking joint.

Measuring Makeup Water

The method used to measure makeup water volume usually involves one of the following:

- A calibrated makeup reservoir (the preferable method)
- A calibrated positive-displacement pump
- A very accurate water meter (not normally recommended)

The test setup can be installed at the end of the pipe, at a service connection, or at a hydrant, as shown in Figure 7-4. Alternatively, a hose with a pressure gauge, pump, and sensitive water meter could be connected from an existing

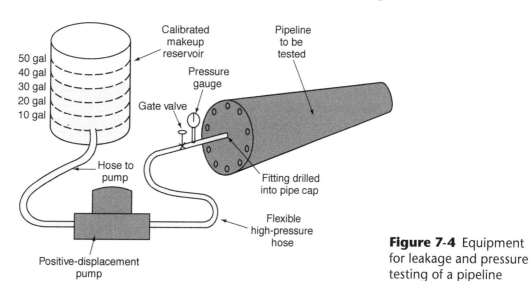

Figure 7-4 Equipment for leakage and pressure testing of a pipeline

operable fire hydrant on the distribution system to the new pipeline being tested. In this case, a double-check valve should also be placed in the connection to prevent any backflow into the distribution system.

Failed Pressure Tests

If a newly installed water main fails the leakage test, it is necessary to find where the excessive leakage is taking place. Steps can be taken to determine if the leak is in a pipe or at a fitting. First, leave the line under normal pressure overnight and repeat the test the next day. If the leakage measured the next day is greater than before, the leak probably is in a pipe joint or a damaged pipe. If the leakage is the same, it is probably in a valve or service connection.

The following is a useful checklist of causes for a failed leak test:

- A hydrant valve may be held open by a piece of rag, wood, or some other foreign object.
- There may be some dirt or foreign material under a coupling or a gasket, or a worker may have forgotten to tighten the bolts on a mechanical joint.
- Fittings, valves, and hydrants in the test section may not have been sufficiently restrained and have moved.
- Corporation stops may not have been tightly closed.
- There may be leakage through a valve at either end of the test section. This cause is particularly likely if there is an old valve at the connection of the new pipeline with the existing water system.
- The packing on a valve may be leaking.
- The test pump may be leaking. This could be the check valve or the gate valve.
- The test may include an overly long section of pipe.
- The saturation time may have been too short. It should be 24 hours.
- There may be a break in the pipe—either a crack or a blowout. A lateral crack may, on occasion, not leak at line pressure but will open up and leak at higher pressure.
- There may be some faulty accessory equipment—possibly a valve, fitting, hydrant, saddle, corporation stop, or relief valve.
- The test gauge may be faulty.
- The test pump suction line may be drawing air.
- Air may be entrained in the test section.

A leak that is large enough to cause a failed test, yet small enough that water does not immediately come to the surface, can be particularly frustrating. In this event, the only solution is first to isolate the section with the leak and make sure the cause is not a leaking valve. Then arrange to continuously subject the line to the highest pressure that can be safely applied, and wait for water to come to the surface. Leak detection equipment can also be used to locate a leak.

Any leaks in the line should be repaired. The line should be retested until the measured leakage is less than or equal to the allowable amount.

Flushing and Disinfection

Any new or repaired water main must be thoroughly flushed, disinfected, and tested for bacteriological quality before it can be put into use for customers. Flushing is primarily necessary to remove any mud and debris that were left in the pipe from the installation.

Flushing

One or more fire hydrants should be used to perform the flushing. A blowoff connection, if one has been installed, can also be used. A velocity of at least 2.5 ft/sec (0.8 m/sec), and preferably 3.5 ft/sec (1.1 m/sec), should be obtained in the pipe. This velocity should be maintained long enough to allow two or three complete changes of water for proper flushing action. Table 7-1 provides information on the rate of water flow necessary to flush various pipe sizes adequately. If the pipeline is large or if there is insufficient water for flushing a new main, the pipe can be cleaned with polypigs.

Disinfection

Chlorine compounds are the most common chemicals used to disinfect large pipelines. Calcium hypochlorite and sodium hypochlorite solutions are generally used for smaller pipelines.

Application Point

The chlorine solution is usually injected through a corporation stop at the point where the new main connects to the existing system. Water utility personnel must ensure that chlorine-dosed water does not flow back into the potable water supply. All high points on the main should be vented to make sure there are no air pockets that would prevent contact between the chlorinated water and portions of the pipe walls.

Chlorine Dosage

The chlorination requirement should normally be in conformance with AWWA Standard C651, *Disinfecting Water Mains* (most recent edition), unless there are other local and state requirements. In general, the rate of application should result in a uniform free chlorine concentration of at least 25 mg/L at the end of the section being treated. Under certain conditions, higher chlorine dosages may be required. It should be kept in mind that too much chlorine for an excessive time can damage mortar linings, brass components, and other fittings.

Table 7-1 Flow rate and number of hydrant outlets required to flush pipelines (40-psi [280-kPa] residual pressure in water main)*

Pipe Diameter		Flow Required to Produce Velocity of Approx. 2.5 ft/sec (0.76 m/sec) in Main		Number of 2½-in. (65-mm) Hydrant Outlets*	Size of Tap, *in. (mm)*		
					1 (25)	1½ (40)	2 (50)
in.	(mm)	gpm	(L/sec)		Number of Taps on Pipe†		
4	(100)	100	(6)	1	1	—	—
6	(150)	200	(13)	1	—	1	—
8	(200)	400	(25)	1	—	2	1
10	(250)	600	(38)	1	—	3	2
12	(300)	900	(57)	2	—	—	2
16	(400)	1,600	(100)	2	—	—	4

*With a 40-psi (280-kPa) pressure in the main and the hydrant flowing to atmosphere, a 2.-in. (65-mm) hydrant outlet will discharge approximately 1,000 gpm (60 L/sec) and a 4.-in. (115-mm) hydrant nozzle will discharge approximately 2,500 gpm (160 L/sec).

†Number of taps on pipe based on no significant length of discharge piping. A 10-ft (3-m) length of galvanized iron (GI) piping will reduce flow by approximately one-third.

Calculating the amount of chlorine and water needed for proper disinfection involves determining the following:

- Capacity of the pipeline
- Desired chlorine dosage
- Concentration of the chlorine solution
- Pumping rate of the chlorine-solution pump
- Rate at which water is admitted to the pipeline

Tables 7-2 and 7-3 provide basic guides to the amount of hypochlorite required for proper disinfection. Additional information on dosage computations is included in Chapter 2 and AWWA Standard C651, *Disinfecting Water Mains* (most recent edition).

Procedures

Three of the most commonly used methods of applying disinfectant are the continuous feed method, the slug method, and the tablet method.

In the **continuous feed method**, water from the distribution system is slowly admitted to the new pipe section while a concentrated chlorine solution is simultaneously forced into the main. Chlorine solution may be injected into the main with a solution-feed chlorinator or by a booster pump.

Table 7-2 Ounces of calcium hypochlorite granules to be placed at beginning of main and at each 500-ft interval

Pipe Diameter*		Calcium Hypochlorite Granules	
in.	(mm)	oz	(g)
4	100	1.7	48
6	150	3.8	108
8	200	6.7	190
10	250	10.5	300
12	300	15.1	430
14 and larger	(350 and larger)		

*Where D is the inside pipe diameter in feet, $D = d/12$.

Table 7-3 Number of 5-g calcium hypochlorite tablets required for chlorine dose of 25 mg/L*

Pipe Diameter		Number of 5-g Calcium Hypochlorite Tablets for Length of Pipe Section ft (m)				
in.	(mm)	13 (4) or less	18 (5.5)	20 (6)	30 (9)	40 (12)
4	(100)	1	1	1	1	1
6	(150)	1	1	1	2	2
8	(200)	1	2	2	3	4
10	(250)	2	3	3	4	5
12	(300)	3	4	4	6	7
16	(400)	4	6	7	10	13

*Based on 3.25 g available chlorine per tablet; any portion of tablet rounded to next higher number.

---

continuous feed method
A method of disinfecting new or repaired mains in which chlorine is continuously added to the water being used to fill the pipe, so that a constant concentration can be maintained.

The concentration used is usually at least 50 mg/L available chlorine. The residual should be checked at regular intervals to ensure that the proper level is maintained. Chlorine application should continue until the entire main is filled and a chlorine residual of at least 25 mg/L can be measured in water being bled from the end of the line. The water should then remain in the pipe for a minimum of 24 hours. During this time, all valves and hydrants along the main must be operated to ensure that they are also properly disinfected.

In the slug method, a long slug of water is fed into the main with a constant dose of chlorine to give it a chlorine concentration of at least 300 mg/L (make sure the pH is within acceptable range). Water is then slowly bled from the end of the line at a rate that will cause the slug to remain in contact with each point on the pipe for at least 3 hours as it passes through the main. As the slug passes tees, crosses, and hydrants, the adjacent valves must be operated to ensure that any dead-end sections of pipe are disinfected. This method is used primarily for large-diameter mains, where continuous feed is impractical.

With the tablet method, calcium hypochlorite tablets are placed in each section of pipe, in hydrants, and in other appurtenances. These tablets are usually glued to the top of the pipe. The main is then slowly filled with water at a velocity of less than 1 ft/sec (0.3 m/sec) to prevent dislodging the tablets and washing them to the end of the main. The final solution should have a residual of at least 25 mg/L. It should remain in contact with the pipe for a minimum of 24 hours. It is also advisable to occasionally bleed a little water from the system to ensure that chlorinated water is distributed within the pipe.

The tablet method cannot be used if the main needs to be flushed before it will be disinfected. If the tablet method is used, workers must take extra care to keep pipe free of dirt, debris, and animals during installation.

Large-diameter pipes or very long pipelines are usually best disinfected with chlorine gas, but special equipment, procedures, and safety precautions are required. Water utility personnel should obtain advice from the manufacturer of the chlorination equipment to be used for the application.

Contact Period

Chlorinated water should normally be left in a pipeline for 24 hours. If unfavorable or unsanitary conditions existed during pipe installation, the period may have to be extended to 48 or 72 hours. If shorter retention periods must be used, the chlorine concentration should be increased, but pH must be monitored. High pH reduces the disinfection effectiveness.

At the end of the contact period, chlorinated water should be flushed to an acceptable location (storm sewer, storage pond, or flood control channel, as allowed by local regulation) until the chlorine residual in water leaving the pipe approaches normal. Highly chlorinated water may kill grass, so it may be necessary to use a fire hose to duct water from a hydrant to the discharge point. Local and state regulatory agencies should be consulted on how to dispose of the highly chlorinated water. The discharge may have to be dechlorinated before being released to the environment.

Bacteriological Testing

After a new pipe has been flushed of chlorinated water and refilled with water from the system, bacteriological tests must be performed. These tests must meet the requirements of the applicable regulatory agencies. Samples must be analyzed by a certified laboratory, and results of the analysis must show that the

slug method
A method of disinfecting new or repaired water mains in which a high dosage of chlorine is added to a portion of the water used to fill the pipe. This slug of water is allowed to pass through the entire length of pipe being disinfected.

tablet method
A method of disinfecting new or repaired water mains in which calcium hypochlorite tablets are placed in a section of pipe. As the water fills the pipe, the tablets dissolve, producing a chlorine concentration in the water.

samples tested negative (no coliform present) before customers are allowed to use the water.

If the results fail to meet minimum standards, the water should be tested again. If the water fails the second test, the entire disinfection procedure must be repeated, including sampling and bacterial testing.

 WATCH THE VIDEO
Water Main Disinfection and Dechlorination (www.awwa.org/wsovideoclips)

Final Inspection

Before a new water main is put into service, as-built plans should be completed and used as the basis for the final inspection. The plans should then be recorded and filed for future use. Detailed records should be completed identifying the types and locations of valve boxes on the new line, as well as the locations of hydrants and other appurtenances.

All valves on the line should be operated and left in the fully open position. The number of turns needed to close and open each valve should be counted and recorded, along with the direction of opening. Each fire hydrant should also be flow-tested to determine its flow capacity and to verify that it is in good operating condition.

Site Restoration

Good restoration requires common sense. It should take whatever form local conditions, owners, and regulatory agencies require. Since restoration is performed at the end of the project, there is often a tendency to give it a low priority. In some cases, this means that some restoration may never get done unless adjacent property owners make strenuous demands. However, operators must remember that the community judges the water utility in great part by what takes place after the work is completed.

Before-and-after photographs or videos of the jobsite are useful in evaluating the quality of restoration. Photographs taken before the job begins show the original condition of trees, shrubs, sidewalks, and fences. They can be particularly useful if someone alleges that the work caused inordinate damage.

The following observations on restoring jobsites to their original condition are generally applicable.

Backfilling Trenches

Properly compacted or settled backfill in trench cuts will reduce settling. If the trench is not well compacted, settlement results, and water utility or street maintenance departments often have to add fill continually to compensate for the settlement, which may extend over several years.

Pavement Repair

Many municipalities require utilities to make temporary repairs of paving cuts by using cold patch. These municipalities assume that additional settlement will occur. They often specify a delay of at least 6 months before a permanent repair is made.

Grass Replacement

A good grade of sod or quick-growing grass placed over a layer of topsoil should be used to replace grass that was disturbed. Sod is usually used to restore lawns and boulevards in urban areas. Depending on the time of year, sod may have to be watered periodically until the roots are established, which may take a month or so. In addition to creating a good image for the water utility, replacing grass is also often necessary to control erosion of the exposed earth.

For larger jobs, grass seed may be sprayed on with a mulch that will help prevent erosion and retain moisture until the grass becomes established. Sprayed seed does not generally require routine watering, but it is slower to grow than sod. Areas that do not grow may require respraying at a later date. Grass matting material is also available to prevent erosion and promote seed growth without much need for maintenance.

Ditches and Culverts

Ditches and culverts in the project area should be checked for proper drainage. Any excessive silt or debris should be removed. A plugged ditch in a heavy storm can flood the surrounding area. Culverts should be checked for any damage caused by the construction, and arrangements should be made for repairs if necessary.

Trees and Shrubs

Damaged roots can cause some species of trees to die. In some cases, the trees might not die for as long as 5 years after the project is completed. A qualified expert should be consulted on the repairs or feeding needed to help save any trees that were damaged. Some locations require that utilities avoid disturbing valuable or historic trees by jacking or boring pipelines rather than performing open-cut trenching adjacent to trees.

Utilities

Ideally, any underground pipes or cables uncovered during construction should be properly supported and protected during the work phase. In some cases, it is impossible to restore these pipes and cables while main installation is in progress. It may be necessary to return later and complete proper restoration.

Curbs, Gutters, and Sidewalks

Concrete and asphalt surfaces removed during excavation should be replaced to match the old surfaces as nearly as possible, including the texture of the finish. In some cases, facilities or private property that are not directly in the line of the work may be damaged during construction. Such damage may be unavoidable, but it may also be due to the carelessness of workers and machine operators. One common example is for sidewalks not directly in the line of work to be damaged by the movement of heavy equipment. If sufficiently damaged, some of these facilities may have to be replaced—even though these replacements weren't originally part of the job plans.

Machinery and Construction Sheds

All construction machinery and structures should promptly be removed from the site when the job is completed. The ground where they were located must then be restored to its original grade and condition.

Watercourses and Slopes

If a stream or wetland must be disturbed during construction, it is important to check with local, state, and federal regulatory officials regarding the need for a permit. Areas that a few years ago may have been considered an unimportant "swamp" may now be considered a "wetland" that must be protected under new environmental protection laws. Some localities now require erosion and sedimentation control during construction.

Slopes should be structurally restored and sodded, seeded, and/or riprapped as necessary. Stream beds should be inspected for excessive mud and debris for a short distance downstream from the actual construction site. Cleanup operations should be conducted promptly to limit the amount of silt and debris sent down the stream.

Roadway Cleanup

Almost any type of construction will track some dirt and dust over the roads being used by construction equipment. Many local governments now have strict laws requiring that roadways be maintained clean and free of dust. If the ground is muddy, arrangements may have to be made to wash or scrape mud from the wheels of trucks before it is tracked onto roadways. During dry weather, dust forms on the roads. It may be necessary to use a power sweeper to clean the roads periodically. On gravel roads, applying calcium will usually help reduce the amount of dust.

Traffic Restoration

When construction is completed, restoration may involve removing signs and filling holes, removing any temporary roads, and putting the area back into its original condition by a process of grading and sodding.

Restoration of Private Property

All construction debris must be removed from the site. Private driveways, walkways, fences, lawns, and appurtenances must be returned to their original condition.

Water Main Installation Safety

During water main installation, the following activities require special precautions:

- Material handling
- Work near trenches
- Traffic control
- Work requiring personal protection equipment
- Chemical handling
- Use of portable power tools
- Vehicle operation

Material-Handling Safety

Back injuries are a common and debilitating type of injury related to material handling. Lifting heavy objects by hand can be done safely and easily if common sense is used and a few basic guidelines are followed. Refer to *Safety Practices for*

Water Utilities (AWWA Manual M3) for detailed descriptions of labor-related safety procedures.

Trench Safety

Trenches can be made safe if proper excavation and shoring rules are followed and if proper equipment is used. Proper trench shoring cannot be reduced to a standard formula. Each job presents unique problems and must be considered under its own conditions. Under any soil conditions, cave-in protection is required for trenches or excavations 5 ft (1.5 m) deep or more. Where soil is unstable, protection may be necessary in much shallower trenches.

Traffic Control Safety

Barricades, traffic cones, warning signs, and flashing lights are used to inform workers and the public of when and where work is going on. These devices should be placed far enough ahead of the work so the public has ample opportunity to determine what must be done to avoid obstructions. If necessary, a flagger should be used to slow traffic or direct it. Everyone involved in the work should wear a bright reflective vest.

Approved traffic safety control devices that should be used for obstructing roadways are described in detail in *Manual on Uniform Traffic Control Devices for Streets and Highways*, prepared by the US Department of Transportation, Federal Highway Administration.

Most states have prepared simplified booklets describing work area protection that should be used during street and utility repairs. The state highway department or department of transportation should be contacted for available information. Water utility operators should be aware that they could be held liable for damages if an accident occurs as a result of utility operations that were not guarded in conformance with state or federal procedures. Figure 7-5 illustrates the directed guarding procedure for utility work in one lane of a low-traffic-volume roadway.

WATCH THE VIDEO
Traffic Zone Safety (www.awwa.org/wsovideoclips)

Personal Protection Equipment

Most utilities must provide a broad range of personal protection equipment for workers. Hard hats, safety goggles, and steel-toed shoes are probably the most widely used safety equipment. The equipment is usually issued to all workers so that they have no excuse for not using proper protective gear. Each person must be responsible for maintaining his or her equipment in good condition and having it available when it is needed.

Gloves are also necessary for protection from rough, sharp, or hot materials. Special long-length gloves provide wrist and forearm protection. Workers should wear rubber gloves when handling oils, solvents, and other chemicals. Gloves should not, however, be used around revolving machinery. The machine might catch onto a glove and injure the worker.

Hard hats, which are necessary whenever an operator is working in a trench or near electrical equipment, have been very successful in reducing serious injuries or deaths due to head injuries. However, metal hard hats should never be used where there is an electrical hazard.

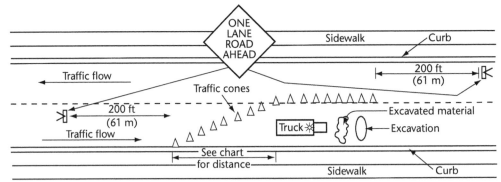

Speed Limit mph (kph)	Lane Width						Number of Cones Required
	10 ft (3 m)		11 ft (3.5 m)		12 ft (3.7 m)		
	Taper Length, ft	(m)	Taper Length, ft	(m)	Taper Length, ft	(m)	
20 (32)	70	(21)	75	(23)	80	(24)	5
25 (40)	105	(32)	115	(35)	125	(38)	6
30 (48)	150	(46)	165	(50)	180	(55)	7
35 (56)	205	(62)	225	(69)	245	(75)	8
40 (64)	270	(82)	295	(90)	320	(98)	9
45 (72)	450	(137)	495	(151)	540	(165)	13
50 (81)	500	(152)	550	(168)	600	(183)	13
55 (89)	550	(168)	605	(184)	660	(201)	13

Figure 7-5 Recommended barricade placement for working in a roadway

Chemical Safety

Chlorine, in the form of chlorine compounds, is the principal chemical used in distribution system operations. Chlorine must be treated, stored, and used carefully to prevent accidents. Even with proper use, accidents can still happen, so the operator should know how to react to them. Depending on the concentration and the length of exposure, chlorine can cause lung irritation, skin irritation, burns, and a burning feeling in the eyes and nose.

Chlorine is most often used in the distribution system in the form of either calcium hypochlorite or sodium hypochlorite. Calcium hypochlorite is available as a dry powder, crystals, or tablets. It is corrosive in small amounts of water and can support combustion. Sodium hypochlorite (bleach) is a strong acid that can cause similar problems. Workers should wear proper protective equipment when handling hypochlorite, including eye goggles, gloves, and a coat.

If chlorine gas is used, supplied-air masks (air packs) should be available, and operators should be trained in their use. Any individuals caught around a chlorine gas leak without a mask should leave the area immediately. They should keep their heads high and mouths closed and avoid coughing and deep breathing.

Portable Power Tool Safety

Electric power tools should always be grounded. A ground-fault interrupter circuit should be used whenever the tool is used outside or near water. The cord should be in excellent condition and should never be used to lift the tool or as a line. The cord or tool should never be left where it could trip someone.

Air tools can be dangerous if the hoses and connections are not correctly maintained. Workers should not point air tools at anyone. They should not clean off any part of their bodies or clothing with compressed air.

Vehicle Safety

Records indicate that the greatest number of accidents in the water utility industry involve vehicles. Workers should be made particularly aware of the potential of accidents while they are operating large trucks and handling heavy construction equipment.

Study Questions

1. If an excavation on a road requires that one of the lanes be closed and the speed limit is 25 mph (40 kph), how many cones are required to divert the traffic?
   a. 6
   b. 9
   c. 13
   d. 15

2. Which of the following is *not* generally a process used for compacting the pipe embedment and backfill placed in a trench?
   a. Tamping
   b. Vibrating
   c. Saturating with water
   d. Sifting

3. Leakage is defined as the volume of water that must be added to the full pipeline to maintain a specified test pressure within a _____ range.
   a. 10-psi (69-kPa)
   b. 5-psi (34-kPa)
   c. 3-psi (21-kPa)
   d. 12-psi (83-kPa)

4. Which of the following is *not* one of the commonly used methods of applying disinfectant?
   a. Stream method
   b. Slug method
   c. Tablet method
   d. Continuous feed method

5. Under any soil conditions, cave-in protection is required for trenches or excavations _____ deep or more.
   a. 15 ft (4.6 m)
   b. 10 ft (3.0 m)
   c. 5 ft (1.5 m)
   d. 2 ft (0.6 m)

6. When disinfecting a new pipeline with calcium hypochlorite tablets or granules, what should the chlorine target be?

7. How deep should the compacted covering layer be for pipe less than 8 in. (200 mm) in diameter? For pipe larger than 8 in. (200 mm) in diameter?

8. How much leakage is allowed for a new pipeline test?

9. How must a pipeline be prepared before a sample is taken for bacteriological testing?

10. What term refers to the process of repeatedly pushing a pressurized water pipe vertically to near the bottom of the loose fill at intervals along the excavation?

Chapter 8
Water Services

Water service lines are the pipes and tubing that lead from a connection on the water main to a connection with the customer's plumbing. Service pipe size varies depending on the pressure at the main, the distance from the main to the building, the quantity of water required, and the residual pressure required by the customer. In most cases, each customer is served through an individual service line. However, there are some exceptions, such as multifamily housing units, which often have only a single service line.

Meter Locations

Good water utility practice is to have a meter on every water service, including schools, churches, parks, public buildings, and the drinking fountain in the town square. The meter's purpose is not simply to keep track of what the customer must pay. A meter also reduces waste by identifying leaking plumbing and provides the best possible accounting of the water used on the system. The location of the meter on a service is, for the most part, governed by both climate and local custom.

Exposed Meters

In areas where there is no danger of freezing, meters can be placed almost anywhere. Some water utilities allow meters to be located in a garage or on the side of a building, as long as the meters are reasonably protected from damage and vandalism.

Meter Boxes

Many water systems install meters in a box, which is usually located in the street right-of-way. Depending on local custom and the size or depth of the box required, meter boxes may also be called meter pits or meter vaults. The following are some of the primary advantages of installing meters outside in boxes:

- In areas where ground frost is nominal, a meter box can be relatively small, inexpensive, and easy to install.
- If the boxes can always be readily located, meter boxes make meter reading easier because meter readers do not have to enter a building. That said, if all meters are located inside buildings and are equipped with remote reading devices, the reading time is generally about the same.
- In areas where most buildings are built without basements, it is often difficult to find a satisfactory inside meter location. Placing all meters outside eliminates the need to find suitable locations.

meter box
A pit-like enclosure that protects water meters installed outside of buildings and allows access for reading the meter.

- Placing meters outside eliminates the need to enter buildings to replace the meters, which is difficult in some inside locations.
- It is common to install meters in a box if there is a long distance between the main and buildings. An example is a rural water system that serves individual farmhouses that may be some distance from the main.

The following are some complications associated with outdoor meter boxes:

- In areas where there is deep frost, meter pits are much more expensive to construct. However, many water systems still require meter pits if there is no appropriate location for a meter in a building.
- Deeper pits often fill with water and must be pumped before the meter can be read. In addition, pit covers are subject to damage if driven over by a heavy vehicle.
- It is sometimes difficult to locate pits that are covered with snow. It may be necessary to remove snow before the meter can be read.

A typical water service with a shallow meter box is illustrated in Figure 8-1. Where meter pits are used in cold climates, the meter can still be raised to near the top of the pit to facilitate reading; meters will not freeze in these locations. If the pit bottom is below the frost line and the cover is tight, the pit will gather enough warmth from the ground to avoid freezing (Figure 8-2). Under extremely cold conditions, pits are sometimes constructed with a double cover to provide added insulation.

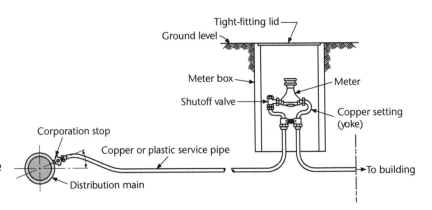

Figure 8-1 Small service connection with shallow meter box

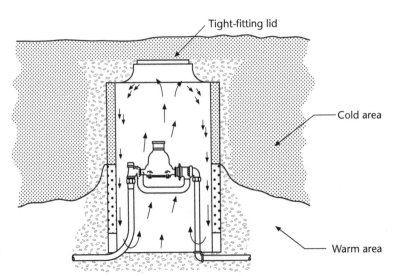

Figure 8-2 Cold-climate meter installation

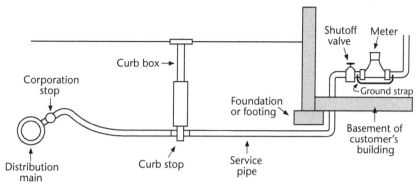

Figure 8-3 Small service connection with the meter located in a basement

Meters Located in Buildings

In areas where most buildings have basements, it is common to locate meters in the basement, as illustrated in Figure 8-3. The service pipe is usually brought up through the floor adjacent to the foundation at a point nearest to the main. The meter should be installed at this point. It is not considered good practice to locate the meter at a distance from the entry point because the property owner might install an illegal connection ahead of the meter.

Water utilities that install meters in basements may be faced with the problem of deciding where to install meters if the buildings only have a crawl space or are constructed on a concrete slab. For these buildings, some utilities require meters to be installed in meter boxes. Other utilities will allow installation in a crawl space or utility closet as long as (1) the meter can easily be accessed by a meter reader or (2) the meter is equipped with a remote reading device.

Service Line Sizes, Materials, and Equipment

This section discusses the sizes and materials of service lines, as well as various additional pieces of equipment.

Service Line Size

The proper service line size required to serve residences and other buildings depends primarily on the following conditions:

- The maximum water demand the customer will require
- The main pressure under peak demand conditions
- Any substantial elevation difference between the main and the highest portion of the building to be served
- The distance between the main and the connection to building plumbing

Single-family residences are most commonly served with a ¾-in. (19-mm) service line. Larger residences, buildings with flushometer toilets, and buildings located far from the main connection should have 1-in. (25-mm) or larger service lines. Apartment buildings, hospitals, schools, and industries all have different use patterns. The principal factor is how much water is likely to be required at one time. Additional details on proper sizing of service lines are included in AWWA Manual M22, *Sizing Water Service Lines and Meters*.

Types of Service Line Pipe and Tubing

The materials that have been used for service line pipe and tubing include lead, galvanized iron, copper, and plastic. All materials used for service lines should comply with the applicable standards (AWWA Standard C800, ANSI/NSF Standard 61).

Lead Pipe

Water services installed on older water systems were generally made of lead. Lead pipe was used as far back as the time of the Roman Empire. It was favored because it is relatively long-lasting and flexible. Joining and attaching fittings to lead pipe involved a molten lead process known as lead wiping, which required considerable expertise by professional plumbers.

Use of lead services gradually decreased as newer materials were found to be less expensive in terms of both material costs and installation. Most older water systems still have many old lead services in use, but these services are gradually being replaced. Most water utilities now have a policy that a leaking lead service cannot be repaired. Instead, it must be replaced with a new service made of another material. The principal reason for such a policy is that evidence points to health risks due to lead leached from plumbing materials. It is therefore recommended to eliminate lead from the system whenever reasonably possible.

Galvanized Iron Pipe

Many older systems allowed the use of galvanized iron pipe in place of lead because of the considerably lower cost. The pipe was usually connected to the corporation stop with a lead "gooseneck" a few feet long to provide flexibility at the connection.

In some soil, galvanized pipe works fairly well. However, it can fail within a year or so under corrosive soil conditions. In addition to general corrosion from contact with soil, galvanic corrosion often causes disintegration of the pipe at the connection with brass fittings unless dielectric fittings are used.

Newer materials have been found to last much longer. They have essentially replaced galvanized pipe for new installations. Some older systems, though, may have many galvanized iron services still in use.

Copper Tubing

Copper tubing became a popular replacement for lead and galvanized iron in service line installations because it is flexible, easy to install, corrosion resistant in most soils, and able to withstand high pressure. It is not sufficiently soluble in most water to be a health hazard, but corrosive water may dissolve enough copper to cause green stains on plumbing fixtures. Allowable copper concentration in drinking water at the tap is regulated by the Lead and Copper Rule. In some cases, corrosion inhibitors must be used to control the copper content. Brass and bronze valves and fittings can be directly connected to the pipe without causing appreciable galvanic corrosion. Copper water service tubing is usually connected by either flare or compression fittings. Although interior copper plumbing is usually connected with solder joints, this method is rarely used for buried service lines.

Plastic Tubing

Of the many types of plastic tubing available, three are generally used for water services: polyvinyl chloride (PVC), polyethylene (PE), and polybutylene (PB). PB is not generally used for new water services since structural problems have occurred in older services. Plastic pipe has a very smooth interior surface, so it has lower friction loss than other types of pipe. It is also very lightweight, making

---

galvanic corrosion
A form of localized corrosion caused by the connection of two different metals in an electrolyte, such as water.

installation easy. The types of tubing generally used for small water services are quite flexible, which allows considerable deflection as the pipe is laid. The pipe is also relatively resistant to both internal and external corrosion.

Plastic tubing must not be used at any location where gasoline, fuel oil, or industrial solvents have contaminated the soil or could do so in the future. Not only will these substances soften the plastic and cause it to eventually fail, but low concentrations will pass through the plastic walls by a process known as permeation. The chemical contamination of the water can cause taste and odor and it may also cause adverse health effects in persons drinking the water. Plastic, because it is nonmetallic, has other disadvantages. It cannot be located with an electronic pipe finder and it cannot be thawed by electric current.

All plastic tubing to be used for potable water must have the NSF International seal of approval printed along the exterior. This ensures that the pipe is made of a material that will not leach products that will lead to tastes, odors, or the formation of harmful chemicals. Plastic pipe is also covered in several AWWA standards in the C900 series. Additional information on specifications and approval for plumbing materials is provided in Appendix B.

Adapters and Connectors

Flare fittings are available for connecting lengths of copper tubing and adapting to iron pipe thread. The same fittings can be used for plastic pipe of the same dimensions.

Compression fittings can be used with either copper or plastic tubing or iron or lead pipe. These fittings seal by means of a compressed beveled gasket. Compression fittings also have a gripping mechanism that positively secures the pipe or tubing. When they are used for PE or PB plastic tubing, it is also recommended that an insert stiffener be placed inside the tube to strengthen the joint. Where stray currents are a problem, dielectric couplings should be used to prevent corrosion.

Corporation Stops

The valve used to connect a small-diameter service line to a water main is called a corporation stop. The plug-type corporation stop has been used for connecting water services to mains for more than 100 years. The corporation stop is also referred to as the corporation cock, corporation tap, corp stop, corporation, or simply corp or stop.

Corporation stops are available in many different sizes and styles, including several combinations of threads and connections for the inlet and outlet ends. The inlet thread on the standard AWWA corporation stop is the one most commonly used. It is commonly known as the Mueller thread. The other thread in general use is the iron-pipe thread. Both threads have approximately the same number of threads per inch. The Mueller thread has a larger diameter and a steeper taper, which gives it greater strength (see AWWA Standard C800, most recent edition).

Corporation-stop outlet ends are available for connection to a variety of service line pipe and tubing materials, including flared copper service connections, iron-pipe threads, increasing iron-pipe threads, lead flange connections, and compression couplings for various materials. Some corporation-stop outlets also include an internal driving thread used for attaching an installation machine adapter. Three plug-type corporation stops are illustrated in Figure 8-4.

A relatively new development is a corporation stop that has a ball valve instead of a plug valve (Figure 8-5). This type of valve operates more easily. It is generally rated for higher pressures than a plug-type corporation valve.

> **corporation stop**
> A valve for joining a service line to a street water main. Cannot be operated from the surface.

Figure 8-4 Plug-type corporation stops with different pipe connections

Courtesy of the Ford Meter Box Company, Inc.

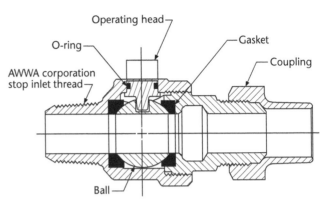

Figure 8-5 Principal parts of a ball-style corporation stop

Courtesy of A.Y. McDonald Mfg. Co., Dubuque, Iowa.

Curb Stops and Boxes

When meters are located outside in a meter box, a shutoff valve for use in temporarily shutting off the service is installed in the box on the inlet side of the meter. When meters are located inside, a shutoff valve called a curb valve or curb stop is installed in the service line. The valve is usually located either between the street curb and the sidewalk or on the property line.

A curb box is a pipe that extends from the curb stop to the ground surface. It allows the valve to be operated with a special key if an operator needs to temporarily discontinue water service to a building. Two styles of curb stops and boxes are in general use in the United States. Arch-pattern base boxes are intended to straddle a curb stop that has no threads on the top (Figure 8-6). The other type is the Minneapolis-style box, which is threaded at the bottom to screw onto threads provided on Minneapolis-style curb stops (Figure 8-7). Water systems generally use one style or the other as a standard.

Each style has some advantages and disadvantages. The arch-pattern base can get sand or soil worked up from the bottom. It can also shift so that it is difficult or impossible to engage the valve handle with a shutoff key. However, if the box is pulled from the ground by construction equipment, the service line won't normally be damaged. With Minneapolis-style curb stops and boxes, the valve key

curb stop
A shutoff valve attached to a water service line from a water main to a customer's premises, usually placed near the customer's property line. It may be operated by a valve key to start or stop flow to the water supply line.

curb box
A cylinder placed around the curb stop and extending to the ground surface to allow access to the valve.

Figure 8-6 Arch-pattern curb box and curb stop

Courtesy of the Ford Meter Box Company, Inc.

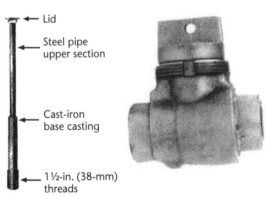

Figure 8-7 Minneapolis-style curb box and curb stop

Courtesy of A.Y. McDonald Mfg. Co., Dubuque, Iowa.

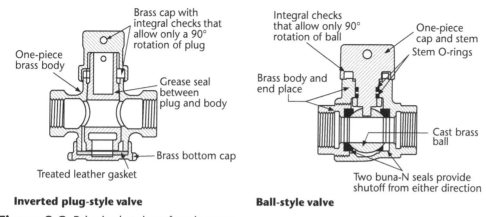

Figure 8-8 Principal styles of curb stops
Courtesy of A.Y. McDonald Mfg. Co., Dubuque, Iowa.

will almost always engage easily. However, if the box is pulled from the ground, it will probably bring the curb stop and service line up with it.

Curb stops are available with a variety of connections for different types of service pipe and tubing. They may have either plug-style or ball-style valves (Figure 8-8).

Water Service Taps

Connections for water services can be made either when the water main pipe is empty (dry taps) or when the pipe is filled with water under pressure (wet taps). Dry taps are usually made only when a new main is installed. Wet tapping is preferred when a service connection is being added to an existing main. It allows the connection to be made without turning off the water or interrupting service to existing customers. There is also less chance of contamination because the pressure in the main tends to expel any foreign matter.

When small taps are made on cast-iron or ductile-iron pipe, the corporation stop is usually screwed directly into a threaded hole in the pipe wall. This process is called direct tapping. Drilling-and-tapping machines make it possible to perform this operation without shutting down the main.

Direct tapping of 6-in. (150-mm) and larger asbestos–cement (A–C) and PVC pipe with ¾-in. (19-mm) and 1-in. (25.4-mm) corporation stops is possible, but it must be done very carefully. The pipe manufacturer's instructions should be followed precisely. It is a good idea to make a practice tap on a piece of scrap pipe before attempting one on a pressurized main. Utility personnel should use a service saddle if they have any doubts about making the installation. AWWA withdrew all of its A–C standards.

Direct Insertion

Tapping a water main and inserting a corporation stop directly into the pipe wall requires a tapping machine. This machine actually performs three operations: drilling, tapping (threading), and inserting the corporation stop. The machine can also be used to remove a corporation stop. AWWA Standard C223, *Fabricated Steel and Stainless Steel Tapping Sleeves*, describes the use of fabricated steel and stainless steel tapping sleeves to provide outlets on pipe. The sleeves are intended for pipe sizes 4 in. (100 mm) through 48 in. (1,200 mm) with branch outlets through 36 in. (900 mm). This standard includes requirements for materials,

dry tap
A connection made to a main that is empty. Compare with wet tap.

wet tap
A connection made to a main that is full or under pressure. Compare with dry tap.

tapping
The process of connecting laterals and service lines to mains and/or other laterals.

dimensions, tolerances, finishes, and testing. It is not intended to apply to tapping sleeves welded to pipe. Fabricated tapping sleeves must be manufactured from steel or stainless steel and are intended for use in systems conveying water. For outlets and main sizes greater than those specified, consult the manufacturer.

Small tapping machines (Figure 8-9) are used for direct installation of ½-in. through 2-in. (13-mm through 50-mm) corporation stops into a main. The hole is drilled into the pipe and the threads are cut by a combination drill and tap. The process of making a tap generally involves the following steps, as shown in Figure 8-10:

1. The water main pipe is excavated and cleaned in the area to be tapped. The drilling-and-tapping machine is clamped in place.
2. The first operation with the machine is to bore a hole into the pipe wall. The water pressure is contained within the sealed body of the machine. After the drill has fully penetrated the pipe, it is advanced further into the pipe to engage the tap. The tap then cuts threads for the corporation stop.
3. The boring bar is retracted. Operators contain the pressurized water from the drilled hole by closing the flapper valve.
4. The combined drill-and-tap tool is removed from the end of the boring bar and replaced with a corporation stop. The corporation stop must be in a closed position before it is inserted.
5. The bar is reinserted into the machine, and the corporation stop is screwed into the threaded hole.
6. The machine can then be removed. The corporation stop is ready to have a service line attached to it.
7. After the service line installation is completed, the corporation-stop valve is opened. The excavation is ready to be backfilled.

Figure 8-9 Drilling-and-tapping machine
Courtesy of Mueller Company, Decatur, Illinois.

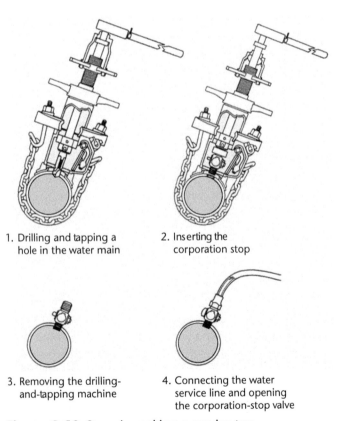

Figure 8-10 Steps in making a service tap
Courtesy of Mueller Company, Decatur, Illinois.

Service Saddles

For taps larger than 1 in. (25 mm) on A–C or PVC pipe or for smaller taps where conditions make it advisable to ensure a perfect tap, a service saddle, also called a service clamp, should be used instead of a direct tap (Figure 8-11). Using a clamp eliminates the chance of the pipe splitting because of the tap, and it avoids the problem of the corporation-stop threads not holding properly in the pipe material. Many water utilities have found it safest to use a clamp regularly for all taps on A–C pipe.

Small Drilling Machines

Small drilling machines may be used to connect service lines 2 in. (50 mm) and smaller, using a service clamp. A manual drilling machine is shown in Figure 8-12. As illustrated in Figure 8-13, the steps in completing a tap are as follows:

1. The pipe is cleaned all the way around and the service clamp is installed.
2. The opened corporation stop is screwed into the service clamp. The drilling machine is then attached to the corporation stop.
3. The drill bit of the drilling machine is extended through the open corporation stop and penetrates completely through the wall of the main.
4. The drill bit is then backed out until it is clear of the corporation-stop valve. The valve is then closed.
5. The drilling machine is removed, and the corporation stop is ready for use.

Self-Contained Taps

Special service connectors for PVC pipe are available that incorporate a built-in tapping tool. Figure 8-14 illustrates a connector of this type.

Tap Location

It is generally agreed that the best location for a tap on a main is at an angle of about 45 degrees down from the top of the pipe. A tap located directly on the top of a main is more likely to draw air into the service, and a tap near the bottom could draw in sediment.

Figure 8-11 Some styles of service saddles
Courtesy of the Ford Meter Box Company, Inc.

Figure 8-12 Manual drilling machine
Courtesy of Mueller Company, Decatur, Illinois.

service saddle
A device attached around a main to hold the corporation stop. Used with mains that have thinner walls to prevent leakage.

1. With the service clamp attached to the main, the corporation stop is threaded into the clamp. The machine is then mounted on the corporation stop, and the stop is opened.

2. The drill penetrates the main without water escaping.

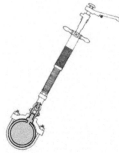

3. The drill bit is retracted, and the corporation stop is closed. The stop now controls the water.

4. The machine is removed, the service line connected, and the corporation stop reopened to activate water service.

Figure 8-13 Using a service clamp to install a corporation stop

Courtesy of Mueller Company, Decatur, Illinois.

Figure 8-14 Service connector made for use with PVC pipe

Courtesy of Dresser Piping Specialties.

The tap should be on the same side of the main as the building to be served. The service pipe should then be supported in a wide S-curve down and into the ditch. Plenty of slack should be provided to allow for earth settlement and pipe expansion and contraction.

 WATCH THE VIDEOS
Tapping Concrete Pressure Pipe, Tapping Ductile-Iron Pipe, Tapping PE Pipe, Tapping PVC Pipe, Tapping Steel Pipe
(www.awwa.org/wsovideoclips)

Leaks and Breaks

Water utilities with old lead and galvanized iron services find that small leaks are relatively common. Some lead pipe will crystallize as it gets old and will crack if flexed. Leaks are most commonly located at the connections to curb stops. These leaks are probably caused by valve movement due to frost heaving or weight applied to the box at the surface. Lead is also prone to leaking after it has been disturbed by adjacent excavation, such as during the installation of a sewer or gas main. Galvanized pipe is more likely to develop a leak after being disturbed. The best policy is not to repair lead or iron services but to replace the entire service, or at least the entire section with the leak.

Copper and plastic pipes rarely have small leaks. The most common problems they encounter are being pulled from a fitting by ground settlement or being cut by adjacent excavation. Copper can sometimes be pulled all the way to the surface by a backhoe without breaking. However, the pipe will probably be kinked in the process, so a new piece will have to be spliced in as a replacement. Plastic pipe is more easily broken or pulled from a connection.

Thawing

In addition to interrupting water service to customers, freezing can also rupture lines and damage meters. The best way to prevent freezing is to ensure that services are buried below the maximum frost line for the area.

Electrical Thawing

Electrical thawing of a service line is possible if the pipe is metallic. A current is run through the pipe, causing heat to be generated that will melt the ice. The current is usually supplied by a portable source of direct current, such as a welding unit.

Electrical thawing can be dangerous and can cause damage to the service line, the customer's plumbing, and electrical appliances. One major concern is that there will be poor conductivity between sections of the service line. Another is that the service line might be in direct contact with other metal pipes or conductors, such as gas pipelines. When these conditions exist, the path of the applied electrical current may be diverted from the service line. The current may then enter adjacent buildings.

Another danger involved with electrical thawing is that the current may damage O-rings, gaskets, and soldered joints on the service. This damage can result in joint failure and leakage. The dangers of stray current and of damage to the service can both be reduced if only low-voltage generators are used and if the voltage and amperage are closely monitored. Only experienced operators should attempt electrical thawing.

Before thawing begins, the property owner and tenant, if any, must be informed of the risks involved and be required to sign a waiver form. This waiver absolves the utility of liability in case of accident or damage. Thawing is a service performed for the customer on the customer's property. The utility must have written confirmation from its insurer stating that adequate liability insurance is in effect to cover possible consequences of the work.

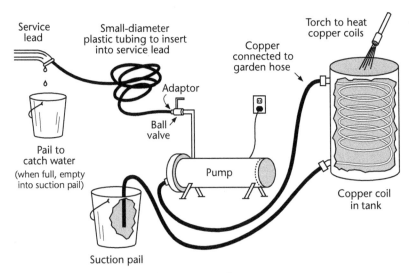

Figure 8-15 System for thawing plastic water line
Adapted from drawing by Randal W. Loeslie, Manager, G.F.-Traill Water District, Thompson, N.D.

Other Methods

Hot-water thawing (Figure 8-15) is becoming more common because it is less dangerous than electrical thawing. In addition, it can be used for plastic pipe. The technique typically involves pumping hot water through a small flexible tube that is fed into the frozen service line.

If a service line or meter is frozen only in the meter box, an effective method of thawing is to circulate warm air from a heat gun or hair dryer. Workers should be careful when using a propane torch for thawing because of the possibility of explosive gases that could accumulate in the vault. The intense, localized heat of a torch can also cause the pipe to expand too rapidly, or it can create steam and rupture a meter or pipe.

After a frozen service has been thawed, it is usually necessary for the customer to leave a faucet running continuously to keep water flowing through the service until the frost in the ground recedes from around the pipe.

Service Line Responsibility

Policies regarding who is responsible for maintaining water service lines vary for different water systems. Most systems have one of the following policies:

- The water utility assumes responsibility for the portion of the service in the street right-of-way (i.e., from the main to the lot line). The portion on private property is the responsibility of the property owner.
- The water utility maintains the service from the main to the curb stop or to the meter if it is in a vault. The remainder is the responsibility of the property owner.
- The entire service is the responsibility of the property owner.

Policies also vary on who does the work of installing the initial service. However, most systems allow the customer's contractor to install the entire service, as long as criteria provided by the water system are followed.

Most water utilities that own tapping equipment prefer to make the taps themselves. This ensures that the work is properly done and that the water system is not likely to be contaminated. The tap is not usually made until the service is installed and inspected. This practice provides a means of ensuring that the service is installed to the satisfaction of the water utility.

The cost of making the tap is then usually billed to the new customer. This bill is often combined with a connection fee, which is also intended to defray other water system costs related to initiating a new service.

Service Line Records

Water service information is often maintained on file cards or computer files, indexed by street address. Important information that should be obtained and recorded before the pipe is covered in the trench includes the following:

- Measurements locating the water main tap
- The type and size of pipe
- Tap size
- Burial depth
- Measurements to the curb stop or meter pit
- Location of the pipe at various points
- Location of the pipe entry into the building
- Date of installation
- Address of the building served

It is particularly important to keep good records of plastic pipe installations because there is no way of electronically locating them later.

Study Questions

1. A corporation stop is used for a
   a. service line.
   b. pump discharge line.
   c. tank inlet.
   d. tank outlet.

2. Compression fittings used with copper or plastic tubing seal by means of
   a. a beveled sleeve.
   b. a compression ring.
   c. a compressed beveled gasket.
   d. compressed O-rings located at either end of the fitting's beveled neck.

3. Water meter pits are usually used in
   a. areas where flooding will most likely occur.
   b. areas where flooding is very rare.
   c. cold climates.
   d. hot climates.

4. Which of the following is *not* an advantage of installing meters outside in boxes?
   a. In areas where ground frost is nominal, a meter box can be relatively small, inexpensive, and easy to install.
   b. If the boxes can always be readily located, meter boxes make meter reading easier because meter readers do not have to enter a building.
   c. In areas where there is deep frost, meter pits are much more expensive to construct.
   d. Placing meters outside eliminates the need to enter buildings to replace the meters, which is difficult in some inside locations.

5. Other utilities will allow installation of a meter in a crawl space or utility closet as long as (1) the meter can easily be accessed by a meter reader or (2)
   a. the meter is equipped with a remote reading device.
   b. the homeowner can easily access and report the meter readings.
   c. the meter can self-report to the utility.
   d. water fees in the area are fixed and readings are unnecessary.

6. What may cause galvanized iron pipe to corrode excessively?

7. What is a curb stop used for?

8. What is the typical service line size for a single-family residence?

9. What is the term for the valve used to connect a small-diameter service line to a water main?

10. What is generally considered the best location for a tap on a main?

Chapter 9
Valves

Numerous types of valves are required in every water system. They are necessary to operate the distribution system as well as to control treatment processes, pumps, and other equipment. The correct size and type of valves must be selected for each use, and most valves require periodic checking to ensure they are operating properly.

Uses of Water Utility Valves

Water utility valves are designed to perform several different functions. The principal uses are as follows:

- Start and stop flow
- Isolate piping
- Regulate pressure and throttle flow
- Prevent backflow
- Relieve pressure

Valves to Start and Stop Flow

Most valves installed in a water system are intended to start and stop flow. They are designed to be either fully open or fully closed under normal conditions. These valves are generally not intended for throttling flow and may be damaged if used in a partially open position for an extended period of time.

Distribution System Isolation Valves

Isolation valves are installed at frequent intervals in distribution piping so that small sections of water main may be shut off for maintenance or repair. The closer the valves are spaced, the fewer the number of customers who will be inconvenienced by having their water turned off while repairs are being made. When a system is laid out in a grid pattern, normal practice is to install at least two valves at each intersection, as illustrated in Figure 9-1. Distribution system valves should normally be maintained in a fully open position.

Hydrant Auxiliary Valves

An isolation valve is usually installed on the section of pipe leading to each fire hydrant. This location allows pressure to be turned off during hydrant maintenance without disrupting service to customers. Fire hydrant design and installation are detailed in Chapter 10.

> **valve**
> A mechanical device installed in a pipeline to control the amount and direction of water flow.
>
> **isolation valve**
> A valve installed in a pipeline to shut off flow in a portion of the pipe, for the purpose of inspection or repair. Such valves are usually installed in the mainlines.

183

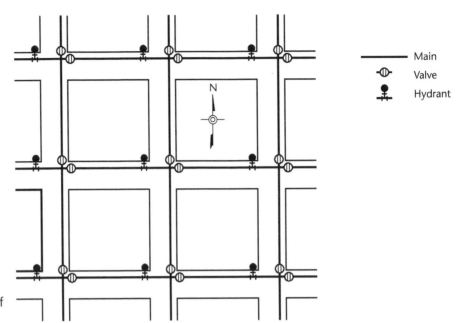

Figure 9-1 Valves installed at intersection of mains in a grid pattern

Pump Control Valves

Each water system pump must normally be furnished with at least two control valves. A discharge valve is essential to stop water from flowing backward through a pump that is not in operation. A valve is also usually installed on the suction side of each pump for use during pump repair.

Water Service Valves

The valves used to connect water services to water mains are usually referred to as corporation stops. They are ordinarily buried, so they are used only when service is being initiated or discontinued.

In service lines, the valves used for temporarily shutting off water to a building are commonly called curb stops. Additional valves are also installed on each side of a water meter for use when the meter is removed for repair. Water service equipment is discussed further in Chapter 8.

Valves for Regulating Pressure and Throttling Flow

Throttling the flow of water requires special valve designs that are durable over a long period of time. The two principal types of throttling valves used in a water system are pressure-reducing valves and altitude valves.

Pressure-Reducing Valves

As illustrated in Figure 9-2, it is sometimes necessary to create two or more separate pressure zones in a distribution system in order to furnish all customers with water at an adequate (but not excessive) pressure. The most common way of accomplishing this is to take water from a higher-pressure zone and reduce the pressure for the lower zone by using pressure-reducing valves. These valves operate automatically to throttle flow and maintain a lower pressure in the lower distribution system zone.

Altitude Valves

Ground-level reservoirs are usually filled through an altitude valve. This type of valve is designed to let the reservoir fill at a regulated rate and to stop flow

---

pressure-reducing valve
A valve with a horizontal disk for automatically reducing water pressures in a main to a preset value.

altitude valve
A valve that automatically shuts off water flow when the water level in an elevated tank reaches a preset elevation, then opens again when the pressure on the system side is less than that on the tank side.

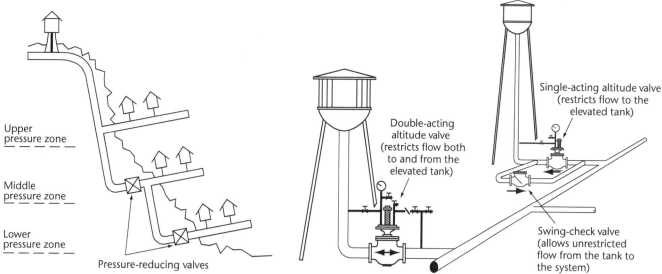

Figure 9-2 Pressure-reducing valve installed on a system that has three pressure zones

Figure 9-3 Altitude valves
Courtesy of GA Industries, Inc.

completely when the tank is full. Altitude valves are also used to control flow to an elevated tank when the tank is not high enough to accept full system pressure. The valve will automatically shut off flow to the tank to keep the tank from overflowing. Single-acting altitude valves allow water to flow in only one direction. Double-acting valves allow flow in both directions. Figure 9-3 illustrates two types of typical water tank installations.

Valves for Preventing Backflow

There are numerous places in a water distribution system where unsafe water may be drawn into the potable water mains if a temporary vacuum should occur in the system. In addition, contaminated water from a higher-pressure source can be forced through a water system connection that is not properly controlled. Several types of valves are available specifically for installation wherever backflow might occur. This equipment is detailed in Chapter 19.

Valves for Relieving Pressure

Special valves are necessary to protect water systems and plumbing from excessive pressure. They are designed to remain tightly shut under normal pressure but will open when excessive pressure occurs.

Pressure-Relief Valves

A rapid increase in pressure in a water distribution system is usually called water hammer. The pressure wave that moves rapidly down a pipe can damage valves, burst pipes, or blow pipe joints apart. A common cause of water hammer is opening or closing a valve too fast. Pressure-relief valves can be installed to release some of the energy created by water hammer.

In household plumbing systems, small pressure-relief valves must be installed on all hot water heaters and boilers on customer services. Their purpose is to allow steam and excess pressure to blow off in the event of equipment overheating. If the excess pressure is not allowed to vent, either the tank or boiler will burst or hot water will be forced back through the water service into the water distribution system. A typical pressure-relief valve is shown in Figure 9-4.

pressure-relief valve
A valve that opens automatically when the water pressure reaches a preset limit to relieve the stress on a pipeline.

Figure 9-4 Pressure-relief valve
Courtesy of Watts Regulator Co.

Figure 9-5 Air-relief valve on a well pump discharge
Courtesy of Henry Pratt Company.

Air-Relief Valves

One of the more common uses of air-relief valves is to automatically vent air that accumulates at high points in transmission pipelines. Air pockets in water mains can substantially reduce the effective area of flow through the pipe. This reduction will result in increased pumping costs and restricted flow to parts of the system.

Air-relief valves must also be installed on the discharge of most well pumps to vent air that has accumulated in the well column while the well is not in use. A typical air-relief installation is shown in Figure 9-5.

Classification of Water Utility Valves

Most valves used in water systems fall into one of the following general classifications (Figure 9-6):

- Gate valves
- Globe valves
- Needle valves
- Pressure-relief valves
- Air-and-vacuum relief valves
- Diaphragm valves
- Pinch valves
- Rotary valves
- Butterfly valves
- Check valves

The designs of various types of valves are covered in several AWWA standards.

air-relief valve
An air valve placed at a high point in a pipeline to release air automatically, thereby preventing air binding and pressure buildup.

 WATCH THE VIDEO
Valves (www.awwa.org/wsovideoclips)

Valves 187

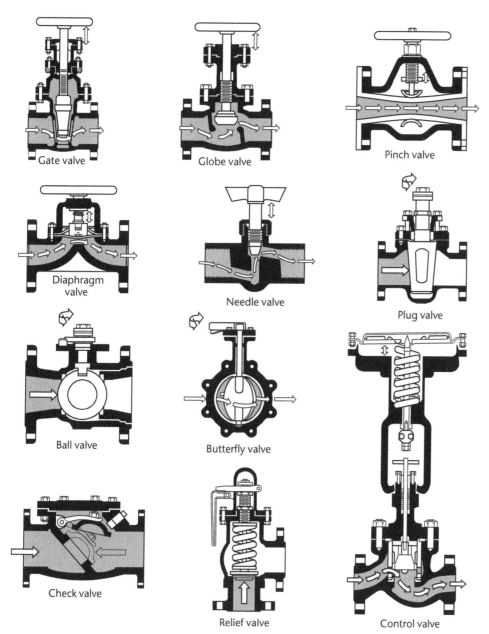

Figure 9-6 Types of water utility valves
Courtesy of the Valve Manufacturers Association of America, Washington DC.

Gate Valves

The most common type of valve found in a water distribution system is the gate valve. The gate, or disk, of the valve is raised and lowered by a screw, which is operated by a handwheel or valve key. When fully open, gate valves provide almost unrestricted flow because the gates are pulled fully up into the bonnet. When closed, the gate seats against the two faces of the valve body and closes relatively tightly unless the faces have become worn or something becomes lodged under the gate.

Gate valves are not designed to be used to regulate or throttle flow. If used for throttling, the gate mechanism will vibrate and eventually be damaged.

Generally, gate valves are not installed where they will need to be operated frequently because they require too much time to operate from the fully open to closed positions. Sizes used in water distribution systems generally range

gate valve
A valve in which the closing element consists of a disc that slides across an opening to stop the flow of water.

from ¾ in. (19 mm) in diameter for water services up to 72 in. (1,830 mm) for transmission mains. To prevent water from leaking past the stem of a gate valve, a seal is provided either by O-rings or conventional packing.

Nonrising-Stem Gate Valves

The most common type of gate valve installed for buried service is the iron-body, bronze-mounted, nonrising-stem, double-disk gate valve (Figure 9-7). The body is made of cast or ductile iron, and the sealing and operating parts are made of bronze. When double-disk valves are closed, the disks are pushed apart by a wedging mechanism as they reach the bottom of the valve. This mechanism pushes the disks outward to seal against the seat rings. The valve stem itself rotates but does not move up and down. The screw mechanism is located within the valve body, so valves can be directly buried. There is no problem with dirt affecting the mechanism.

Rising-Stem Gate Valves

In situations where a valve is not buried and the water system operator will need to know from inspection whether the valve is open or closed, a rising-stem valve with an outside screw and yoke is often used (Figure 9-8). Outside screw-and-yoke valves are frequently used in treatment plants and pumping stations.

Horizontal Gate Valves

Gate valves of the double-disk type, in sizes 16 in. (400 mm) or larger, are often designed to lie horizontally. In this position, the valve-operating mechanism does not have to lift the weight of the gate to open the valve (Figure 9-9). These valves may also be used where a large main is not buried very deeply and a vertical valve would extend above ground level.

Horizontal valves are equipped with special tracks in the valve body and bonnet. Rollers on the disks ride in the tracks throughout the length of travel to support the weight of the disks. Bronze scrapers are also provided to move ahead of the rollers in both directions to remove any foreign matter from the tracks. In rolling-disk valves, the disks themselves serve as rollers.

nonrising-stem valve
A gate valve in which the valve stem does not move up or down as it is rotated.

horizontal valve
A gate that valve that is designed such that the valve-operating mechanism does not have to lift the weight of the gate to open the valve.

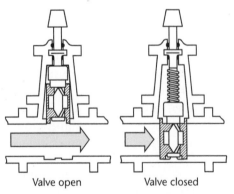

Figure 9-7 Operation of nonrising-stem gate valve

Figure 9-8 Valve with an outside screw and yoke
Courtesy of Mueller Company, Decatur, Illinois.

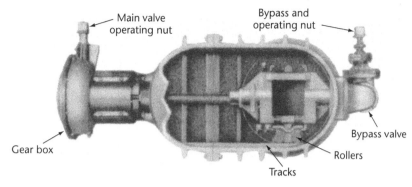

Figure 9-9 Horizontal gate valve
Courtesy of Mueller Company, Decatur, Illinois.

Because of the weight of the disk and the internal friction in the valve, large gate valves are furnished with a geared operator. Several turns of the operating nut may therefore be required to make one turn of the operating shaft. It may take 30 minutes or more to fully open or close a large gate valve. If the valve is old, it usually requires two operators to turn the valve key by hand. Most water utilities with large valves now use a portable power-operated valve operator for operating valves. These units greatly reduce the time and effort required in operating valves.

Bypass Valves

Horizontal thrust against the disk of a closed gate valve increases rapidly with valve size. For example, the thrust against a closed valve that is under 75 psi (517 kPa) water pressure is approximately 8,480 lbf (37,700 N) for a 12-in. (305-mm) valve. For the same pressure, the thrust is 15,080 lbf (67,000 N) for a 16-in. (406-mm) valve, and 76,340 lbf (339,000 N) for a 36-in. (914-mm) valve. These thrusts cause a lot of friction on the valve guides, which makes large gate valves extremely difficult to open. For this reason, bypass valves may be needed. Bypass valves are smaller-diameter valves that will allow water to pass around the larger valve. This reduces the differential pressure across the closed disk and makes the main valve easier to open and close. Bypass valves also provide for low-volume flow without the need for opening the main valve.

Tapping Valves

As illustrated in Figure 9-10, a tapping valve is a specially designed gate valve used to connect a new water main to an existing main under pressure. One end

Figure 9-10 Tapping valve
Courtesy of Mueller Company, Decatur, Illinois.

bypass valve
A small valve installed in parallel with a larger valve. Used to equalize the pressure on both sides of the disc of the larger valve before the larger valve is opened.

tapping valve
A special shutoff valve used with a tapping sleeve.

of the valve is a flange designed to attach to a tapping sleeve, and the other end is usually a regular mechanical joint pipe bell. Tapping valves have a slightly larger inside diameter to allow the tapping machine cutter to pass. Details of tapping procedures are included in Chapter 6.

Cutting-in Valves

As illustrated in Figure 9-11, a cutting-in valve has one oversized end connection designed to be used with a cutting-in sleeve. The valve and sleeve used together greatly facilitate installation of a new valve in an existing main. Main pressure must be shut off for a period of time to make the installation. This approach is a relatively inexpensive way of adding a valve in a distribution system.

Inserting Valves

When it is necessary to install a valve in a water main without shutting off pressure, a special inserting valve can be used (Figure 9-12). Water utilities normally have an inserting valve installed by a company that has the special equipment required to make the installation and specializes in this work.

Resilient-Seated Gate Valves

New designs of gate valves are available that use a resilient seat of rubber or synthetic material. When the valve is closed, the seat provides a seal between the valve body and the disk. The design offers easier operation and leak-tight shutoff. Designs vary, but in general the resilient material is bonded to either the valve body or the disk. Figure 9-13 shows a cutaway of a resilient-seated valve.

Slide Valves

Where the pressure to be regulated is relatively low and tight shutoff is not important, knife gate or sluice gate valves may be used. These valves use a relatively thin gate or blade that slides up and down in a recess, as illustrated in Figure 9-14. These valves, called slide valves, may be square, oblong, or round.

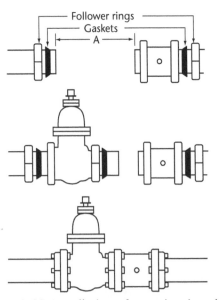

Figure 9-11 Installation of a cutting-in valve
Courtesy of US Pipe and Foundry Company.

---

cutting-in valve
A specially designed gate valve used with a sleeve that allows the valve to be placed in an existing main.

inserting valve
A shutoff valve that can be inserted by special apparatus into a pipeline while the line is in service under pressure.

resilient-seated valve
A gate valve with a disc that has a resilient material attached to it to allow leak-tight shutoff at high pressure.

slide valve
A gate valve that uses a relatively thin gate or blade that slides up and down in a recess to stop low-pressure flows where tight shutoff is not important.

Valves **191**

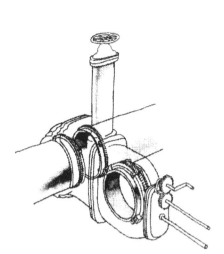

Figure 9-12 Inserting valve

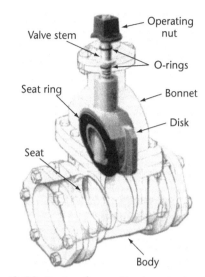

Figure 9-13 Parts of a resilient-seated gate valve
Courtesy of Mueller Company, Decatur, Illinois.

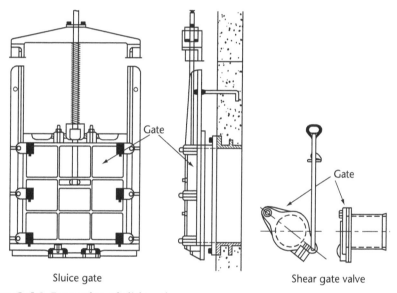

Figure 9-14 Examples of slide valves

Globe Valves

The globe valve principle is commonly used for water faucets and other household plumbing. As illustrated in Figure 9-6, the valves have a circular disk that moves downward into the valve port to shut off flow. Although they seat very tightly, globe valves produce high head loss when fully open. They are not suited for distribution mains where head loss is critical.

Needle Valves

Needle valves are similar to globe valves except that a tapered metal shaft fits into a metal seat when the valve is closed (see Figure 9-6). Needle valves are

globe valve
A valve having a round, ball-like shell and horizontal disc.

needle valve
A valve that is similar to a globe valve except that a tapered metal shaft fits into a metal seat when the valve is closed; available only in small sizes and are primarily used for precise throttling of flow.

available only in small sizes and are primarily used for precise throttling of flow. A common water utility practice is to install needle valves on the hydraulic lines to valve actuators. The needle valves then allow the opening and closing speeds of the actuator to be adjusted.

Pressure-Relief Valves

Pressure-relief valves are similar to globe valves, but their disks are normally maintained against the seat by a spring. The tension on the spring can be adjusted to allow the valve to open if the desired preset pressure is exceeded.

Air-and-Vacuum Relief Valves

As illustrated in Figure 9-15, air-and-vacuum relief valves consist of a float-operated valve that will allow air to escape as long as the float is down. When water fills the container, the float rises and closes the valve. If trapped or entrained air subsequently enters the unit, the float drops long enough to vent the air and then recloses. If a vacuum occurs in a pipeline, the relief valve will admit air to prevent collapse or buckling of the pipe.

Diaphragm Valves

Diaphragm valves operate similarly to globe valves. A manually operated diaphragm valve is illustrated in Figure 9-6. In other types of diaphragm valves, water pressure exerted on a diaphragm is used to assist in closing the valve. Figure 9-16 illustrates the principal parts of an altitude valve, which is the most common type of diaphragm valve used in a water system. Figure 9-3 illustrates how altitude valves are connected to control water to an elevated tank.

Pinch Valves

Pinch valves are closed by pinching shut a flexible interior liner. This type of valve is available only in relatively small sizes, but it is particularly useful for throttling the flow of liquids that are corrosive or might clog other types of valves. The cross section of a pinch valve is illustrated in Figure 9-6.

air-and-vacuum relief valve
A dual-function air valve that (1) permits entrance of air into a pipe being emptied, thus preventing a vacuum, and (2) allows air to escape in a pipe while being filled or under pressure.

pinch valve
A valve that is closed by pinching shut a flexible interior liner.

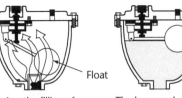
During the filling of the line, air entering the valve body will be exhausted to atmosphere. When the air is expelled and water enters the valve, the float will rise and cause the orifices to be closed.

The large and small orifices of the air-and-vacuum valve are normally held closed by the buoyant force of the float.

While the line is working under pressure, small amounts of trapped or entrained air are exhausted to atmosphere through the small orifice.

Air is permitted to enter the valve and replace the water while the line is being emptied.

Figure 9-15 Operation of an air-and-vacuum relief valve
Courtesy of GA Industries, Inc.

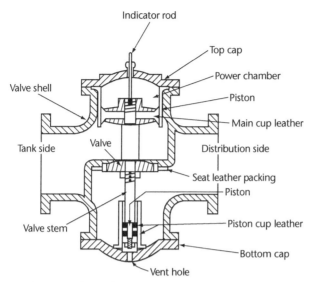

Figure 9-16 Principal parts of an altitude valve

Rotary Valves

The two principal types of rotary valves used in water systems are plug valves and ball valves. The movable element in a plug valve (Figure 9-6) is a cylinder-shaped or cone-shaped plug that has a passageway or port through it. It requires a one-quarter turn to move from fully open to fully closed. Small plug valves are used as curb stops and corporation stops. Large plug valves are most often used where pressure is relatively high and positive shutoff is required, such as on the discharge of a high-lift pump. Plug valves may also be used for throttling flow without damage.

Ball valves (Figure 9-6) operate similarly to plug valves, except that the operating element is a sphere with a circular hole bored through it. Both plug and ball valves are widely used for water service connections.

Butterfly Valves

Butterfly valves (Figure 9-17) consist of a body in which a disk rotates on a shaft to open or close the valve. In the fully open position, the disk is parallel to the axis of the pipe and the flow of water. In the closed position, the disk seals against rubber or synthetic elastomer bonded either on the valve seat of the body or on the edge of the disk. Because the disk of a butterfly valve stays in the water path in the open position, the valve creates a somewhat higher resistance to flow (i.e., a higher pressure loss) than does a gate valve. However, butterfly valves have the advantage of operating easily and relatively quickly. Water pressure on both halves of the disk is relatively equal as the valve is operated.

The laying length (i.e., the length of the valve body along the pipe axis) of butterfly valves is quite short. In the open position, the disk may extend beyond the valve body into the piping on either side. For instance, a 16-in. (400-mm) diameter flanged-end butterfly valve has an 8-in. (200-mm) laying length. A gate valve for the same pipe diameter is 17 in. (430 mm) long.

Wafer-type butterfly valves require even less laying length because they have no flanges. Instead they are sandwiched between flanges on the adjoining piping.

plug valve
A valve in which the movable element is a cylindrical or conical plug.

ball valve
A valve consisting of a ball resting in a cylindrical seat. A hole is bored through the ball to allow water to flow when the valve is open. When the ball is rotated 90°, the valve is closed.

butterfly valve
A valve in which a disc rotates on a shaft as the valve opens or closes. In the fully open position, the disc is parallel to the axis of the pipe.

Figure 9-17 Butterfly valve with electric actuator
Courtesy of Tyco Valves & Controls.

Wafer-type valves are also considerably less expensive because they are easier to manufacture.

The lug-wafer design is basically the same as the wafer design except for the addition of lugs around the body. The lugs are drilled and tapped to allow the valve to be attached directly to one of the flanges for easy pipeline modification or cleaning.

The cost of large butterfly valves is substantially less than the cost of large gate valves. Butterfly valves are also easier and quicker to operate. There are, however, disadvantages to using butterfly valves in a distribution system. The main disadvantage is that if it should ever be necessary to clean a main by the use of pigs or swabs, the operation would be blocked by the valve disks. In addition, when butterfly valves are used in a distribution system, they must not be operated too quickly or serious water hammer could result.

The short laying length of butterfly valves makes them particularly useful in piping for pumping stations and treatment plants, where space is often limited. Although butterfly valves are primarily designed for on–off service, they can be used in some situations for throttling under low-flow and low-pressure conditions. They should not be used for high-pressure throttling, however, because the disk will vibrate and eventually be damaged.

Check Valves

Check valves are designed to allow flow in only one direction. They are commonly used at the discharge of a pump to prevent backflow when the power is turned off. Foot valves are a special type of check valve installed at the bottom of the pump suction so that the pump will not lose its prime when power is turned off. If allowed to close unrestrained, a check valve may slam shut, creating water hammer that can damage pipes and valves. A variety of devices are available to minimize the slamming of check valves. These devices include external weights, restraining springs, and automatic slow-closing motorized drives. Various designs of check valves commonly used in water systems are illustrated in Figure 9-18. Backflow-prevention devices are also a type of check valve. They are described in more detail in Chapter 19.

check valve
A valve designed to open in the direction of normal flow and close with reversal of flow. An approved check valve has substantial construction and suitable materials, is positive in closing, and permits no leakage in a direction opposite to normal flow.

Slanting disk check valve

Cushioned swing check valve

Rubber flapper swing check valve

Double door check valve

Foot valve

Figure 9-18 Five types of check valves
Reprinted with permission of APCO/Valve & Primer Corp.

Valve Operation

Water system valves can be operated either manually or through a power actuator. The method used depends primarily on how the valve is used and how frequently it must be operated.

Manual Operation

Manual operation of an exposed valve, such as in a treatment plant, is usually with a handwheel, chainwheel, or floor stand operator, as shown in Figure 9-19. Buried distribution system valves are usually fitted with a 2-in. (50-mm) square operating nut for manual operation from the surface. Figure 9-20 illustrates a water main valve-operating key. Service-line valves are equipped with tee heads and require a slotted key for operation.

Power Actuators

Valves may be operated by any of several different types of power actuators. Each has some advantages and disadvantages in terms of speed of operation, reliability, and ease of control.

actuator
A device, usually electrically or pneumatically powered, that is used to operate valves.

Figure 9-19 Handwheel operator
Courtesy of M&H Valve and Fire Hydrant.

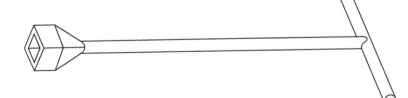

Figure 9-20 Valve key for water main valves

Electric Actuators

Electric actuators use a small electric motor to rotate the valve stem through a gear box. The actuator is turned on by a switch or by remote signal. It continues to operate until turned off by a limit switch when the valve is fully open or closed. Some actuators are fitted with controls that can be adjusted to leave the valve partially open for use in throttling. An electric valve actuator is shown attached to a butterfly valve in Figure 9-17. The handwheel on the unit is for emergency operation of the valve by hand in the event of a power failure.

Hydraulic Actuators

Valves in plants and pumping stations are frequently operated by hydraulic cylinders using either plant water pressure or hydraulic fluid. The fluid is admitted to the cylinders through electric solenoid valves to operate the valve in each direction.

Although water-operated valve cylinders are widely used, there are two warnings that should be observed. Under some water quality conditions, deposits may build up within the cylinders. The cylinders will have to be cleaned periodically to maintain smooth operation. The other problem is that if system pressure is used to operate the valves, there may be no means of operating the valves in the event of low system pressure. Some water plants guard against this by having all valve actuators operated by a separate pressure system that has a backup pressure tank.

Pneumatic Actuators

Pneumatic actuators operate much like hydraulic actuators, except that compressed air is used as the operating force. The installation requires a source of compressed air with sufficient reserve capacity to provide continued operation after a power failure. Although pneumatic actuators can be used on isolation valves, they are more common on control valves.

Actuator Operating Speed

It is usually advisable for valve actuators to operate slowly to prevent water hammer in the system. However, if there is a power failure, the pump discharge valves must close fast enough to prevent water from flowing backward through the pumps. The operating speeds of hydraulic and pneumatic operators are usually individually adjustable for each direction by means of needle valves that control the flow of fluid or air to the cylinder.

Valve Storage

Whenever possible, valves should be stored indoors. If outside storage is necessary, gears, power actuators, cylinders, valve ports, and flanges should be protected from the weather. Valves stored outside in freezing climates should be closed and have their disks in a vertical position. If the disks are in a horizontal position, rainwater will collect on top of the disk and may seep into the valve body, freeze, and crack the casting.

Valve Joints

Iron and brass valves 2 in. (50 mm) and smaller are generally furnished with female (inside) iron pipe thread. The most common joint types furnished with larger metallic valves are flanged, mechanical, and push-on. Plastic valves are also

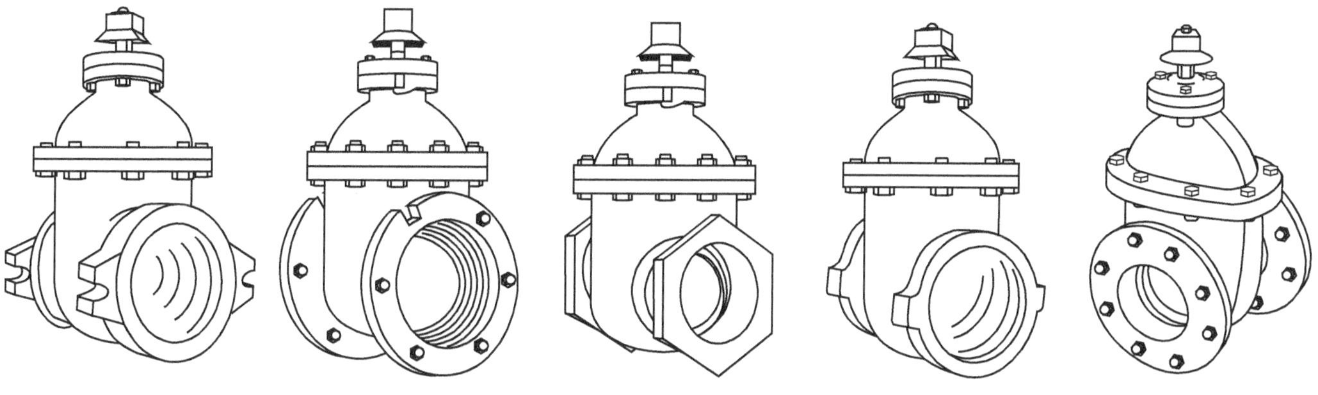

Valves 197

| Tyton® joint connection | Mechanical-joint connection | Screw-end connection | Ringtite® joint connection | Flanged connection |

Figure 9-21 Common types of valve couplings

commonly furnished with solvent-weld joints (see Chapter 5). Figure 9-21 illustrates valves with typical joints.

Valve Boxes and Vaults

Buried valves are made accessible for operation either by a valve box placed over them or by a valve vault built around them. Since valves are placed at various depths, valve boxes are made in two or more pieces that telescope to adjustable depths underground. Figure 9-22 shows a typical valve box installation.

When cast-iron valve boxes are used, they should rest above the valve so the weight of traffic passing over them will not be transferred to the valve or the pipe. The bottom flared edge of the box may require extra support. Valve boxes must be installed vertically and centered around the operating nut of the valve to ensure that the valve wrench can engage the nut.

Valve vaults are also often installed around valves to provide access for operation and maintenance. Round vaults are usually constructed in the field. They are formed of either rounded concrete blocks or precast concrete segments (Figure 9-23).

valve box
A metal, concrete, or composite box or vault set over a valve stem at ground surface to allow access to the stem so that the valve can be opened and closed. A cover for the box is usually provided at the surface to keep out dirt and debris.

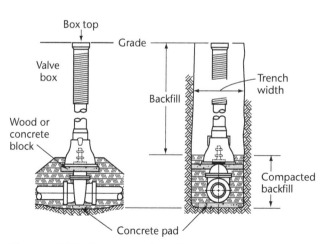

Figure 9-22 Valve box installation
Courtesy of the Ductile Iron Pipe Research Association.

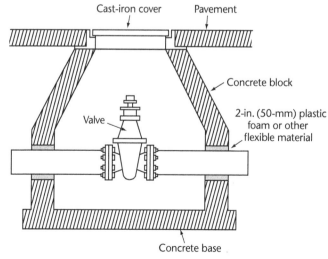

Figure 9-23 Valve vault

When valves that have exposed gearing or operating mechanisms must be installed below grade (underground), a vault must be installed around them. When a vault is constructed around a valve, the operating nut must be accessible for operation with a valve key from the surface through the access cover or through a special opening in the top slab.

Care must be taken when constructing valve vaults to ensure that the weight of the vault will not rest on the water main. Valve vault drains should not be connected to storm drains or sanitary sewers because the water system could become contaminated if the sewer should back up and flood the vault. Vaults are usually drained to underground absorption pits.

Valve Records

It is extremely important that accurate records be maintained on each valve in a distribution system. Records of each valve's location must include measurements from at least three different permanent reference objects. This method makes it possible to locate the valve even if some of the reference objects are changed or moved over a period of years. The record should also include the make, size, date of installation, and type of valve to facilitate obtaining the proper parts if the valve must be repaired.

Study Questions

1. Which type of valve should be installed at a dead-end water main?
    a. Vacuum valve
    b. Air valve
    c. Blowoff valve
    d. Pressure-relief valve

2. Which type of valve operates similar to a diaphragm valve?
    a. Vacuum relief valve
    b. Globe valve
    c. Pressure-relief valve
    d. Butterfly valve

3. Which type of valve can go from fully open to fully closed with a quarter turn?
    a. Plug valve
    b. Needle valve
    c. Globe valve
    d. Pinch valve

4. Foot valves are a special type of
    a. relief valve.
    b. control valve.
    c. check valve.
    d. plug valve.

5. Which type of valve would be best to use to precisely throttle flow?
   a. Globe valve
   b. Butterfly valve
   c. Rotary valve
   d. Needle valve

6. Which type of valve is used to isolate a pump on the suction side?
   a. Butterfly valve
   b. Globe valve
   c. Gate valve
   d. Ball valve

7. When fully opened, which type of valve will have the highest head loss?
   a. Gate valve
   b. Plug valve
   c. Globe valve
   d. Ball valve

8. Which type of valve will prevent the collapse of a pipe?
   a. Pressure-relief valve
   b. Needle valve
   c. Pinch valve
   d. Air-and-vacuum relief valve

9. List five types of valve used in water systems.

10. What type of valve is installed at frequent intervals in distribution piping so that small sections of water main may be shut off for maintenance or repair?

11. What are the two principal types of throttling valves used in a water system?

12. What are the three types of power actuators commonly used to operate valves?

13. Records of each valve's location must include measurements from at least how many different permanent reference objects?

Chapter 10
Fire Hydrants

Fire hydrants are one of the few parts of a water system visible to the public, so keeping them well maintained can help a water utility project a good public image. In smaller towns, hydrants are so seldom used for fighting fires that it may be easy to forget how important it is to keep them well maintained for quick and reliable service when needed. A hydrant that does not operate when needed can result in a loss of life and property.

Fire Hydrant Uses

In addition to fighting fires, hydrants have other uses. This section discusses the different uses of fire hydrants.

Fire Fighting

The major purpose of fire hydrants is for public fire protection. The water utility is usually responsible for keeping hydrants in working order, although the fire department assumes this responsibility in some communities. Water system operators should be aware that they can be held liable for damages if property is destroyed by fire because neglected hydrants fail to operate.

Following are guidelines on where hydrants should be located:

- Hydrants should not be located too close to the buildings they are intended to protect. Fire fighters will not position their fire trucks where a building wall could fall on them if the building should collapse during a fire.
- Hydrants should preferably be located near street intersections. This way, hose can be strung to fight a fire in any of several directions.
- Hydrants should be placed back far enough from a roadway to minimize the danger of them being struck by vehicles. However, hydrants should be close enough to pavement to ensure a secure connection between the pumper and hydrant without the risk of the truck getting stuck in mud or snow.
- In areas of the country with heavy snow, hydrants must be located where they are least likely to be covered by plowed snow or struck by snow-removal equipment.
- Hydrants that are very low or in a hole will be difficult for fire fighters to use. They are more quickly covered by snow and may be difficult to find in winter and more difficult to dig out for use. Hydrants should generally be high enough that valve caps can be removed with a standard wrench, without the wrench hitting the ground.

fire hydrant
A device connected to a water main and provided with the necessary valves and outlet nozzles to which a fire hose may be attached. The primary purpose of a fire hydrant is to fight fires, but it is also used for washing down streets, filling water-tank trucks, and flushing out water mains.

- Some residents like to screen hydrants by planting bushes around them, but this may make it difficult for fire fighters to find the hydrant in an emergency. Residents should be politely told that the hydrant must be kept exposed for proper protection of their property.
- Some residents are bothered by a hydrant's bright color. They may take it upon themselves to repaint the one in front of their house a darker color to match the vegetation. If the change in color could make it harder to find the hydrant in an emergency, the darker color should not be allowed.

Miscellaneous Fire Hydrant Uses

Other frequent uses for fire hydrants include the following:

- Flushing water mains
- Flushing sewers
- Filling tank trucks for street washing, tree spraying, and other uses
- Providing a temporary water source for construction jobs, such as for mixing mortar and settling dust

Restricting Miscellaneous Uses

Authorized uses of hydrants for miscellaneous purposes should be rigidly controlled and generally discouraged. *Unauthorized* use should be absolutely prohibited. One reason for discouraging miscellaneous uses is that the water is often not paid for. A more important reason is that inexperienced persons can unknowingly or carelessly cause a hydrant to be damaged or become inoperable.

Inexperienced persons can damage hydrants by

- failing to note that a hydrant has not properly drained after being shut off;
- not shutting the main valve completely, resulting in slight leakage;
- leaving attachments on the hydrant when it is not in use; and
- using a pipe wrench for operation, which will damage the top nut.

Allowing Miscellaneous Uses

Several points should be observed in allowing miscellaneous uses of hydrants. First, it is preferable to require permits for anyone outside of the utility or local government staff to use hydrants. When the permit is issued, the water utility can check that the person or firm has the proper equipment and understands the proper operation of hydrants. At the same time, it is possible to charge a fee for the water that will be used, either through a flat permit fee or by metering the water taken from hydrants. Figure 10-1 illustrates a hydrant meter used for metering water use at a construction site.

When the utility issues a permit, it should specify whether certain hydrants are not to be used. Types and locations of hydrants that should be excluded from use if possible include the following:

- Very old hydrants (they may not seat properly when closed)
- Hydrants on dead-end mains (their use may stir up sediment)
- Hydrants on busy streets (they may disrupt traffic)
- Hydrants in locations with a high groundwater table (they may not drain properly)

Everyone who is allowed to use hydrants should be instructed in their proper operation. These people must be explicitly told that only appropriate operating

Figure 10-1 Fire hydrant meter
Courtesy of Neptune Technology Group Inc.

wrenches and hose connections with a throttling valve are to be used. They should also be instructed to close hydrants slowly, check for draining, and guard against any possibility of creating a cross-connection.

The importance of restricting hydrant use should be explained to police and fire department personnel and other municipal staff. These groups should be asked to cooperate by immediately reporting any suspected unauthorized use.

System Problems Caused by Hydrant Operation

Problems can arise if the distribution system is not designed to handle the demands placed on it by hydrants during a fire. For example, if the mains are not large enough to provide adequate fire flow, a fire department pumper can create a negative pressure in the main. This in turn can cause any cross-connections on the system to siphon nonpotable water into the distribution system. It can also cause hot water heaters to drain back through service lines and damage customer meters.

Standard practice is to install hydrants only on mains 6 in. (150 mm) or larger. Larger mains are often necessary to ensure that the residual pressure during fire flow remains greater than 20 psi (140 kPa).

Whenever hydrants are operated, the increase in flow causes the water to move faster through the mains. The faster flow scours any sediment that may have accumulated. This material can produce discolored and cloudy water. It will usually result in customer complaints unless the public has been notified of the problem beforehand and instructed on how to deal with it.

Hydrants should always be closed slowly. Stopping flow quickly can cause a water hammer condition that could move the hydrant backward if it is not firmly blocked. It could also lead to other distribution system damage.

Types of Fire Hydrants

The types of hydrants generally available are classified as follows:

- Dry-barrel hydrants
- Wet-barrel hydrants
- Warm-climate hydrants
- Flush hydrants

> **fire flow**
> The rate of flow, usually measured in gallons per minute (gpm) or liters per minute (L/min), that can be delivered from a water distribution system at a specified residual pressure for fire fighting. When delivery is to fire department pumpers, the specified residual pressure is generally 20 psi (140 kPa).

Dry-Barrel Hydrants

As illustrated in Figure 10-2, **dry-barrel hydrants** are equipped with a main valve and a drain in the base. The barrel is filled with water only when the main valve is open. A small drain valve connected to the operating stem opens as the main valve is closed, allowing water to drain from the barrel.

Dry-barrel hydrants are primarily designed for use in freezing climates, but they are also used in warmer parts of the country. An advantage of dry-barrel hydrants is that there is no flow of water from a broken hydrant.

Wet-Top Hydrants

One type of dry-barrel hydrant is the **wet-top hydrant**, which is constructed so that the threaded end of the main rod and the operating nut are not sealed from water when the main valve is open (Figure 10-3).

Dry-Top Hydrants

The other type of dry-barrel hydrant is the **dry-top hydrant**. As shown in Figure 10-4, the threaded end of the operating stem is sealed from water in the barrel when the hydrant is in use. This design reduces the possibility of the threads becoming fouled by sediment or corrosion.

---

dry-barrel hydrant
A hydrant for which the main valve is located in the base. The barrel is pressurized with water only when the main valve is opened. When the main valve is closed, the barrel drains. This type of hydrant is especially appropriate for use in areas where freezing weather occurs.

wet-top hydrant
A dry-barrel hydrant in which the threaded end of the main rod and the revolving or operating nut are not sealed from water in the barrel when the main valve of the hydrant is open and the hydrant is in use.

dry-top hydrant
A dry-barrel hydrant in which the threaded end of the main rod and the revolving or operating nut is sealed from water in the barrel when the main valve of the hydrant is in use.

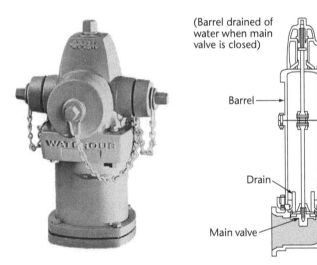

Figure 10-2 Dry-barrel hydrant
Courtesy of Waterous Company.

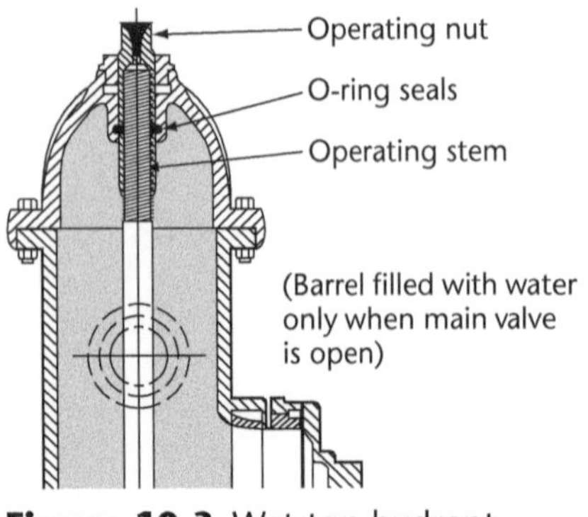

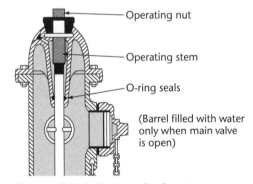

Figure 10-3 Wet-top hydrant
Courtesy of Mueller Company, Decatur, Illinois.

Figure 10-4 Dry-top hydrant
Courtesy of Mueller Company, Decatur, Illinois.

Valve Types

Dry-barrel hydrants are also classified according to the type of main valve. The primary types of valves that have been used, as illustrated in Figure 10-5, are described as follows:

- In a standard compression hydrant, the valve closes with the water pressure against the seat to aid in providing a good seal.
- In a slide gate hydrant, the valve is a simple gate valve, similar to one side of an ordinary rubber-faced gate valve.
- In a toggle (Corey) hydrant, the valve closes horizontally, and the barrel extends well below the branch line.

Breakaway Hydrants

Early dry-barrel hydrants were generally constructed with a solid barrel extending from the base to the top, as shown in Figure 10-6. Many of these hydrants are still in use today. The principal disadvantage with this design is that when the barrel is broken, usually by being struck by a vehicle, it is necessary to excavate down to the base and replace the entire barrel. The operating stem is usually also bent and must be replaced. The parts and labor for repairing a broken hydrant of this type are quite expensive. If there is deep frost in the ground, the repair will be particularly difficult or may have to be deferred until spring.

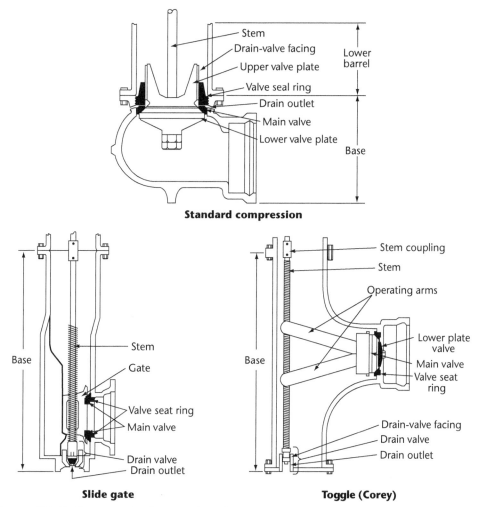

Figure 10-5 Common hydrant main valve types

breakaway hydrant
A two-part, dry-barrel post hydrant with a coupling or other device joining the upper and lower sections. The coupling and barrel are designed to break cleanly when the hydrant is struck by a vehicle, preventing water loss and allowing easy repair.

wet-barrel hydrant
A fire hydrant with no main valve. Under normal, nonemergency conditions, the barrel is full and pressurized (as long as the lateral piping to the hydrant is under pressure and the gate valve ahead of the hydrant is open). Each outlet has an independent valve that controls discharge from that outlet. The wetbarrel hydrant is used mainly in areas where temperatures do not drop below freezing. The hydrant has no drain mechanism.

warm-climate hydrant
A fire hydrant with a two-piece barrel that is the main valve located at ground level.

flush hydrant
A fire hydrant with the entire barrel and head below ground elevation. The head, with operating nut and outlet nozzles, is encased in a box with a cover that is flush with the ground line. Usually a dry-barrel hydrant.

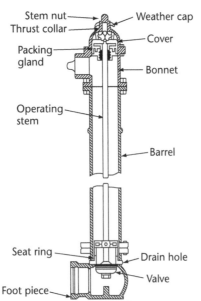

Figure 10-6 Dry-barrel hydrant

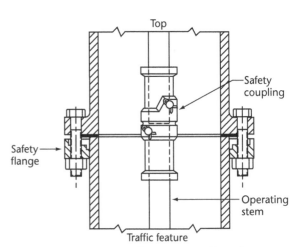

Figure 10-7 Detail of one type of breakaway flange and stem coupling
Courtesy of American Cast Iron Pipe Company.

The breakaway hydrant (or "traffic" hydrant) design is now in general use. It has a two-part barrel with a flanged coupling just above the ground line (Figure 10-7). The flange is designed to break on impact without further damage to the barrel. The operating stem is also in two pieces, with a coupling designed to break on impact. In most cases, workers can repair a hydrant that has been struck by simply replacing the breakaway flange and stem coupling at nominal cost for parts and labor. No excavation is required. The design of dry-barrel hydrants is covered in AWWA Standard C502, *Dry-Barrel Fire Hydrants* (most recent edition).

Wet-Barrel Hydrants

A wet-barrel hydrant is completely filled with water at all times (Figure 10-8). The hydrant itself has no main valve. Instead, each nozzle is equipped with a valve. Wet-barrel hydrants cannot be used in climates where temperatures fall below freezing. Another disadvantage of this type of hydrant is that large amounts of water will flow from a broken hydrant until a repair crew can shut it off. The design of wet-barrel hydrants is covered in AWWA Standard C503, *Wet-Barrel Fire Hydrants* (most recent edition).

Warm-Climate Hydrants

Warm-climate hydrants have a two-part barrel. The main valve is located at the ground line, and the lower barrel is always full of water and under pressure. The main valve controls flow from all outlet nozzles, and there is no drain mechanism.

Flush Hydrants

The entire standpipe and head of a flush hydrant are below ground. As shown in Figure 10-9, the operating nut and outlet nozzles are encased in a box with a removable cover that is at the ground surface level. Flush hydrants are used on aprons and taxiways at airports, pedestrian malls, and other locations where post-type hydrants are considered unsuitable. Flush hydrants are usually of the dry-barrel type.

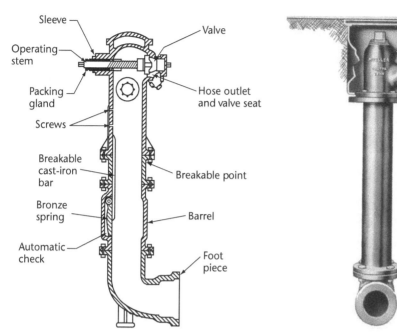

Figure 10-8 Wet-barrel hydrant

Figure 10-9 Flush hydrant
Courtesy of Mueller Company, Decatur, Illinois.

 WATCH THE VIDEO
Hydrants—Types and Parts (www.awwa.org/wsovideoclips)

Hydrant Parts

This section discusses the components of fire hydrants.

Upper Section

The upper section of a hydrant is often called the nozzle section or head. Figure 10-10 shows some of the principal parts. The operating nut is usually a five-sided nut so that it cannot be operated with a regular socket wrench. Special wrenches are used for hydrant operation (Figure 10-11). Opening a hydrant usually involves turning the operating nut counterclockwise. An arrow and the word *open* should be cast in relief on the top of the hydrant to designate the direction of opening.

The top cover or closure on the hydrant upper barrel is usually referred to as the bonnet. The bonnet may or may not be pressurized. The upper barrel is a gray cast-iron or ductile-iron section that carries water from the lower barrel to the outlet nozzles.

The outlet nozzles are threaded bronze outlets on the upper barrel. They are used (1) to connect hose lines for direct use of main pressure or (2) to connect a suction hose from the hydrant to the pumper truck. Most water systems use hydrants with two 2.5-in. (64-mm) nozzles and one 4.5-in. (114-mm) nozzle. The smaller-diameter outlet nozzles are for connecting to fire hoses. The larger nozzle, called a pumper outlet or steamer outlet, is used for connecting to a pumper suction hose. Outlet-nozzle threads are normally National American Standard threads (*Standard for Fire Hose Connections* is available from the National Fire

---

upper section
The upper part of the main hydrant assembly, including the outlet nozzles and outlet-nozzle caps. The upper section is usually constructed of gray cast iron. Also known as nozzle section or head.

operating nut
A nut, usually pentagonal or square, rotated with a wrench to open or close a valve or hydrant valve. May be a single component or it may be combined with a weather shield.

bonnet
The top cover or closure on the hydrant upper section. It is removable for the purpose of repairing or replacing the internal parts of the hydrant.

outlet nozzle
A threaded bronze outlet on the upper section of a fire hydrant, providing a point of hookup for hose lines or suction hose from hydrant to pumper truck.

lower section
The part of a dry-barrel hydrant that includes the lower barrel, the main valve assembly, and the base.

lower barrel
The section of a hydrant that carries the water flow between the base and the upper section. Usually buried in the ground with the connection to the upper section approximately 2 in. (50 mm) above ground line.

main valve
In a dry-barrel hydrant, the valve in the hydrant's base that is used to pressurize the hydrant barrel, allowing water to flow from any open outlet nozzle.

base
The inlet structure of a fire hydrant. An elbow-shaped piece that is usually constructed as a gray cast-iron casting. Also known as the shoe, inlet, elbow, or foot piece.

travel-stop nut
A nut, used in dry-barrel hydrants, that is screwed on the threaded section of the main rod. It bottoms at the base of the packing plate, or revolving nut, and terminates downward travel (opening) of the hydrant valve.

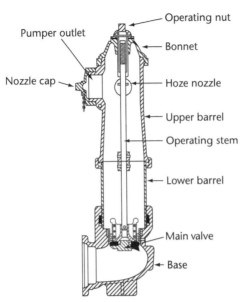

Figure 10-10 Main parts of a typical fire hydrant
Courtesy of US Pipe and Foundry Company.

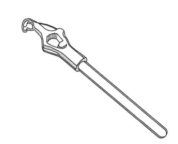

Figure 10-11 Adjustable hydrant wrench

Protection Association; see Appendix C). However, some utilities use special threads. Both utility and fire department personnel must be aware of the thread type used so that it will be compatible with the firefighting equipment.

The outlet-nozzle caps are cast-iron covers that screw onto outlet nozzles. These caps protect the nozzles from damage and unauthorized access. Caps are usually furnished with a nut the same size as the hydrant operating nut. They should be removed only with a hydrant wrench.

Lower Section

The lower section of a hydrant includes the lower barrel, main valve, and base. The lower barrel is made of static-cast or centrifugally cast gray or ductile iron. It carries water flow between the base and the upper barrel. It should be buried in the ground so that the connection to the upper barrel is approximately 2 in. (50 mm) above the ground line.

The main valve assembly includes the operating stem, resilient valve gasket, and other attached parts. The base is also known as the shoe, inlet, elbow, or foot piece. It is usually cast iron. Many hydrant mechanisms include a travel-stop nut to prevent the main rod and valve from moving down too far. If a hydrant doesn't have such a nut, a stop must be built into the middle of the base. The valve assembly will then rest against the stop.

Auxiliary Valves

An auxiliary valve should be installed on every hydrant. This device enables each hydrant to be individually turned off for maintenance or repair. Some water systems may have old hydrants that were installed without auxiliary valves. It is advisable to install valves on these hydrants as time becomes available so that repairs can be made without shutting down an entire section of the distribution system.

The type of auxiliary valve most often used is directly connected to the hydrant by a flanged connection, as shown in Figure 10-12. One advantage of this arrangement is that the valve cannot separate from the hydrant. Another is that

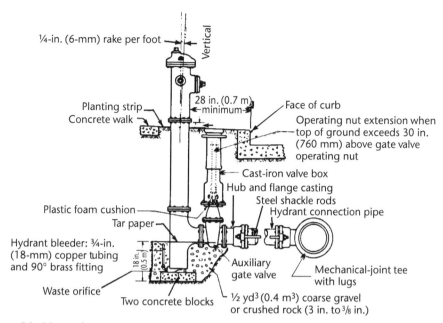

Figure 10-12 Hydrant auxiliary valve installation

all valves are at a standard location in relation to the hydrant. The valves are then easily located in an emergency.

Inspection and Installation

This section discusses the steps required before a hydrant can be put into service—namely, inspection, installation, and testing.

Inspection of New Hydrants

Hydrants should be inspected at the time of delivery in order to verify that they meet specifications and were not damaged during shipment. Points that should be checked include the following:

- direction to open the hydrant
- size and shape of the operating nut
- depth of bury (distance to the main connection below the ground surface)
- size and type of inlet connection
- main valve size
- outlet-nozzle sizes and configuration
- nozzle thread dimensions

The hydrant should be cycled to the fully open and fully closed positions to ensure that no internal damage or breakage has occurred during shipment and handling. All external bolts should be checked for proper tightness.

After inspection, the hydrant valves should be closed and the outlet-nozzle caps replaced to prevent foreign matter from entering. Stored hydrants should be protected from the weather whenever possible. If they must be stored outside in cold weather, they should be placed with the inlet down to prevent water from entering and freezing.

Installation Procedures

The following are important concerns during the installation of a fire hydrant:

- Location
- Footing and blocking
- Drainage
- Paint color

Location

Hydrants should generally be set back at least 2 ft (0.6 m) from the curb. This will place the auxiliary valve box behind the curb and locate the hydrant beyond normal vehicle overhang. In rural areas or where there is no curb, hydrants must be protected from traffic with a larger setback. The pumper outlet nozzle should always face the street. If the location is particularly vulnerable to traffic, guard posts should be installed.

Footing and Blocking

Hydrants must be set on a firm footing that will not rot or settle. A flat stone or concrete slab is ideal. A carpenter's level should be used to ensure that hydrants are set plumb (i.e., exactly vertical).

Hydrants must also be securely blocked or restrained from movement in the direction opposite the main connection (usually the back of the hydrant). If they are not, they may move and open a joint if there is a water hammer in the system. The excavation for a hydrant should, if possible, be made in a manner that will preserve undisturbed earth a short distance behind the hydrant base. After the hydrant is placed, a concrete block can be poured between the base and the undisturbed earth. Installers must carefully place the concrete so as not to block the drain hole in the base.

When it is not possible to obtain good support for a block behind a hydrant, restraint must be provided by rods or fittings. A common practice is to use two shackle rods ¾ in. (19 mm) in diameter to tie a hydrant valve to the tee in the main.

Drainage

Unless the soil surrounding a hydrant is very porous, special provisions must be made to carry off the drainage from dry-barrel hydrants. The accepted method is to excavate an area around the base of the hydrant to be large enough to permit the placement of approximately ⅓ yd^3 (0.25 m^3) of clean stone. The top of this layer should be slightly above the drain openings. To prevent the drainage stone from being clogged with dirt, it should be covered with heavy polyethylene or tar paper before backfill is placed.

If hydrants must be installed where the barrel will not fully drain because the water table is above the hydrant base, it is advisable to plug the hydrant drains. These hydrants must then be specially marked so that the barrel will be pumped dry with a hand pump after each use.

Hydrant Painting

Hydrants should be painted with colors that are easily visible both day and night. Red, orange, and yellow are usually the most visible colors. Some water utilities paint the operating nut or portions of the nozzles with reflective paint to make the hydrants readily visible at night.

Table 10-1 Standard hydrant color scheme to indicate flow capacity

Hydrant Class	Color†	Usual Flow Capacity at 20 psig (140 kPa [gauge])*	
		gpm	(L/min)
AA	Light blue	1,500	(5,680)
A	Green	1,000–1,499	(3,785–5,675)
B	Orange	500–999	(1,900–3,780)
C	Red	Less than 500	(Less than 1,900)

*Capacities are to be rated by flow measurements of individual hydrants at a period of ordinary demand. See AWWA Standard C502, *Dry-Barrel Fire Hydrants*, most recent edition, for additional details.

†As designed in Federal Standard 595B, General Services Administration, Specification Section, Washington DC.

Color coding hydrants to indicate their capacity helps the fire department select the most productive hydrants for fighting a fire. Table 10-1 shows a commonly used color scheme for painting hydrant tops and/or caps to indicate the expected flow rate, as determined by hydrant flow tests. Another scheme for color coding might be based on water main size.

Testing

New hydrants installed at the same time as a new main should not be pressure-tested along with the main. All hydrant auxiliary valves should be closed during the main pressure test. After the main test is completed—and preferably before hydrant excavations have been backfilled—the auxiliary valves can be opened and the hydrants tested as follows:

1. Open the hydrant fully and fill it with water. Then close all outlets.
2. Vent air from the hydrant by leaving one of the caps slightly loose. After all air has escaped, tighten the cap before proceeding.
3. Apply a pressure up to a maximum of 150 psig (1,000 kPa [gauge]) by using a pressure pump connected to one of the nozzles. If it is not practical to apply higher pressure, system pressure will suffice.
4. Check for leakage at flanges, outlet nozzles, and the operating stem.
5. If leakage is noted, repair or replace components until the condition has been corrected.

It is not uncommon for outlet nozzles and connecting bolts to become loose as a result of rough handling in shipment, storage, and installation. Loose flange bolts and nozzles should be tightened.

Utility personnel can test dry-barrel hydrants for proper drainage by placing the palm of one hand over the outlet-nozzle opening immediately after the main valve is shut. The barrel usually drains fast enough that a noticeable suction can be felt. If there is a question of whether the barrel has completely drained, a string with a small weight should be dropped in through one nozzle to see if it comes out wet.

After backfilling is complete, hydrants should be operated so that any foreign material is flushed out. When nozzle caps are replaced, they should be made tight enough to prevent removal by hand, yet not excessively tight. New hydrants should promptly be painted the color preferred by the water system.

Operation and Maintenance

This section discusses common operation and maintenance procedures for hydrants that have been placed in service.

Hydrant Operation

Hydrants are designed to be operated by one person using a 15-in. (380-mm) wrench. The use of a longer wrench or piece of pipe added to a standard wrench (cheater) is not good practice. If one person cannot operate a hydrant with a standard wrench, the hydrant should be repaired or replaced. To prevent damage to the operating nut, wrenches not specifically designed for hydrant operation should not be used.

If the main valve in a dry-barrel hydrant is not fully closed, there may be leakage through the drain valves. This leakage will eventually saturate the surrounding soil and prevent proper draining of the barrel. In soil that can easily wash away, the leakage can undermine the hydrant footing or blocking.

The main valve of a dry-barrel hydrant should always be completely opened when in use. This will ensure that the drain valve is completely closed. In addition, throttling the main valve may cause damage to the valve seat and rubber. If flow from a hydrant must be throttled, a nozzle cap fitted with a gate valve should be used, with the hydrant valve fully opened.

Hydrant Maintenance

Regularly scheduled inspection of hydrants is necessary to ensure satisfactory operation. All hydrants should be inspected at least annually. In freezing climates, each hydrant should be inspected in the autumn to make sure no standing water is in the barrel. Some water systems do this by "**stringing**" the hydrant—dropping a weighted string down the barrel to see if it comes out wet.

Water systems that have a problem with sediment accumulating in their mains usually perform an annual system flush and hydrant check at the same time. If a hydrant is found to be inoperable and cannot be repaired immediately, a barricade or barrel should be placed over it. The fire department should then be notified that the hydrant is out of service.

Inspection Procedures

Following are some general inspection procedures:

- Check that there is nothing near the hydrant that will interfere with its operation.
- Ensure that there is nothing attached to the outlet nozzles.
- Visually check the hydrant to be sure it is not leaning. Look for other indications that it may have been struck by a vehicle.
- Remove one outlet-nozzle cap from dry-barrel hydrants and check for water or ice standing in the barrel.
- Use a listening device to check for seat leakage on dry-barrel hydrants. Visually inspect each valve on wet-barrel hydrants.
- On dry-barrel hydrants, replace the outlet-nozzle cap and open the hydrant to a fully open position while venting air from the barrel. Check the ease of operation. If the stem action is tight, repeat the action several times until it is smooth and free.

---

stringing (hydrants)
The practice of dropping a weighted string down the barrel of a hydrant to check if the barrel has fully drained.

- While the hydrant is under pressure, check for leakage at joints, around outlet nozzles and caps, and at packing or seals. If leakage is observed, it might be necessary to (1) tighten or recaulk outlet nozzles, (2) lubricate and tighten compression packing, or (3) replace O-rings, seals, and gaskets. If leakage cannot be corrected with the tools at hand, record the nature of the problem for prompt attention by the repair crew.
- Wet-barrel hydrants require that a special outlet-nozzle cap be used to test and operate each valve.
- Remove an outlet-nozzle cap and attach a section of hose (or a flow diverter if necessary) to direct flow into the street. Open the hydrant and allow it to flow for a short time to remove foreign material from the interior and lateral piping.
- On dry-barrel hydrants, close the main valve to the position at which the drains open. Allow flow through the drains under pressure for about 10 seconds to flush the drain. Then close the main valve completely.
- Remove all outlet-nozzle caps and inspect them for thread damage from impact or cross threading. Clean and lubricate outlet-nozzle threads. Use caps to check for easy operation of threads. Be sure outlet-nozzle cap gaskets are in good condition.
- Lubricate operating-nut threads in accordance with the manufacturer's instructions. Some hydrants require oil in the upper operating-nut assembly. This lubrication is critical and must not be overlooked during routine maintenance.
- Check that the barrel drains properly on each dry-barrel hydrant. You should feel suction rapidly being created when your hand is placed over an outlet nozzle during drainage.
- Use a listening device to check again for seat leakage on dry-barrel hydrants. Visually check wet-barrel hydrants.
- If dry-barrel hydrants do not drain, pump out any residual water. Check with a listening device to make sure the main valve is not leaking.
- Check outlet-nozzle cap chains for free action on each cap. If you observe binding, open the chain connection around the cap until the action is free. This prevents kinking when the cap is removed under emergency conditions.
- On traffic hydrants, check the breakaway device. Inspect couplings, cast lugs, special bolts, and other parts for damage due to impact or corrosion.

Hydrant Repair

Any condition that cannot be corrected during the regular inspection should be recorded and reported for subsequent action by a repair crew. Before any repair is started, the fire department must be notified that the hydrant will be out of service.

Turning off the hydrant auxiliary valve is usually the first step in repairing the internal parts of a hydrant. The hydrant is then disassembled according to procedures supplied by the manufacturer. In addition to any worn or damaged parts, all gaskets, packing, and seals should be replaced while the hydrant is apart, regardless of whether they appear worn or not.

When the repair is completed, the auxiliary valve should be opened. The hydrant should then be thoroughly tested for proper operation and leaks. The fire department may then be notified that the hydrant is back in service. The date and details of the repair should be promptly recorded on the hydrant record card.

Flow Testing

As a municipality expands, the increased demand on the water distribution system causes higher velocities in mains. This results in increased head loss in the piping. At the same time, because of corrosion, scale, and sediment in the pipes, the carrying capacity of old piping in the system may slowly decrease. It is therefore necessary to check the capacity of the system periodically to determine (1) the need for additional feeder or looping mains and (2) the need to clean and line existing pipes.

Hydrants should periodically undergo flow testing to provide data on distribution system flow and fire flow capabilities. Testing will also identify other problems, such as a valve that has inadvertently been left closed. Flow capability should also be retested after any major changes are made in the distribution system piping.

WATCH THE VIDEO
Hydrants—Installation and Repair (www.awwa.org/wsovideoclips)

Hydrant Records

A meaningful inspection and maintenance program requires that a record card, sheet, or computerized record be kept on each hydrant. Some initial information that should be recorded at the time each hydrant is installed includes the make, model, and location. The record should also list any other pertinent information that may be of use in the future when repairs are required. Most water systems assign a number to each hydrant to assist in record keeping.

Each time a hydrant is inspected, an entry should be made in the record indicating the date of inspection and the condition of the hydrant. This record must be accurately maintained. It can serve as proof of inspection in the event the water utility is ever charged with negligence because of a hydrant failing to operate properly in an emergency. A record should also be kept of all repair work that is performed.

Hydrant Safety

In addition to the general safety precautions detailed in other chapters for installing piping and protecting the public, special precautions must be taken to prevent injury and damage to private property during hydrant flushing. Following are several safety concerns that should be taken into account:

- Remember, the force and volume of water from a full hydrant stream are sufficient to seriously injure workers or pedestrians.
- If traffic is not adequately controlled, drivers trying to avoid a hydrant stream might stop quickly or swerve. An accident might result.
- If the temperature is below freezing, water that is allowed to flow onto pavement may freeze and cause accidents.
- If flow is diverted with a hose to a sewer, care must be taken not to create a cross-connection.
- If flow is diverted with a hose, the end of the hose must be securely anchored. A loose hose end can swing unpredictably and could cause serious injury.

Figure 10-13 Flow diffuser
Courtesy of Pollardwater.com.

During flow tests, the hydrant nozzle must be unobstructed, so the only way of protecting property is to choose the nozzle that will do the least damage. Barricades should be provided to divert traffic. Other precautions may have to be taken to minimize property damage and prevent personal injury.

Steps can be taken to divert flushing flow to prevent property damage. Flow diffusers or a length of fire hose should be used where necessary to direct the flow into a gutter or drainage ditch (Figure 10-13). A rigid pipe connected to a hydrant outlet and turned at an angle to divert flow down a gutter is not considered a good idea. The torque produced by the angular flow could be enough to twist or otherwise damage the hydrant.

Study Questions

1. It is standard practice to install fire hydrants on mains with diameters of _____ or larger.
   a. 6 in. (150 mm)
   b. 8 in. (200 mm)
   c. 10 in. (250 mm)
   d. 12 in. (300 mm)

2. Most water systems use hydrants with two _____-diameter nozzles and one _____-diameter nozzle.
   a. 2.0-in. (51-mm); 3.0-in. (76-mm)
   b. 2.0-in. (51-mm); 4.0-in. (102-mm)
   c. 2.5-in. (64-mm); 3.5-in. (89-mm)
   d. 2.5-in. (64-mm); 4.5-in. (114-mm)

3. Miscellaneous use of fire hydrants
   a. is never authorized.
   b. is at the public's discretion on an as-needed basis.
   c. is generally discouraged but may be authorized in a controlled context.
   d. is a useful way of testing hydrants indirectly.

4. Which of the following is *not* a common classification of fire hydrants?
   a. Flow hydrant
   b. Warm-climate hydrant
   c. Wet-barrel hydrant
   d. Dry-barrel hydrant

5. The lower barrel should be buried in the ground so that the connection to the upper barrel is approximately _____ above the ground line.
   a. 12 in. (300 mm)
   b. 1 in. (25 mm)
   c. 6 in. (150 mm)
   d. 2 in. (50 mm)

6. What type of fire hydrant is used in freezing climates?

7. What is an auxiliary valve used for?

8. Why is a fire hydrant operating nut five sided?

9. In what type of fire hydrant are the entire barrel and head below ground elevation?

10. In the standard hydrant color scheme, what color are class A hydrants?

Chapter 11
Water Storage

Water storage is essential for meeting all of the domestic, industrial, and fire demands of most public water systems. Water may be stored before and/or after treatment. The primary subject of this chapter is distribution storage, which refers to the storage of treated water ready for use by customers.

Water Storage Requirements

The type and capacity of water storage required in a distribution system vary with the size of the system, the topography of the area, how the water system is laid out, and various other considerations. There are four primary types of water storage structures:

- Hydropneumatic tanks
- Ground-level reservoirs
- Buried reservoirs
- Elevated tanks

Purposes of Water Storage

Water storage in the distribution system may be required for the following purposes:

- Equalizing supply and demand
- Increasing operating convenience
- Leveling out pumping requirements
- Decreasing power costs
- Providing water during power source or pump failure
- Providing large quantities of water to meet fire demands
- Providing surge relief
- Increasing detention times
- Blending water sources

Equalizing Supply and Demand

As illustrated in Figure 11-1, the demand for water normally changes throughout the day and night. If treated water is not available from storage, the wells or water treatment plant must have sufficient capacity to meet the demand at peak hour (the busiest water-use hour). This high capacity is not generally practical or economical. The peak-hour demand in Figure 11-1 is approximately 175 percent of the average

217

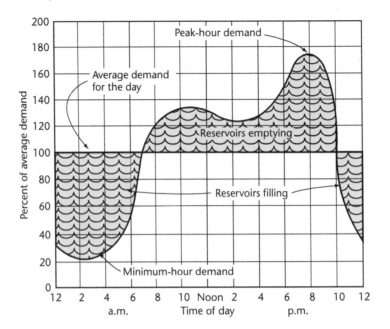

Figure 11-1 Daily variation of system demand

demand for the day. This means that without storage, the plant capacity would have to be almost double the size that is needed to meet average demand.

With adequate storage, water can be treated or supplied to the system at a relatively even rate over a 24-hour period. During midday, when the demand increases, the excess requirement can be made up from storage. Figure 11-1 shows that from 10 p.m. to 7 a.m., the demand on the system is below the average rate. During this time, the reservoirs are being filled. From 7 a.m. to 10 p.m., the demand is greater than the supply. The reservoirs are then being used to feed water back into the system.

A water system that purchases water from another water system is usually offered these three storage options:

1. The purchasers can draw water as needed, in which case they may need no storage of their own. They will, however, have to pay a premium price for the water, because the supplying system must supply the necessary storage or adequate plant capacity to meet demands.

2. The purchasers can take water at a relatively uniform or maximum rate throughout the day. In this case, the water cost is usually substantially less, but the purchasing system will have to provide enough storage of its own to supply peak demands.

3. The purchasers can draw all or most of their water at night, when the treatment plant usually has excess capacity. The water purchased during off-peak hours is usually much less expensive. The rate charged for the water is often called a "dump rate." Under this plan, the purchasing system must furnish adequate storage to supply daytime demand for at least one day.

Increasing Operating Convenience

In some situations, storage is provided to allow a treatment plant to be operated for only one or two shifts, thereby reducing personnel costs. In this situation, storage provides the water required through the night when the plant is shut down.

Leveling Out Pumping Requirements

The demand for water is continually changing in all water systems. Demand depends on the time of day, day of the week, weather conditions, and many other factors. If there is no storage at all, the utility must continually match the changing demand by selecting pumps of varying sizes. Frequent cycling of pumps (i.e., turning them on and off frequently) causes increased wear on pump controls and motors. It also increases electrical costs.

If a water system has adequate elevated storage, pump changes can be minimized. Water will then flow from the tank if a sudden increase in consumer use is greater than what the operating pump can provide. Likewise, water in excess of the current demand will refill the tank when there is a reduction in water use on the system. The pumping rate will not need to be changed immediately.

However, if a water system has only underground storage, the utility must meet system water demands by constantly adjusting the pumping rate, either from the water source or from a storage facility. In this situation, the utility cannot take advantage of the force of gravity to move water through the system.

Decreasing Power Costs

Storage also allows pumping costs to be reduced if there are special rate structures for electrical power. Many electric power utilities have off-peak rates that are lower at certain times—usually during the night. A system can often achieve substantial savings in power costs if most large pumps are operated during off-peak hours. Water utilities can pump water to elevated storage during the night and do a minimum amount of pumping during the day.

Elevated storage can also help a water system avoid using large pumps to meet occasional peak demands, which usually provides a savings in electricity costs as well.

Providing Water During Source or Pump Failure

There are times when power failure, mechanical breakdown, or maintenance work will prevent use of source water pumps or even an entire treatment plant. In addition, sudden extreme increases in demand can be caused by main breaks, broken hydrants, or similar problems. Elevated storage will automatically provide additional water to the system during such emergencies. However, if only pumped storage is available, the utility must be able to begin pumping immediately in order to maintain system pressure.

Another situation in which a relatively large amount of storage may be required involves a water system that depends on a single, long transmission main as its only water source. These systems usually provide storage equal to at least one average day's water use to allow for possible repair or maintenance of the transmission main.

Providing Water to Meet Fire Demands

A major purpose of distribution system storage is to meet fire demands. Although fire demand may not occur very often, the rate of water use is usually much greater than for domestic peak demands. Water systems are usually designed to meet fire demand in addition to normal customer needs. Fire demand can account for as much as 50 percent of the total capacity of a storage system.

Providing Surge Relief

As pumps turn on and off and valves are opened and closed, tremendous pressure changes, known as water hammer, can surge through the distribution system. Excessive water hammer can seriously damage pipes and appurtenances. Elevated storage tanks greatly assist in absorbing the shocks of water hammer by allowing the shock wave to travel up the riser and into the upper tank section.

---

elevated storage
In any distribution system, storage of water in a tank supported on a tower above the surface of the ground.

fire demand
The required fire flow and the duration for which it is needed, usually expressed in gallons (or liters) per minute for a certain number of hours. Also used to denote the total quantity of water needed to deliver the required fire flow for a specified number of hours.

Increasing Detention Times

The time that water spends in storage after disinfectants are added, but before the water is delivered to the first customer, can be counted toward the disinfectant contact time required for surface water treatment. Storage reservoirs located at the discharge of a treatment plant provide system storage and help meet disinfection requirements at the same time.

Blending Water Sources

Some water systems use water from two or more sources, with each source having different water quality. An example is a system that uses water from both ground and surface sources. Blending different sources together in a reservoir will often improve the quality of marginally acceptable water.

Both industrial and domestic customers dislike day-to-day changes in hardness, taste, temperature, or chemical composition. Blending water from different sources in a reservoir allows water of relatively uniform quality to be furnished to the system.

Capacity Requirements

Distribution storage capacity is based on the maximum water demands in the different parts of the system. It varies for different systems and can be determined only by qualified engineers after a careful analysis and study of the system. The storage capacity needed for fire protection should be based on the recommendations of fire underwriters' organizations. Because so many variables are involved, operators should contact the Insurance Services Office or the fire insurance rating office in their state to obtain information.

The amount of emergency storage required is also based on both the reliability of the water source and the availability of backup equipment and standby power sources. A system having two power lines from different substations is usually considered to have a relatively secure power source in an emergency. It is even more secure to have engine generators and/or engine-driven pumps that will supply at least average system demands. A water system should also have backup pumps, so that maximum demands can be met even if major pumping units should fail.

Types of Treated-Water Storage Facilities

Water storage tanks can be classified based on type of service, configuration, and type of construction material.

Type of Service

Water storage tanks can be used for either operating storage or emergency storage. An **operating storage** tank generally "floats" on the system. In other words, the tank is directly connected to the distribution piping, and the elevation of the water in the tank is determined by the pressure in the system. Water flows into the tank when water demand is low, and it empties from the tank when demand exceeds supply.

Emergency storage is designed to be used only in exceptional situations, such as during high-demand fires. A tank installed by an industry for use with the industry's own fire protection sprinkler system is an example of emergency storage. Because emergency storage water is not constantly being turned over, it stagnates, loses all residual chlorine, and can become contaminated. It is not

---

operating storage
A tank supplying a given area and capable of storing water during hours of low demand, for use when demands exceed the pumps' capacity to deliver water to the district.

emergency storage
Storage volume reserved for catastrophic situations, such as a supply line break or pump-station failure.

usable as a potable water source unless it is further treated. Because the water in emergency storage tanks is not circulated, the tanks must be provided with heating equipment in colder climates to prevent the water from freezing.

In establishing storage for primarily standby or seasonal use, one must consider how to keep the stored water from degrading in quality. Water stored for a period of time can, in some circumstances, develop a foul taste or odor or become unsafe.

Some operators of water systems with reservoirs that are primarily designed to meet summer demand have found it best to drain some of the reservoirs during the winter. Otherwise, a procedure must be established to regularly remove some of the water from each reservoir and replace it with fresh water every day or two.

Configuration

Distribution storage facilities can be either located at the ground level or elevated. These facilities are classified as tanks, standpipes, or reservoirs.

Elevated Tanks

Elevated tanks generally consist of a water tank supported by a steel or concrete tower. In general, an elevated tank floats on the distribution system. Occasionally, system pressure could become so high that the tank would overflow. In these cases, an altitude valve must be installed on the tank fill line to keep the tank from overflowing.

Standpipes

A tank that rests on the ground and has a height that is greater than its diameter is generally referred to as a standpipe. In most installations, only water in the upper portion of the tank will furnish usable system pressure, so that most larger standpipes are equipped with an adjacent pumping system that can be used in an emergency to pump water to the system from the lower portion of the tank.

Standpipes combine the advantages of elevated storage with the ability to store a large quantity of water. Note that a relatively large amount of water must regularly be circulated through the tank to keep the water fresh and to prevent freezing.

Reservoirs

The term reservoir has a wide range of meanings in the water supply field. For raw water, reservoirs are generally ponds, lakes, or basins that are either naturally formed or constructed for water storage. For storage of finished water, the term is usually applied to a large aboveground or underground storage facility. Reservoirs may also be referred to as ground-level tanks.

Distribution system reservoirs are usually used where very large quantities of water must be stored or when an elevated tank is objectionable to the public. If a reservoir can be located on a high rise of ground, it can float on the system. This way, all or most of the water is directly available to the system without the need for pumping.

When a ground-level or buried reservoir is located at a low elevation on the distribution system, water is admitted through a remotely operated valve. A pump station is provided to transfer the water into the distribution system. The valves and pumps are usually controlled from the treatment plant or a central control center.

Completely buried reservoirs are often used where an aboveground structure is objectionable, such as in a residential neighborhood. In some cases, the land over a buried reservoir can be used for recreational facilities such as a ball field or tennis court.

---

elevated tank
A water distribution storage tank that is raised above the ground and supported by posts or columns.

standpipe
A ground-level water storage tank for which the height is greater than the diameter.

reservoir
(1) Any tank or basin used for the storage of water. (2) A ground-level storage tank for which the diameter is greater than the height.

ground-level tank
In a distribution system, storage of water in a tank whose bottom is at or below the surface of the ground.

The initial cost of a ground-level reservoir is generally less than providing equal elevated storage. However, the water must regularly be circulated in and out of a ground-level reservoir to keep it fresh. Thus, there is a continuing cost for power and the maintenance of pumping equipment. The cost of constructing a completely buried reservoir will be higher than for one at ground level because of the considerable amount of excavation required.

> **WATCH THE VIDEO**
> Water Storage Tanks—Types (www.awwa.org/wsovideoclips)

Hydropneumatic Systems

For very small water systems that cannot afford to install an elevated or ground-level tank, a hydropneumatic system will often provide adequate continuity of service for domestic use. As illustrated in Figure 11-2, a steel pressure tank is kept partially filled with water and partially filled with compressed air. Some tanks also have a flexible membrane that separates the air and water.

The compressed air maintains water pressure when use exceeds the pump capacity. The tank also reduces the frequency with which the pump needs to be turned on and off. It provides water for a limited time in the event of pump failure.

Type of Construction Material

Over the years, reservoirs have been constructed from a variety of materials. Materials of construction and surfaces in contact with potable water should comply with all applicable standards (ANSI/AWWA D100, D102, D103, D104, D110, D115, D120, D130, and ANSI/NSF Standard 61). Early reservoirs were constructed with earth-embankment techniques. Today, steel and concrete are the most widely used construction materials.

Earth-Embankment Reservoirs

Early distribution storage facilities were often earth-embankment reservoirs that were built on higher ground, if available, or near the service area. These reservoirs were usually constructed partly by excavation and partly by embankment. They were paved with stone riprap, brick, or concrete on the slopes and bottoms and were almost always open at the top.

hydropneumatic system

A system using an airtight tank in which air is compressed over water (separated from the air by a flexible diaphragm). The air imparts pressure to water in the tank and the attached distribution pipelines.

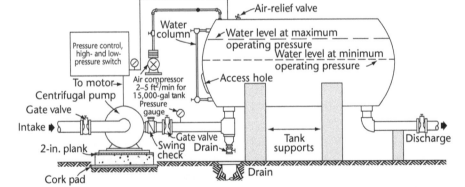

NOTE: Use special rubber hose fitting between pump and pressure tank for quiet operation.

Figure 11-2 Hydropneumatic water pressure system

This material is used with permission of John Wiley & Sons, Inc., from *Environmental Engineering and Sanitation*, 4th ed., Joseph A. Salfvato, © 1992 by John Wiley & Sons, Inc.

Water Storage 223

These installations were effective and valuable in system operation at the time, but problems included leakage, freezing, contamination by birds and animals, and algae growth. State and federal regulations now prohibit the storage of potable water in open reservoirs.

Steel Tanks

Most elevated tanks and standpipes, as well as many ground-level tanks, are constructed out of steel. The thickness of the steel used in constructing a tank varies with the pressure exerted on the tank wall. The upper walls may be relatively thin, but the lower walls of a tall tank may have a thickness of 2 in. (50 mm) or more.

Early steel tanks were constructed of panels that were riveted together, but this method was gradually replaced by welding starting around 1930. Welded tanks are much easier to maintain because of their smooth surface. Welding also allows tanks to be designed in new, more pleasing designs that would not be possible with riveted construction. Standards for constructing potable water tanks from welded steel are covered in AWWA Standard D100, *Welded Carbon Steel Tanks for Water Storage* (most recent edition).

Figure 11-3 shows an example of the unusual types of designs that can be achieved with welded steel construction. Most newer elevated tanks are of the single-pedestal design, which is more visually pleasing and easier to maintain because of its smooth surfaces.

Bolted tanks are also available in reservoir and standpipe configurations. The tanks are constructed of uniformly sized panels that are assembled with gaskets or a sealant to achieve a watertight seal at the bolted joints. The panels are factory coated for long-term corrosion protection. The coating systems presently available include hot-dipped galvanized, fused glass, thermoset liquid epoxy, and thermoset dry epoxy types. Figure 11-4 shows a typical bolted tank. Guidelines for

Figure 11-3 A 1,000,000-gal single-pedestal spheroidal elevated tank painted to look like a hot-air balloon
Courtesy of CB&I.

Figure 11-4 Typical bolted steel tank
Courtesy of Engineered Storage Products Company.

bolted steel tanks are found in AWWA Standard D103, *Factory-Coated Bolted Carbon Steel Tanks for Water Storage* (most recent edition).

Concrete Tanks and Reservoirs

The first concrete reservoirs were open, but later designs were roofed with wood or concrete. A properly ventilated reservoir will generally not have sanitation problems. Concrete reservoirs may be constructed using several different techniques.

Cast-in-place concrete tanks are constructed much the same as the basement for a house. However, much more reinforcing is needed so the tank can resist the weight of the water. In addition, more care is needed to reduce leakage. The shape of a cast-in-place concrete reservoir is generally square or rectangular.

It is difficult to prevent some cracking of cast-in-place concrete tanks, but the cracks can usually be repaired with new types of flexible caulking materials. Otherwise, membrane liners can be added to prevent water loss.

Circular, prestressed concrete tanks are constructed by first installing an inner concrete core wall that establishes the reservoir's circular form. Steel wire is then wrapped around the core wall under tension (Figures 11-5 and 11-6). The steel wrapping is then protected by layers of hydraulically applied concrete (gunite).

The walls of prestressed tanks can be made thinner and with less reinforcing steel than those of cast-in-place concrete tanks. However, the tanks must be built only by qualified contractors. Prestressed tanks are constructed in accordance with AWWA Standard D110, *Wire- and Strand-Wound, Circular, Prestressed Concrete Water Tanks* (most recent edition).

Hydraulically applied concrete-lined reservoirs consist of an earth embankment that is covered with reinforced, hydraulically applied concrete. This process is similar to that used for constructing many swimming pools. Reservoirs can be constructed at relatively low cost with this method, but they are usually shallow and difficult to cover.

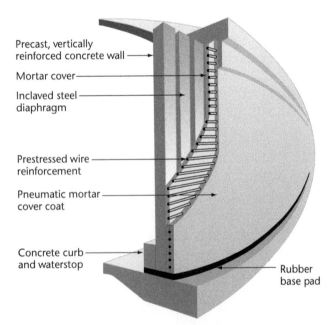

Figure 11-5 Sectional view of a prestressed concrete tank

Courtesy of Preload Inc.

Figure 11-6 Circumferential wire prestressing operation

Courtesy of Preload Inc.

Location of Distribution Storage

The location of distribution storage facilities is closely associated with the system hydraulics and water demands in various parts of the system. It is also affected by the availability of appropriate land and by public acceptance of the structure.

Elevated Storage

Some concerns linked to the location of elevated storage include the quality of the system hydraulics, pumping and transmission costs, and aesthetic considerations.

Relationship to System Hydraulics

When elevated storage is placed adjacent to the pump station, as illustrated in Figure 11-7A, the head loss to the farthest portion of the system may be excessive through normal-sized piping. As a result, additional transmission mains may have to be added to the system to provide enough pressure to remote areas.

To avoid the cost of increased main size, the elevated tank can be located beyond the service area (Figure 11-7B), so that the pressure will be significantly improved with existing main size. In this situation, however, there must be adequate main capacity to the remote location to refill the tank during off-peak periods.

An even better solution is shown in Figure 11-7C. Here the tank is located in the part of the service area that originally had the lowest pressure. In this case, the pressure is slightly improved over the situation before the tank was installed.

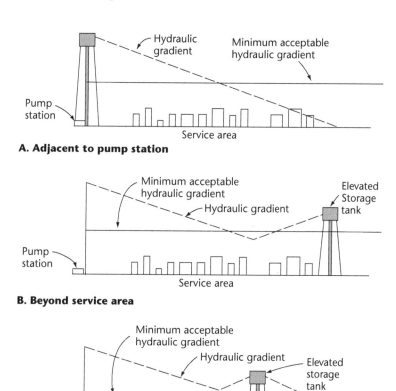

Figure 11-7 Different locations of elevated storage

Courtesy of *Public Works Magazine*.

In addition, smaller-diameter mains can be connected to the tank because the flow from the tank is split into two directions.

It is often desirable to provide several smaller storage units in different parts of the system rather than one large tank of equal capacity at a central location. Smaller pipelines are used for decentralized storage. In addition, other things being equal, the smaller units do not require the flow-line elevation or pumping head to be as high. Of course, the cost of building and maintaining several smaller tanks must be compared with the cost of larger water mains that would accomplish the same objective.

An elevated tank should be located on the highest point of ground that also meets other hydraulic criteria. A tank at a higher elevation will not need to be as tall as a tank at a lower elevation. Eliminating 50–75 ft (15–23 m) in the height of a tank can provide a considerable cost savings.

Minimizing Pumping and Transmission Costs

As illustrated in Figure 11-8, there can be a very substantial pressure loss when water is pumped over a long distance. The situation worsens drastically under high-demand conditions. The location of storage near the load center, as shown in Figure 11-8, would allow the pumping station to operate near average-day conditions most of the time.

Aesthetic Concerns

Although existing elevated tanks rarely bother anyone, property owners will usually object to a new tank being built near their homes. Although many designs are very pleasing and colors can be selected to minimize the visual impact, objections are usually still made. Another complaint is that an elevated tank might cause television interference.

As a result, the best hydraulic location and the most economical design are not always the deciding factors in the location of an elevated tank. In some cases, the only acceptable location will be in an industrial area or public park. In cases where public feeling is very strong, a water utility may have to construct ground-level storage, which is more aesthetically acceptable.

Ground-Level Storage

Ground-level storage is generally used where a large quantity of water must be stored or where topography or community objections do not permit the economical location of an elevated tank. A relatively large parcel of land is required to

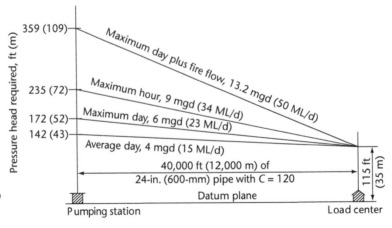

Figure 11-8 Sufficient pumping head must be provided when there is no storage

accommodate both the reservoir and an accompanying pump station. If the distribution system has several pressure zones, ground-level storage and booster pumping are often located at the pressure zone boundary. This way, water from the lower zone flows into the reservoir and is pumped to the zone with higher pressure.

Water Storage Facility Equipment

This section discusses the types of equipment associated with elevated and ground-level storage facilities.

Elevated Storage Tanks

Elevated storage tanks (Figure 11-9) are often constructed of steel. The tanks contain a variety of equipment options, including inlet and outlet pipes, overflow piping, drain connections, monitoring devices, valving, air vents, access hatches, ladders, protective coatings, cathodic protection, and obstruction lighting.

Inlet and Outlet Pipes

The same pipe is generally used as both the inlet and outlet pipe on an elevated tank. This pipe is called a **riser**. In cold climates, any exposed risers on multicolumn tanks are generally 6 ft (1.8 m) in diameter or larger to allow for some freezing around the edge and for expansion when water turns to ice. In extremely cold climates, an exposed riser may have to be insulated or heated. An advantage of the single-pedestal support design is that heat can be provided within the support column to keep the temperature around the riser pipe above freezing. In locations where freezing temperatures do not occur, a small riser pipe with just enough capacity to carry maximum flow may be used.

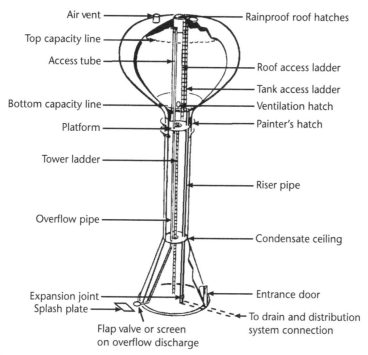

Figure 11-9 Principal accessories for an elevated storage tank
Courtesy of CB&I.

riser
The vertical supply pipe to an elevated tank.

Overflow Pipe

An overflow pipe is necessary on all tanks to safeguard the tank in the event that the water-level controls fail. The overflow pipe should be brought down from the maximum tank level to a point within about 1 ft (0.3 m) of the ground surface. It should discharge to a splash plate or drainage inlet structure to prevent soil erosion at the tank foundation. The overflow pipe should never be directly connected to a sewer or storm drain.

The overflow pipe discharge should be covered by a weighted flap valve or screen to keep out bugs and animals, but it must positively open or break away so that the discharge is unobstructed if overflow takes place.

Drain Connection

An elevated tank must be furnished with a drain connection to empty the tank for maintenance and inspection. A commonly used method is to install a fire hydrant on the main connection to the distribution system, with a valve on either side of the hydrant tee (Figure 11-10). Utility personnel can empty the tank by draining it through the hydrant after closing one of the valves. After draining is complete, the valve positions can be reversed so that the hydrant can be used for tank maintenance work.

If the water drained from a tank is discharged onto the ground, the utility must make provisions to carry away the considerable flow created without eroding or flooding the surrounding area. If the drained water is discharged to a storm sewer, the design must be such that a potential cross-connection is not created.

Monitoring Devices

The water level in a tank can be measured either by a pressure sensor at the base of the tank or by a level sensor inside the tank. The level is then usually indicated on an instrument at the site, and the data are transmitted to a central location. In many systems, water level information is used to manually or automatically operate pumps that maintain system pressure. The level indicator is usually also furnished with alarms to alert the operator of an unusually high or low water level in the tank.

Valving

An elevated tank must be furnished with a valve at the connection to the distribution system. This valve is used to shut the tank off for maintenance and inspection. When a tank floats on the system, this valve is left open for water to pass in and out of the tank as pressure fluctuates.

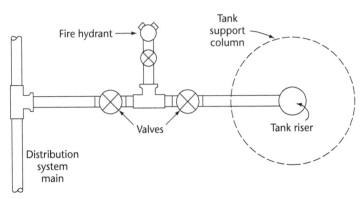

Figure 11-10 Fire hydrant installed for draining an elevated tank

If a tank is not tall enough to accept full system pressure without overflowing, an altitude valve is installed on the distribution system connection. This valve automatically shuts off flow to the tank when the water level approaches the overflow point. Flow out of the tank, however, is usually unrestricted.

Air Vents

Air vents must be installed to allow air to enter and leave the tank as the water level falls and rises. If the vent capacity is insufficient to admit air when water is draining from the tank, the vacuum created could collapse the tank.

Vents should be screened to keep out birds and animals that might contaminate the water. A screen with ¼-in. (6-mm) mesh openings is required by most state regulations. Insects seldom contaminate stored water, and insect screening is not recommended because it may become clogged or covered with ice. Some vent designs have flap valves that will operate to relieve excess pressure or vacuum if the screen should become blocked.

Access Hatches

Water storage tanks must have hatches installed for both entry and ventilation during maintenance and inspection. Hatches on the roof of a tank must be installed with rims under the cover. These rims are designed to prevent surface runoff from entering the tank.

The hatch at the bottom of a large-diameter wet riser must be designed to withstand the pressure of the water column. Some designs use a large exterior flange with multiple bolts. Others have a hatch with an interior cover that is held in place by a crossbar and large bolt. All hatches and access holes must be securely locked when not in use to prevent vandalism and reduce the security risk. Figure 11-11 illustrates how all access to a single-column tank is through a secure doorway.

Ladders

Multicolumn tanks generally have three ladders. The tower ladder runs up one leg of the tower from the ground to the balcony. The bottom of most ladders begins about 8 ft (2.5 m) from the ground to deter unauthorized persons from climbing the tank. On other tanks, the ladder is installed with the base at the ground surface but with a heavy metal shield locked in place over the bottom of the ladder to prevent unauthorized entry.

The second ladder, or tank ladder, runs from the balcony to the roof. The third ladder, called the roof ladder, is installed on the top of the roof from the tank ladder to the roof access hatch. Tanks located in warm climates may also have an inside ladder for the convenience of those working inside the tank. However, in climates where ice may form in the tank, an inside ladder cannot be permanently installed because it would be destroyed by the moving ice.

Pedestal tanks such as those illustrated in Figures 11-9 and 11-11 are furnished with similar ladders. One extends from the ground to the bottom of the tank, another goes up through the access tube to the top of the tank, and another is located inside the tank if icing is not a problem.

All ladders and safety devices including safety cage, safety cable, or safety rail, must meet Occupational Safety and Health Administration (OSHA) requirements. Ladders may also need to have rest platforms or roof-ladder handrails.

 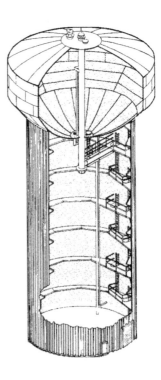

Figure 11-11 Exterior view and cutaway drawing of a fluted column elevated tank
Courtesy of CB&I.

Coatings

Steel, when exposed to the environment, oxidizes and deteriorates, particularly when the environment includes both oxygen and moisture. It is therefore necessary to protect both the interior and exterior surfaces of steel tanks from corrosion. Several types of paints or coatings can be used for this purpose.

Interior coatings must be able to withstand the following conditions:

- Constant immersion in water
- Varying water temperatures
- Alternate wetting and drying periods
- Ice abrasion
- High humidity and heat
- Chlorine and mineral content in the water

The exterior tank coating must endure similar conditions. In addition, it must maintain a good appearance over a reasonable period of time. A recent concern that has emerged concerning older tanks is that many were originally coated with lead-based paint. If the old coating is to be removed by sand blasting, it is important to observe state and federal regulations that restrict the levels of both lead and silica that can be released into the atmosphere. Special blast abrasive and methods of containing the lead must be used if a lead coating is to be removed. In addition, at the conclusion of the work, the removed material must be tested. It may have to be disposed of as a potentially hazardous waste.

Water Storage 231

It is important that the new coating be effective and safe. It must not cause taste or odor problems. A reliable manufacturer should be consulted in selecting a coating system that meets NSF International standards (see Appendix B).

Inspecting the work of a painting contractor is both dangerous and beyond the knowledge of most water system operators. It is generally a good idea to employ a qualified third-party firm to inspect the work. The inspectors selected should be trained and experienced in tank painting. They must be able to climb and use rigging to reach the work areas for inspection. Competing painting contractors should not be used as inspectors because of possible conflicts of interest.

Cathodic Protection

In addition to a good inside coating system, cathodic protection can further reduce corrosion of the interior walls below the water line of steel tanks. Cathodic protection reverses the flow of current that tends to dissolve iron from the tank surface and cause rust and corrosion. If electrodes are placed in the water and a direct current is impressed on them, the electrodes will corrode (or "be sacrificed") instead of the tank.

In warm climates, the electrodes can be suspended from the tank roof and left in place until they have corroded to the point of requiring replacement (Figure 11-12). In climates where ice is formed in the tank, the ice will quickly pull down suspended electrodes. A submerged system can be installed that resists damage by icing (Figure 11-13). Although the anodes will generally last up to 10 years, it is recommended that cathodic protection systems be inspected annually to ensure they are operating properly.

Although cathodic protection systems can be operated on bare steel or in a tank with a badly deteriorated coating, a large amount of power will be used and the anodes will quickly disintegrate. The usual approach is to install a good coating system and a cathodic system. This combination extends the useful life of the coating and anodes with minimal power consumption. Details on the installation and operation of cathodic protection for steel water tanks are covered in

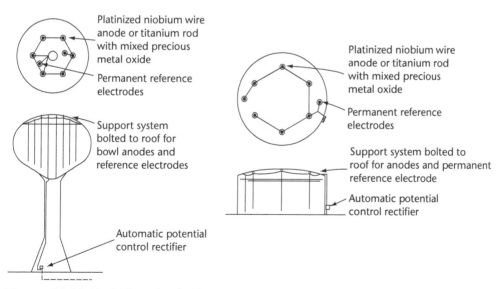

Figure 11-12 Typical methods of suspending cathodic protection anodes in steel tanks where icing conditions do not exist

Courtesy of Corrpro Waterworks, a Corrpro Company, 1055 West Smith Road, Medina, Ohio 44256.

cathodic protection
An electrical system for preventing corrosion to metals, particularly metallic pipe and tanks.

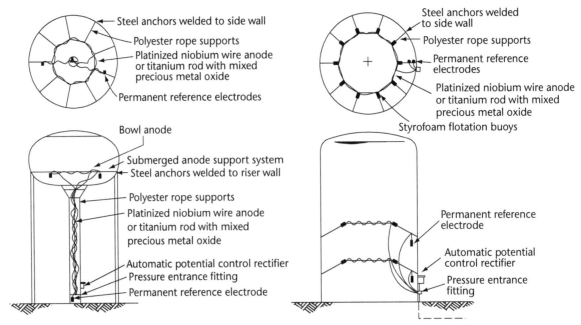

Figure 11-13 Typical methods of suspending cathodic protection anodes in steel tanks where icing conditions exist
Courtesy of Corrpro Waterworks, a Corrpro Company, 1055 West Smith Road, Medina, Ohio 44256.

AWWA Standard D104, *Automatically Controlled, Impressed-Current Cathodic Protection for the Interior of Steel Water Tanks* (most recent edition).

Obstruction Lighting

Depending on the height and location of an elevated tank, the Federal Aviation Administration (FAA) may require the installation of obstruction lights or strobe lights to warn aircraft in the vicinity. In particularly hazardous locations, orange and white obstruction painting may also be required. It is even possible that the FAA will not allow the construction of an elevated tank in some locations, so an application for FAA clearance should be made before a site location is finalized.

Ground-Level Storage Facilities

The equipment required for a ground-level storage facility includes inlet and outlet pipes; overflow pipes, vents, and hatches; drains; and corrosion protection.

Inlet and Outlet Pipes

Ground-level and buried storage tanks usually have a single pipe that serves as both inlet and outlet. Separate inlet and outlet pipes may be used if there is a need to improve the circulation of water within the tank. Improved circulation helps the quality of the water remain uniform and reduces freezing. The opening of the inlet–outlet pipe is usually located a short distance above the floor. Alternatively, a silt stop may be installed to keep sediment on the floor from being drawn into water leaving the tank.

Overflow Pipes, Vents, and Hatches

The same general comments for overflows, vents, and hatches on elevated tanks also apply to ground-level tanks. However, for ground-level tanks, this equipment is generally even more accessible to vandals. In particular, if the area over or around a buried reservoir is used for a play area or other public use, all exposed equipment must be especially sturdy and designed with child safety in mind.

Drains

If the location and topography are appropriate, it is convenient to have a gravity drain on a ground-level tank. However, this should not be done if a possible cross-connection will be created. If a drain cannot be installed, and if there is a pump station at the tank, the usual procedure for emptying the tank is to pump most of the water to the system. It is not generally wise to pump all the way to the outlet opening because any sediment on the bottom will be disturbed and pumped into the system as cloudy water. The remaining water in the tank must then be removed by portable pumps.

Corrosion Protection

Steel ground-level tanks should be painted and protected by cathodic protection in the same manner as elevated steel tanks. Aboveground steel tanks are generally set on a concrete ring-wall foundation. The section of steel wall immediately above the foundation is a particularly vulnerable point for rusting. Care should be taken to keep this area free of weeds and dirt all the way around the tank.

Concrete tanks are not generally painted, but occasionally there is a demand to paint the exterior of an aboveground tank for aesthetic reasons. An exterior coating may also be used for corrosion protection in areas near the ocean with salty air or where there are industrial gases in the air. The utility must take extreme care in selecting the proper time and correct painting system. If the paint does not stick properly and begins to flake or peel, removing the coating could be extremely costly. There may also be a need to coat the interior or to place a membrane liner in an older concrete tank. This serves as a means of protecting concrete that is disintegrating.

Operation and Maintenance of Water Storage Facilities

This section discusses some of the normal concerns in the daily operation of water storage facilities.

Cold-Weather Operation

In cold climates, freezing water in tanks can be a serious problem. Freezing is more likely in systems using surface water sources because the water entering the tank may, at times, be just above freezing temperature. Even though groundwater is usually somewhat warmer, icing problems in storage tanks can also occur during extremely cold weather.

If ice forms inside a tank, it can damage the paint, cathodic protection system, ladders, and overflow pipes that are exposed. If it is expected that ice may form in a tank, there should be nothing exposed in the interior that can be damaged.

Freezing in ground-level tanks is not usually serious because of the way in which the tanks are operated. As long as some water is pumped from the tank to the system and new, warmer water is filled back in every day, there may only be nominal freezing around the walls.

Multicolumn elevated tanks constructed in areas with freezing temperatures are furnished with greatly oversized risers. It is expected that some ice will form around the walls in very cold weather. The water circulating in and out of the tank will normally be able to pass through the riser. However, if the riser is allowed to freeze solid, the upper tank may then also freeze. If this happens, severe structural damage can result because water expands as it turns to ice.

It is impossible to keep ice from forming around the walls of an elevated tank under extremely cold conditions. This ice is not normally of concern. What is most important is to prevent thick ice from forming on the water surface. If ice is allowed to become so thick that it will resist the pressure of the water, the tank will become unusable. The tank could also be damaged by the expansion of the ice.

To keep ice from getting too thick, the utility must make a special effort in extremely cold weather to provide frequent changes in the water level. This will both discharge cold water and bring in warmer water. It will also break up any surface ice that has formed. If automatic controls normally maintain an almost constant water level in a tank, they should be overridden in very cold weather to provide pressure fluctuations. If a tank should become frozen, it is possible to use a steam generator or electric heaters to thaw it, but this process can be very costly.

Basic Maintenance

It is recommended that all storage structures be completely inspected and cleaned every 3–5 years. Consult your local regulatory requirements regarding storage tank maintenance.

Elevated Tanks

Elevated tanks should periodically be drained, cleaned, inspected, repaired (if necessary), and painted. Because of the specialized nature of the work and the many dangers involved in working in high and confined places, this work should be performed by a competent contractor.

Ground-Level Tanks

Ground-level tanks can generally be drained, cleaned, and inspected by water system workers as long as careful attention is given to the dangers of working in confined spaces and other potentially dangerous conditions. The surfaces of the walls and floor should be cleaned thoroughly with a high-pressure water jet or by sweeping or scrubbing. All water and dirt must then be flushed from the tank. If workers notice pitting of steel tank walls or disintegration of concrete, it is usually best to obtain professional advice on how to prevent further damage.

If painting is necessary, it is extremely important that adequate measures be taken to exhaust the paint fumes. Paint fumes in a confined space can cause injury or even death.

Disinfection

After cleaning and/or painting are complete, water storage tanks must be disinfected before being placed in service. Either (1) liquid sodium hypochlorite solution or (2) calcium hypochlorite granules or tablets may be used. Alternate methods of chlorination are listed in AWWA Standard C652, *Disinfection of Water-Storage Facilities* (most recent edition). The state water supply agency should also be consulted for any specific requirements.

In the first method, the volume of the entire tank is chlorinated, so that the water will have a free chlorine residual of at least 10 mg/L after the proper detention time. The detention time is 6 hours if the disinfecting water is chlorinated before entering the tank, and 24 hours if the water is mixed with hypochlorite in the tank. The chlorine level must be reduced to acceptable levels before the water is used or discharged.

The second method involves spraying or painting all interior tank surfaces with a solution of 200 mg/L available chlorine. This procedure requires special precautions and should be done only by trained, experienced, and properly equipped personnel.

In the third method, 6 percent of the tank volume is filled with a solution of 50 mg/L available chlorine for at least 6 hours. Then the tank is completely filled and the solution held for 24 hours. An advantage of this method is that if the test results are satisfactory, the chlorine levels of the water are generally acceptable for immediate use of the water in the distribution system.

The amount of chlorine needed can be determined from Table 11-1. The amount of chemical required when the chlorine solution must be neutralized before being discharged is shown in Table 11-2. Check with the local sewer department before discharging any highly chlorinated wastewater into a sanitary sewer. Consult with the state or local environmental agency (regulatory agency) before discharging highly chlorinated water anywhere. The agency may have special disposal requirements.

After the chlorination procedure is completed, and before the tank is placed in service, water from the full tank must be tested for bacteriological safety (coliform test). If the tests show the sample to be bacteriologically safe according to

Table 11-1 Amounts of chemicals required to give various chlorine concentrations in 100,000 gal (378.5 m^3) of water*

Desired Chlorine Concentration in Water, mg/L	Liquid Chlorine Required		Sodium Hypochlorite Required						Calcium Hypochlorite Required 65% Available Chlorine	
			5% Available Chlorine		10% Available Chlorine		15% Available Chlorine			
	lb	(kg)	gal	(L)	gal	(L)	gal	(L)	lb	(kg)
2	1.7	(0.77)	3.9	(14.7)	2.0	(7.6)	1.3	(4.9)	2.6	(1.2)
10	8.3	(3.76)	19.4	(73.4)	9.9	(37.5)	6.7	(25.4)	12.8	(5.8)
50	42.0	(19.05)	97.0	(367.2)	49.6	(187.8)	33.4	(126.4)	64.0	(29.0)

*Amounts of sodium hypochlorite are based on concentrations of available chlorine by volume. For either sodium hypochlorite or calcium hypochlorite, extended or improper storage of chemicals may have caused a loss of available chlorine.

Table 11-2 Amounts of chemicals required to neutralize various residual chlorine concentrations in 100,000 gal (378.5 m^3) of water

Residual Chlorine Concentration, mg/L	Chemical Required, lb (kg)							
	Sulfur Dioxide (SO_2)		Sodium Bisulfite ($NaHSO_3$)		Sodium Sulfite (Na_2SO_3)		Sodium Thiosulfate ($Na_2S_2O_3 \cdot 5H_2O$)	
1	0.8	(0.36)	1.2	(0.54)	1.4	(0.64)	1.2	(0.54)
2	1.7	(0.77)	2.5	(1.13)	2.9	(1.32)	2.4	(1.09)
10	8.3	(3.76)	12.5	(5.67)	14.6	(6.62)	12.0	(5.44)
50	41.7	(18.91)	62.6	(28.39)	73.0	(33.11)	60.0	(27.22)

the requirements of the state regulatory agency, the tank can be placed in service. If the samples test unsafe, further disinfection must be performed and repeat samples taken until two consecutive samples test safe.

Inspection

Water storage facilities must be inspected periodically to find any structural problems and correct them before they become serious. Cracks or holes in tanks can cause a loss of water, contamination of the water supply, and possible eventual total failure of the structure.

Tanks should be inspected for corrosion and cracks on both the inside and outside. This requires draining the tank to check the surfaces and the operation of the cathodic protection equipment. Overflows and vents should be examined to make sure that they are not blocked and that screens are clean and in place. AWWA Manual M42, *Steel Water Storage Tanks* (most recent edition), includes maintenance procedure information.

If an altitude valve is used, it should be checked to make sure it is allowing water into the tank and stopping the flow when the tank is full. It should also be checked for speed of opening and closing to make sure it does not cause water hammer. Inspectors should check level sensors and pressure gauges by changing the water level in the tank. Controls should be examined to see if they are operating at the desired pressures.

Water storage tanks should be checked frequently for vandalism and signs of forced entry. Access doors and their locks should be kept in good repair. Ladders must provide safe access for authorized personnel, but the entry should be locked to exclude unauthorized access. Security must be provided to prevent unauthorized access and protect the water from possible contamination.

Aviation warning lights, if installed, should be maintained so they will provide adequate warning to aircraft. Lights should be checked regularly to make sure they are working and clean enough that they do not have reduced light output. FAA regulations require that the bulbs be replaced before they reach 75 percent of their normal life expectancy.

Records

Most water systems have a relatively small number of distribution storage tanks, so the record-keeping system does not have to be extensive. Basic information to be recorded should include the tank location, dates of inspection, conditions noted during the inspection, dates on which maintenance was performed, and notes on the maintenance performed.

If maintenance is performed by outside contractors, a copy of the contracts, reports, and lists of repairs that were made should be maintained in a separate file for each structure. The phone numbers of equipment suppliers and contractors equipped to make repairs should also be kept available for use in an emergency.

Water Storage Facility Safety

Only trained and experienced operators should be allowed to work on elevated and ground-level storage tanks and standpipes. This is hazardous work and dangerous for untrained workers. Special precautions are also needed for work on or in tanks. In these confined working areas, workers must guard against slipping and falling from dangerous heights.

In addition to the safety precautions discussed in Chapter 7, the following guidelines apply to work in and around storage facilities:

- The security of ladders must be checked frequently. Required safety cages or safety cable equipment must be provided.
- Workers must be provided with boots and clothing for working in wet and slippery conditions.
- Workers performing disinfection must be provided with special protective goggles and gloves.
- Special fans or other ventilation equipment must be provided inside tanks while work is being done there.
- Adequate light must be provided inside tanks for workers to perform their work properly and safely. Special care must be taken to use waterproof wiring and light units to prevent shocks in a wet environment.

Study Questions

1. The height of water in three differently shaped tanks is 22.4 feet. Which tank will have the highest psi at the bottom?
    a. The square tank
    b. The rectangular tank
    c. The cylindrical tank
    d. It will be the same in all three tanks.

2. Which of the following is the proper detention time for disinfecting a water storage tank that is filled with already chlorinated water such that the free chlorine residual is 10 mg/L after the proper detention time is completed?
    a. 4 hours
    b. 6 hours
    c. 8 hours
    d. 24 hours

3. Which of the following is the proper detention time for disinfecting a water storage tank with water that is mixed with hypochlorite already in the tank such that the free chlorine is 10 mg/L after proper detention time is complete?
    a. 6 hours
    b. 8 hours
    c. 12 hours
    d. 24 hours

4. In what type of water storage system does water generally "float" on the system?
    a. Elevated storage
    b. Demand storage
    c. Emergency storage
    d. Operating storage

5. The _____ may require the installation of obstruction lights or strobe lights on an elevated tank, depending on its height and location, to warn aircraft in the vicinity.
   a. Occupational Safety and Health Administration
   b. American Water Works Association
   c. Federal Aviation Administration
   d. Federal Communications Commission

6. What is the result of utilizing storage to minimize pumping?

7. What is the term for a tank that rests on the ground and has a height that is greater than its diameter?

8. What is the term for a vertical supply pipe to an elevated tank?

9. Ponds, lakes, or basins are all examples of what type of raw-water storage container?

Chapter 12
Electrical and Instrumentation-and-Control Systems

Electricity and Magnetism

For purposes of explanation, electricity is often classified as either static or dynamic. Both forms of electricity are composed of large numbers of electrons, and both forms interact with magnetism. The study of the interaction of electricity and magnetism is called electromagnetics.

Static Electricity

Static electricity refers to a state in which electrons have accumulated but are not flowing from one position to another. Static electricity is often referred to as electricity at rest. Once given the opportunity, it is ready to flow. An example of this phenomenon is often experienced when you walk across a dry carpet and then touch a doorknob; you may notice a spark at your fingertips and feel a sudden shock. Static electricity is often caused by friction between two bodies, as by rotating machinery—conveyor belts and the like—moving through dry air. Another common example of static electricity is the natural buildup of static electricity between clouds and the earth that results in an electrical discharge—a lightning bolt. In this case, the friction occurs between the air molecules. Static electricity is prevented from building up by properly bonding equipment to ground or earth.

Dynamic Electricity

Dynamic electricity is electricity in motion. Motion can be of two types—one in which the electrical current flows continuously in one direction (direct current), and a second in which the electrical current reverses its direction of flow in a periodic manner (alternating current).

Direct Current

Current that flows continuously in one direction is referred to as direct current (DC). In this case, the voltage remains at a fixed polarity, or direction, across the path through which the current is flowing. Direct current is developed by batteries and can also be developed by rotating-type DC generators.

Alternating Current

Electrical current that reverses its direction in a periodic manner—rising from zero to maximum strength, returning to zero, and then going through similar variations of strength in the opposite direction—is referred to as alternating current (AC). In this case, the voltage across the circuit varies in potential force

---

electromagnetics
The study of the combined effects of electricity and magnetism.

static electricity
A state in which electrons have accumulated but are not flowing from one position to another.

direct current (DC)
Current that flows continuously in one direction.

alternating current (AC)
Electrical current that reverses its direction in a periodic manner.

in a periodic manner similar to the variations in current. Alternating current is generated by rotating-type AC generators.

Induced Current

A change in current in one electrical circuit will induce a voltage in a nearby conductor. If the conductor is configured in such a way that a closed path is formed, the induced voltage causes an induced current to flow. This phenomenon is the basis for explaining the property of a transformer. Since changes of current are a requirement for a transformer to operate, it is obvious that DC circuits cannot use transformers, whereas AC circuits are adaptable to the use of transformers. The function of a transformer is of great value because it allows the voltage of the induced current to be increased or diminished.

WATCH THE VIDEO
Electricity and Power (www.awwa.org/wsovideoclips)

Electromagnetics

The behavior of electricity is determined by two types of forces: electrical and magnetic. A familiar example of the electrical force is the static force that attracts dust to phonograph records in dry climates. The magnetic force holds a horseshoe magnet to a piece of iron and aligns a compass with magnetic north. Wherever there are electrons in motion, both forces are created. These electromagnetic forces act on the electrons themselves and on the matter through which they move. Under certain conditions, the motion of electrons sends electromagnetic waves of energy over great distances. Figure 12-1 shows an electrical current being produced when a coil of wire is moved through a magnetic field.

Electromagnetic energy is responsible for the transmission of a source of power. Because of electromagnetic energy, power can be transmitted across the air gap between the stator and rotor of a motor. Radio waves are electromagnetic. Radar is electromagnetic. So is light, which unlike most forms of electromagnetic waves, can be seen. The doorbell is an electromagnetic device; so is the solenoid valve on your dishwasher. In fact, much of the electrical equipment with which you are familiar is electromagnetic; your clock motor, the alternator on your car, and the electric typewriter use electromagnetics.

You will find it worthwhile to make your own list of equipment and appliances that are all-electric and that are electromagnetic. You will be surprised at how much you already know about this subject.

In fact, because electricity is such a part of everyday life, even persons who have not studied electricity know more about it today than did early scientific

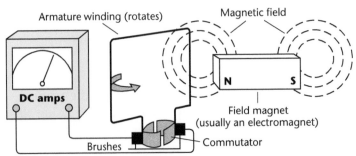

Figure 12-1 Current produced in a generator when a coil of wire moves through a magnetic field

Electrical and Instrumentation-and-Control Systems **241**

observers. Electricity is often taken for granted, but it can be dangerous and should be respected. Every operator should become familiar with safety rules and practices pertaining to electricity.

Electrical Measurements and Equipment

The following paragraphs introduce the basic terminology used to describe electricity and electrical equipment. To help illustrate the new concepts, the behavior of electricity is compared to the behavior of water (Figure 12-2).

Molecule of water ↔ *electron of electricity.* The smallest unit of water is a molecule, whereas the smallest unit of electricity is an electron.

Flow rate (gpm) ↔ *current (A).* The rate at which water flows through a pipe is expressed as gallons per minute (gpm) or liters per minute (L/min). It could be expressed as molecules per minute, but the numbers involved would be inconveniently large. Similarly, the rate at which electricity flows through a conductor (called current) could be expressed as electrons per second, but the numbers involved would be too large to be practical. Therefore, the unit of electrical current commonly used is the ampere (A), which represents a flow rate of about 6,240,000,000,000,000,000 electrons per second (also termed 1 coulomb per second). The flow rate of liquid expressed in gallons per minute is, therefore, analogous to the flow rate (current) of electricity expressed in amperes.

Pressure (psi) ↔ *potential (V).* Liquid flow requires a certain head or pressure. The liquid will tend to flow from the high pressure to the low pressure. Electrical potential is similar to head or pressure. Electrical current will always tend to flow from high potential to low potential. Electrical potential is expressed in terms of volts (V). Whereas liquid pressure is generally expressed in terms of pounds per square inch with reference to atmospheric pressure, electrical potential is generally expressed in terms of voltage with respect to ground or earth, with the general assumption that ground or earth is at zero potential.

Pressure drop ↔ *voltage drop.* The flow of liquid through a pipe is accompanied by a pressure drop as a result of the friction of the pipe. This pressure drop is often referred to as friction head loss. Similarly, the flow of electricity through a wire is accompanied by a voltage drop as a result of the resistance of the wire.

Friction ↔ *resistance.* The flow of water through a pipe is limited by the amount of friction in the pipe. Similarly, the flow of electricity through a wire is limited by resistance.

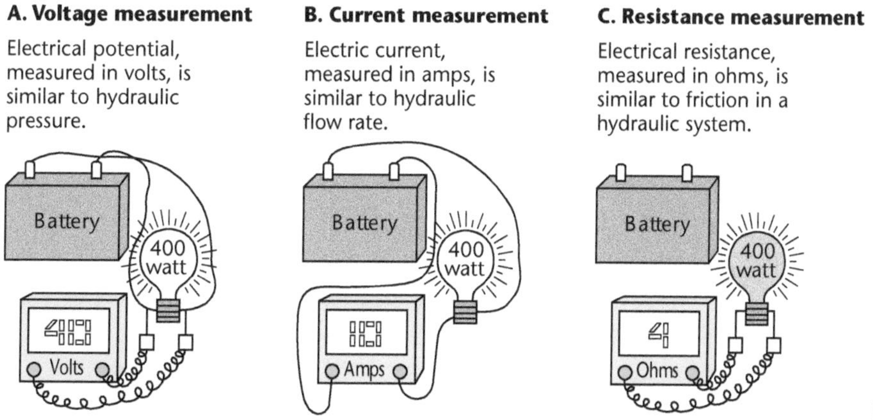

A. Voltage measurement
Electrical potential, measured in volts, is similar to hydraulic pressure.

B. Current measurement
Electric current, measured in amps, is similar to hydraulic flow rate.

C. Resistance measurement
Electrical resistance, measured in ohms, is similar to friction in a hydraulic system.

NOTE: Ohm's law states that voltage drop E across an electrical component equals current I flowing through the component multiplied by resistance R of the component: $E = IR$.

Figure 12-2
Representation of Ohm's law

Pump ↔ generator. A pump uses the energy of its prime mover, perhaps a gasoline engine, to move water; the pump creates pressure and flow. Similarly, a generator, powered by a prime mover such as a gasoline engine or a water turbine, causes electricity to flow through the conductor; the generator creates voltage and current.

Turbine ↔ motor. A water-driven turbine, like those found in hydroelectric power stations, takes energy from flowing water and uses it to turn the output shaft. An electric motor uses the energy of an electrical current to turn the motor shaft.

Turbine-driven pump ↔ motor-driven generator. A turbine running off a high-pressure, low–flow-rate stream of water (or hydraulic fluid) can be used to drive a pump to move water at low pressure, but at a high flow rate. (Such an arrangement might be useful for dewatering, for example, although it is not common.) Similarly, a motor operating from a high-voltage, low-current source can be used to run a generator having a low-voltage, high-current output. Note that both the water system and the electrical system could be reversed, taking a source with a low pressure (voltage) and high flow rate (current) and creating an output with a high pressure (voltage) and low flow rate (current). The motor-driven generator (also called a dynamotor) is sometimes used to convert DC battery power to a higher-voltage, lower-current AC power.

Turbine-driven pump ↔ AC transformer. For most electrical applications, the motor-driven generator is replaced with a transformer, which does exactly the same thing—it transforms low-voltage, high-current electricity into high-voltage, low-current electricity; or, it can transform high-voltage, low-current electricity into low-voltage, high-current electricity. The transformer will work only with AC power. It is a very simple device that uses the electromagnetic properties of electricity.

Reservoir ↔ storage battery. Liquids can be stored in tanks and reservoirs, whereas direct-current electricity can be stored in batteries and capacitors.

Flooding ↔ short circuits. The washout of a dam or the break of a water main can cause various degrees of flooding and damage, depending on the pressure and, more important, on the quantity of water that is released. Similarly, an electrical fault current, referred to as a short circuit, can cause excessive damage, depending on the voltage and, more important, on the quantity of electrical energy that is released. This latter item, the quantity of electrical energy that might be released, is referred to as the short-circuit capability of the power system. It is important to recognize that the short-circuit capability determines the physical size and ruggedness of the electrical switching equipment required at each particular plant. In other words, for nearly identical plants, the electrical equipment may be quite different in size because of the different short-circuit capabilities of the power system at the plant site.

Instantaneous transmission of pressure ↔ electricity. When water is let into a pipe that is already filled, the water let into the pipe at one end is not the same as that which promptly rushes out of the other end. The water let in pushes the water already in the pipe out ahead of it. The pressure, however, is transmitted from end to end almost instantaneously. The action of electricity is much the same. The electrical current in a wire travels from end to end at high speeds, very close to that of light. However, individual electrons within the wire move relatively slowly and over short distances. When a light switch is turned on, the power starts the electrons on their way.

Instrumentation-and-Control Systems

Instruments allow a distribution system operator to monitor and control flow rates, pressures, and water levels. The main categories of instrumentation and control are as follows:

- *Primary instruments* measure flow, pressure, level, and temperature.
- *Secondary instruments* respond to and display the information from the primary instruments.
- *Control systems* either manually or automatically operate equipment such as pumps and valves.

Primary Instrumentation—Monitoring Sensors

The sensors that measure process variables are called primary instrumentation because they are basically necessary to obtain the information required to operate the monitoring system.

Flow Sensors

The various types of flow meters used in a water distribution system are discussed in Chapter 15. The types of meters primarily used for measuring flow in mains are differential pressure meters (such as venturi meters) and velocity meters (such as propeller meters). For monitoring flow at remote locations and for control purposes, meters must be provided with either a pulse or an electrical output that is proportional to the flow rate.

Pressure Sensors

Pressure sensors are commonly used to measure suction and discharge pressure on pumps and at points on the distribution system. The three common types of direct-reading pressure gauges that have been used for many years are illustrated in Figure 12-3. The bellows sensor uses a flexible copper can that expands and contracts with varying pressure. The helical sensor has a spiral-wound tubular element that coils and uncoils with changes in pressure. The Bourdon tube uses a semicircular tube with a C shape that opens under increasing pressure.

Most pressure sensing for operating electronic recorders and controls today is done with pressure transducers that produce an electrical current in proportion to pressure. Transducers are very accurate, require essentially no maintenance, and can be adjusted to register pressure over either a narrow or a wide range.

Level Sensors

Several methods of measuring the depth of water in a tank or the elevation of a water surface are illustrated in Figure 12-4. One of the oldest types uses a float mechanism attached by a wire to a pulley on a shaft, which is rotated as the water level rises and falls.

The diaphragm element type of sensor operates on the principle that the confined air in the tube compresses in relation to the head of water above the diaphragm. Changes in pressure are detected by a transducer and electronically translated into water depth.

Bubbler tubes were once widely used for determining water depth, particularly in dirty water. A constant flow of air must be maintained in the tube, which is suspended in the water. Bubbler tubes work on the principle that the pressure

---

primary instrumentation
Instruments required to operate the monitoring system by obtaining information relating to water flow, pressure, level, and temperature.

bubbler tube
A level-sensing device that forces a constant volume of air into the liquid for which the level is being measured.

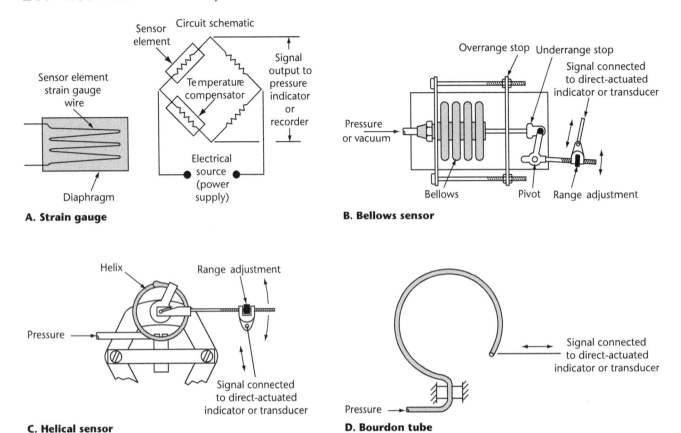

Figure 12-3 Types of pressure sensors

required to discharge air from the tube is proportional to the head of water above the bottom of the tube. Although the bubbler assembly is essentially maintenance free, maintaining a continuous air supply presents some maintenance problems.

Ultrasonic units bounce radio signals off the water surface and translate the return times into distances. Knowing the elevation of the water surface, one can then convert the figure into water depth.

Transducers are now widely used for measuring water depth. The pressure of the water over the unit can be translated directly into feet of head or pressure. In a similar fashion, a transducer can be installed at the base of a water tower to indicate the elevation of water in the tower.

Electronic probes may also be used for sensing water depth. An insulated metallic probe suspended in the water has an electronic circuit that detects a change in capacitance between the probe and water and electronically converts the information into depth. If a probe is installed in a nonmetallic tank, a second probe is required.

A variable-resistance level sensor consists of a wound resistor inside a semiflexible envelope. As the liquid level rises, the flexible outer portion of the sensor presses against the resistor and gradually shorts it out. The resistance is then converted to a liquid-level output signal.

Temperature Sensors

Two types of temperature sensors that are commonly connected to instrumentation are thermocouples and thermistors. Thermocouples use two wires of

thermocouple
A sensor, made of two wires of dissimilar metals, that measures temperature.

Electrical and Instrumentation-and-Control Systems   **245**

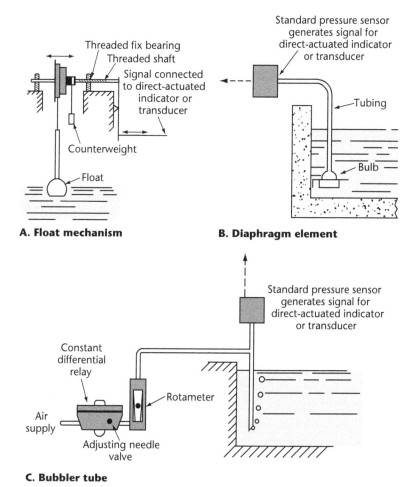

Figure 12-4 Types of level sensors

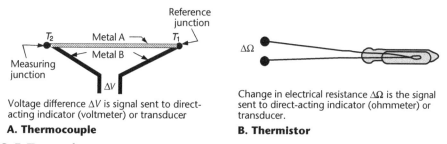

Figure 12-5 Types of temperature sensors

different material, as illustrated in Figure 12-5. The wires are joined at two points. One is called the sensing point and the other is the reference junction. Temperature changes between the two points cause a voltage to be generated, which can be read directly or amplified.

Thermistors are also called resistance temperature devices (RTDs). They use a semiconductive material, such as cobalt oxide, that is compressed into a desired shape from the powder form and then is heat treated to form crystals to which wires are attached. Temperature changes are reflected by a corresponding change in resistance as measured through the wires.

thermistor
A semiconductor type of sensor that measures temperature.

voltage
(1) A measure of electrical potential (electrical pressure), measure in volts. One volt will send a current of 1 ampere through a resistance of 1 ohm. (2) In telemetry, a type of signal in which the electromotive force (measured in volts) varies as the parameter being measured varies.

current
(1) The flow rate of electricity, measured in amperes. (2) In telemetry, a signal whose amperage varies as the parameter being measured varies.

resistance
A characteristic of an electrical circuit that tends to restrict the flow of current, similar to friction in a pipeline. Measured in ohms.

power
A measure of the amount of work done per unit time by an electrical circuit, expressed in watts.

D'Arsonval meter
An electrical measuring device, consisting of an indicator needle attached to a coil of wire, placed within the field of a permanent magnet. The needle moves when an electric current is passed through the coil.

wattmeter
An instrument for measuring real power in watts, stated as kilowatt-hours (kW·h).

Electrical Sensors

It is frequently necessary to monitor the status of operating electrical equipment. The variables that may be measured are voltage (in volts), current (in amperes), resistance (in ohms), and power (in watts). The measurement of volts, amperes, and ohms can all be made with the same instrument, the D'Arsonval meter. Electrical current passing through the meter's coil creates a magnetic field, which reacts with the field of the permanent magnet and causes the indicator needle to move. The parameter measured by the meter depends on how it is connected in the circuit. The D'Arsonval meter is now rapidly being replaced by digital equipment that indicates the electrical value directly (Figure 12-6).

To measure power, an instrument must combine the measurements of volts and amperes (watts = volts × amps, assuming the power factor is 1). The electrical energy used is measured and totalized by a wattmeter that reads in kilowatt-hours (kW·h). One thousand watts (1 kW) drawn by a circuit for 1 hour results in an energy consumption of 1 kW·h. Meters that register kilowatt-hour usage are essentially totalizing wattmeters (Figure 12-7).

A rotating disk can be seen through the glass front on older kilowatt-hour meters, and the rotations can be timed with a stopwatch to determine power consumption. Newer meters have other types of dial and digital readouts. A local power utility representative can be consulted on how to read power consumption on a particular meter.

Equipment Status Monitors

Some of the more common equipment status monitors are vibration sensors, position and speed sensors, and torque sensors. Vibration sensors are one of the most commonly used types in the water industry and are often mounted on pumps and motors that are located in unsupervised locations. The sensors are usually wired into the power circuit to disconnect power to the motor and activate an alarm if the vibration in the equipment exceeds a specified value.

Process Analyzers Because of increasingly stringent water quality requirements, many water systems are installing online monitoring equipment to continuously analyze water quality. Typical equipment of this type includes turbidity, pH, and chlorine residual monitors; particle counters; and streaming current meters. These monitors provide continuous analyses of water quality parameters and will

Figure 12-6 Digital multimeter
Courtesy of Fluke Corporation.

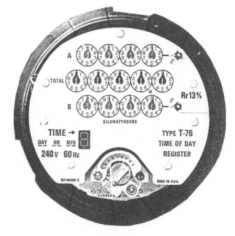

Figure 12-7 Totalizing wattmeter

activate alarms if values go above or below preset limits. They may also feed into a computer control system that will automatically make corrective adjustments of the operating system.

Turbidity Monitors Online turbidity monitors are critical in water treatment plant operation, and they are required by water quality regulations. However, they are also useful for distribution system monitoring. By carefully placing these analyzers, they can be effectively used to detect changes in water quality.

pH Monitors These analyzers provide another sensitive measure of changes within the distribution system. The pH of water entering the distribution network can change unacceptably because of water reactions of pipe walls and storage tank coatings. Also, if pH modification is being used to provide corrosion protection, this can be monitored to ensure effectiveness.

Disinfectant Residual Monitors Chlorine and chloramine residuals (and chlorine dioxide) can be reliably measured by online analyzers within the distribution system. Strategically placing these monitors near storage tanks and in areas of concern can provide early warning of disinfectant decay. This information can be used by operators to adjust water flow and respond to avoid the possibility of problems developing.

Other Analyzers Used in Distribution Systems Several other types of analyzers are commonly used to monitor water quality in distribution systems. These are often employed along with those previously mentioned to provide a "suite" of analyses to better understand the following important information:

- Conductivity
- Ultraviolet absorbance
- Temperature
- Dissolved organic carbon
- Online gas chromatography or mass spectrometry

Secondary Instrumentation

Secondary instruments display information provided by sensors. The display may be mounted adjacent to the sensor, in a nearby control room, or in a distant control center.

Signal Transmission

Because of improvements in electronic circuitry, most new equipment operates with electrical transmission using either current or voltage. The most common is 4–20 milliamps (mA) DC.

Receivers and Indicators

Receivers and indicators convert signals from sensors for use by the water system operator or to be fed into the control system. Alternative methods include the following:

- Direct-reading value of the parameter (e.g., gpm, volts, pressure)
- Recording of the information (as on a strip chart)
- Total accumulated value since the unit was last reset (e.g., 5,000 gal)
- Some combination of the above methods

---

secondary instrumentation
Instruments that display information provided by sensors.

receiver
(1) The part of a meter that converts the signal from the sensor into a form that can be read by the operator; also called the receiver–indicator. (2) In a telemetry system, the device that converts the signal from the transmission channel into a form that the indicator can respond to.

indicator
The part of an instrument that displays information about a system being monitored. Generally either an analog or digital display.

Two types of instrument display are analog and digital. An example of an analog display is a dial indicator. The values range smoothly from the minimum to the maximum, and it is easy to see the relative position of the reading to the entire range. An analog display makes it easy to estimate readings that fall between the primary divisions on the dial.

A digital display shows decimal numbers. The numbers may be a mechanical readout or an electronic display like a digital watch. Digital indicators are generally more accurate than analog displays because they are not subject to the errors associated with mechanical systems. They are also easier to read correctly. A disadvantage is that there is no way of estimating the exact value when it is between the divisions provided on the display.

Telemetry

When the distance between a sensor and the indicator is relatively short, the information can be transmitted between them by using variations in current or voltage. But if there is an appreciable distance between them, telemetry must be used because the signal must be a type that will not vary in spite of variations in the wiring or radio signal.

Early telemetry equipment used audio tones or electrical pulses. Most new systems are digital and use a binary code. The sensor signal feeds into a transmitter that generates a series of on–off pulses that represent the exact numerical value of the measured parameter. For example, off–on–off–on represents the number 5. The receiver then translates the code to number or letter readings.

The transmitting device in a digital system is called a remote terminal unit (RTU) and the receiver is called the control terminal unit (CTU).

Multiplexing

Multiplexing refers to sending signals from more than one sensor over the same transmission line. Several methods of multiplexing are available.

Tone-frequency multiplexing sends several signals over one wire or radio signal by having tone-frequency generators in the transmitter and sending each parameter at a different frequency. Filters in the receiver then sort out the signals and send them to the proper indicator. As many as 21 frequencies can be sent over a single voice-grade telephone line.

Scanning equipment transmits the value of each of several parameters one at a time in a set sequence. The receiver decodes the signal and displays each one in turn. Scanning can be used with all types of signals and all types of transmission. Scanning and tone-frequency multiplexing can be combined to allow even more signals over a single line. For example, a four-signal scanner combined with a 21-channel, tone-frequency multiplexer would yield 84 signal channels.

Polling is another method of sending several different signals over a single line. In this system, each instrument has a unique address (identifying number). A system controller, usually located at the central control center, sends out a message requesting a specific piece of equipment to transmit its data.

The controller can poll the instruments as often as necessary, which may be every few minutes to every hour or so. In more sophisticated units, the controller regularly polls each piece of equipment to determine whether there is any new information. If the status report indicates there is new information, the instrument is instructed to send its data. Some systems also provide for key instruments to interrupt other transmissions to send urgent new data.

---

telemetry
A system of sending data over long distances, consisting of a transmitter, a transmission channel, and a receiver.

remote terminal unit (RTU)
A computer terminal used to monitor the status of control elements, monitor and transmit inputs from instruments, and respond to data requests and commands from the master station.

control terminal unit (CTU)
The receiving device in a digital telemetry system.

multiplexing
The use of a single wire or channel to carry the information for several instruments or controls.

tone-frequency multiplexing
A method of sending several signals simultaneously over a single channel.

scanning
A technique of checking the value of each of several instruments, one after another.

polling
A technique of monitoring several instruments over a single communications channel with a receiver that periodically asks each instrument to send current status.

Duplexing also allows an operator to send control signals back to the site of a transmitting sensor. This can be done using a single transmission line in one of three ways:

1. Full duplex allows signals to pass in both directions at the same time.
2. Half-duplex allows signals to pass in both directions but only in one direction at a time.
3. Simplex allows signals to pass in only one direction.

Transmission Channels

Four types of transmission channels are regularly used by water utilities for transmitting telemetry signals. These channels include a privately owned cable, such as a wire between two buildings on the same property; a leased telephone line; a radio channel; and a microwave system. A system using space satellites is also available but is presently quite expensive, and equipment is available that can send signals over a cellular phone.

A leased telephone line is usually the least expensive transmission channel and is generally quite reliable and free of interference. Most modern telemetry transmitters are designed to operate over voice-grade lines. Radio channels may be in the VHF (very high frequency) or UHF (ultra-high frequency) bands. Both radio and microwave systems generally require line-of-sight paths with no obstructions such as buildings.

Control Systems

Control systems enable the control of equipment in a variety of ways, from completely manual to completely automatic.

Direct Manual Control

Under complete manual control, each piece of equipment is adjusted by the water system operator directly turning it on and off (e.g., turning the handwheel on a valve). Manual control has the advantage of low initial cost and little complicated equipment to maintain. It may require more work for the operator, however, and proper operation of the equipment depends completely on the operator's expertise and judgment. If the equipment to be operated is at different locations, the operator must go to each location to perform the operation.

Remote Manual Control

Remote manual control still requires the operator to initiate each adjustment, but it is not necessary to go to the equipment location. Instead, the operator has a remote station, such as a switch or push button, which turns the equipment on and off. Examples of actuators for remote operation are solenoid valves, electric relays, and electric motor actuators. Proper operation of the equipment is still dependent on the judgment of the plant operator.

Semiautomatic Control

Semiautomatic control combines manual control by the plant operator with automatic control of specific pieces of equipment. For example, a circuit breaker will disconnect automatically in response to an overload but must then be reset manually.

duplexing
A means by which an operator sends control signals back to the site of a transmitting sensor using a single transmission line.

full duplex
Capable of sending and receiving data at the same time.

half-duplex
Capable of sending or receiving data but not both at the same time.

simplex
Related to a telemetry or data transmission system that can move data through a single channel in only one direction.

control system
A means of controlling of equipment in a variety of ways.

manual control
A type of system control in which personnel manually operate the switches and levers to control equipment from the physical location of the equipment.

remote manual control
A system in which personnel in a central location manually control equipment at a distant site.

semiautomatic control
A system equipment in which many actions are taken automatically but some situations require human intervention.

Automatic Control

Automatic control systems turn equipment on and off or adjust operation in response to signals from sensors and analytical instruments. The plant operator does not have to exercise any control. A simple example is the thermostat on a heating system. The two general modes of operation for automatic control are on–off differential and proportional control.

On–off differential control turns a piece of equipment either full on or off in response to a signal. For example, the same signal that activates a service pump can turn on a chlorinator. If there is any need to adjust the chlorine feed rate, it must be done manually by the plant operator.

Proportional control can adjust the operation of a piece of equipment in response to a signal in several ways.

Feedforward proportional control measures a variable and adjusts the equipment proportionally (Figure 12-8). An example is the adjustment of chlorinator feed rate from a flowmeter signal. The faster the water flows through the meter, the more chlorine that is fed. As long as the chlorine demand of the water remains constant, this method of operation is satisfactory.

Feedback proportional control measures the output of the process and reacts backward to adjust the operation of the piece of equipment. In the illustration in Figure 12-9, the chlorine residual analyzer is set by the operator to maintain a specific chlorine residual. It then adjusts the feed rate of the chlorinator to maintain the residual in spite of variations in both chlorine demand and changes in the water flow rate. This is also called closed-loop control because it is continuously self-correcting. The principal problem with this control system is that, if there are wide variations in water flow rate, the system will spend a lot of time seeking the correct value. If the flow rate increases, the

> **on–off differential control**
> A mode of controlling equipment in which the equipment is turned fully on when a measured parameter reaches a preset value, then turned fully off when it returns to another preset value.
>
> **proportional control**
> A mode of automatic control in which a valve or motor is activated slightly to respond to small variations in the system, but activated at a greater rate to respond to larger variations.
>
> **feedforward proportional control**
> A control system that measures a variable and adjusts the equipment proportionally.
>
> **closed-loop control**
> A form of computerized control that automatically adjusts for changing conditions to produce the correct output, so that operator intervention may be minimized.

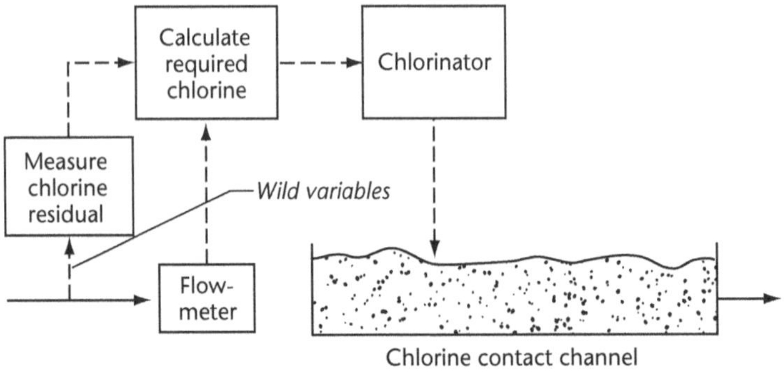

Figure 12-8 Feedforward control of chlorine contact channel

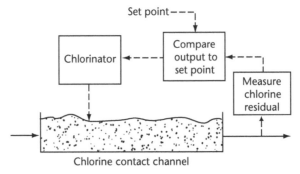

Figure 12-9 Feedback control of chlorine contact channel

analyzer will detect a low residual and will increase the feed rate. But it will probably overfeed for a short time, then underfeed again, and so on until it finds the correct feed rate.

The ultimate automatic control for this situation is a combination of both feedforward and feedback control. The chlorinator is set up to primarily adjust in response to changes in flow rate, but the analyzer then monitors the residual and makes minor adjustments in the feed rate as necessary to maintain the selected residual in the finished water.

Supervisory Control and Data Acquisition

Supervisory control and data acquisition (SCADA) has become a principal method of control in the water industry. A SCADA system consists of four basic components. The first two components have already been discussed—sensors (RTUs) to monitor the variables and telemetering to send the information to a central location.

The third component is a central control location that has equipment for monitoring the operation and sending back commands. The fourth component is equipment for reviewing the operation, giving commands to the remote equipment, and recording information for historical purposes. This component generally consists of a video screen, a keyboard, and computer data storage equipment.

When a water system has everything controlled from a single computer, it is called *centralized computer control*. Early SCADA systems were of this type, having one big computer to control everything on the water system. Two problems with this type of operation are the great dependency on the computer and on the telemetry links to the remote locations. For example, if a remote pumping station is completely controlled from a central computer and there is loss of the telephone telemetry line, the station would shut down and be unusable. The only way to operate it is to have someone operate it manually from the station.

As smaller, more powerful, and less expensive computers became available, SCADA systems called *distributed computer control* were developed (Figure 12-10). It is now possible to have local computer control of subsystems and

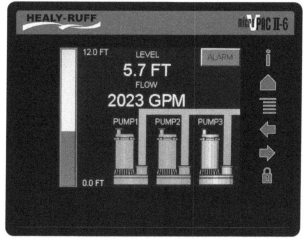

Figure 12-10 Remote supervision of a reservoir
Courtesy of ICS Healy-Ruff.

supervisory control and data acquisition (SCADA)
A methodology involving equipment that both acquires data on an operation and provides limited to total control of equipment in response to the data.

individual pieces of equipment. The trend has extended to the development of "smart equipment," which adjusts itself and monitors its own operation.

State regulatory agencies typically have mixed views about operating water system equipment completely by computer control. On the plus side, the computer eliminates the possibility of human error—the computer cannot fall asleep on the job. On the negative side, breakdown of a computer system or an error in the software could result in loss of pressure or contamination of the system. State agencies will always want to see that an automated system is furnished with all possible monitoring and reporting capabilities and is programmed to fail-safe in the event of an emergency. See AWWA's Standard G100 for more detailed recommendations regarding unstaffed water treatment plant operation.

WATCH THE VIDEO
SCADA (www.awwa.org/wsovideoclips)

Study Questions

1. Which basic electrical unit is used to measure a material's opposition to the flow of electricity?
   a. Ampere
   b. Ohm
   c. Volt
   d. Joule

2. All sensors that respond to liquid pressure will perform poorly if _____ enter(s) the sensor.
   a. air
   b. corrosive chemicals from water treatment processes
   c. corrosive chemicals from piping
   d. iron bacteria

3. SCADA systems consist of what distinct components?
   a. Remote terminal units (RTUs), communications, and human–machine interface (HMI)
   b. Sensing instrument, RTUs, communications, and HMI
   c. Sensing instrument, RTUs, communications, master station, and HMI
   d. RTUs, communications, master station, and HMI

4. What is the electronic standard range?
   a. 4 to 20 mA DC
   b. 4 to 20 mA AC
   c. 0% to 100%
   d. 0 to 1 binary

5. The D'Arsonval meter is
   a. an amperometric meter.
   b. a type of pH meter.
   c. an analog (uses a needle) meter.
   d. a digital (number displays on unit) meter.

6. Which of the following is *not* an example of a secondary instrument?
   a. Telemetry device
   b. Multiplexing device
   c. pH monitor
   d. Indicator

7. Secondary instrumentation transmits information of what type?

8. What are the two types of temperature sensors commonly connected to instrumentation?

9. In what type of control system is each piece of equipment adjusted by the water system operator directly turning it on and off?

10. What type of automatic control system measures a variable and adjusts the equipment proportionally?

Chapter 13
Motors and Engines

Motors

Electric motors are used to power 95 percent of the pumps used in water supply operations. Internal-combustion engines are primarily used for standby service, although some utilities operate engines during peak demand periods to reduce electrical costs.

The alternating current (AC) electricity furnished by electric utilities for water utility use is generally in the form of three-phase current. This is then usually reduced to single-phase current within the plant for operating lights and small equipment. Principal motor components are labeled in Figure 13-1.

 WATCH THE VIDEO
Motors (www.awwa.org/wsovideoclips)

Single-Phase Motors

Single-phase motors are typically used only in fractional-horsepower sizes, but if there is a special requirement, they can be furnished with ratings up to 10 hp (7.5 kW) at 120 or 240 V. A single-phase motor has no power to bring it up to speed (starting torque), so it must be started by some outside device.

A starting winding is usually built into the motor to provide initial high torque. Then as the motor comes up to speed, a centrifugal switch changes

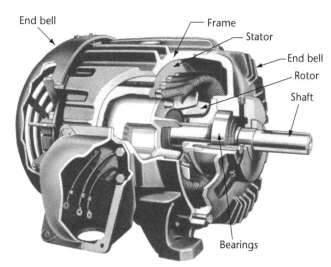

Figure 13-1 Motor components

single-phase
Alternating current (AC) power in which the current flow reaches a peak in each direction only once per cycle.

connections to the running winding. Single-phase motors are of the following three basic types:

1. *Split-phase motors* use a rotor with no windings. They have a comparatively low starting torque, so they require a comparatively low starting current.
2. *Repulsion–induction motors* are more complex and expensive than split-phase motors and also require a higher starting current.
3. *Capacitor-start motors* have high starting torque and high starting current. They are used in applications where the load can be brought up to speed very quickly and infrequent starting is required.

Three-Phase Motors

Motors used in water treatment or distribution systems that are more powerful than ½ hp (0.4 kW) are generally three phase and may be operated at 230, 460, 2,300, or 4,000 V. The three main classes used are squirrel-cage induction, synchronous, and wound-rotor induction.

- A squirrel-cage induction motor is the simplest of all AC motors. The rotor windings consist of a series of bars placed in slots in the rotor and connected together at each end (creating the appearance of a squirrel cage). The stator windings located in the frame are connected to the power supply, and the current flowing through them induces a rotating magnetic field. Simple starting controls are usually adequate for most of the normal- and high-starting-torque applications of these motors.

- A synchronous motor has power applied to the windings in such a way that a revolving magnetic field is established. The rotor is constructed to have the same number of poles as the stator and they are supplied with direct current so the rotor's magnetic field is constant. A slip-ring assembly (commutator) and graphite brushes are used to connect power to the rotor. Synchronous motors are used where the motor speed must be held constant and, because the motor has a power factor of 1.0, in areas where the power company has a penalty for low power factor conditions.

- A wound-rotor induction motor has a stator similar to a squirrel-cage motor, except that the resistance of the rotor circuit can be controlled while the motor is running, which varies the motor's speed and torque output. The starting current required for a wound-rotor motor is seldom greater than the full-load operating current. In contrast, squirrel-cage and synchronous motors generally have starting current requirements between 5 and 10 times their full-load current.

Motor Temperature

Motors convert electrical energy into mechanical energy and heat. About 5 percent of the energy is lost in heat and it must be removed quickly to prevent the motor temperature from rising too high. Motors are designed for an external (ambient) temperature of 104°F (40°C), so ventilation air should never have a higher temperature. The useful life of a motor is considerably shortened by being run at high temperatures. Care must be taken not to obstruct air flow around motors.

Mechanical Protection

The design of the motor housing must be considered in relation to where the motor will be located. A motor powering a pump inside a building can generally

---

three-phase
Alternating current (AC) power in which the current flow reaches three peaks in each direction during each cycle.

squirrel-cage induction motor
The most common type of induction electric motor. The rotor consists of a series of aluminum or copper bars parallel to the shaft, resembling a squirrel cage.

stator
The stationary member of an electric generator or motor.

synchronous motor
An electric motor in which the rotor turns at the same speed as the rotating magnetic field produced by the stator. This type of motor has no slip.

commutator
A device that is part of the rotor of certain designs of motors and generators. The motor unit's brushes rub against the surface of the spinning commutator, allowing current to be transferred between the rotor and the external circuits.

wound-rotor induction motor
A type of electric motor, similar to a squirrel-cage induction motor but easier to start and capable of variable-speed operation.

be of the simplest design because it will be in a clean environment. A motor to be installed outside must have protection from rain, dust, and wind-driven particles.

Motor housing designs commonly available include open, drip-proof, splash-proof, guarded, totally enclosed, totally enclosed with fan cooling, explosion-proof, and dust-proof.

Motor Control Equipment

Motor Starters

Small motors are usually started by directly connecting line voltage to the motor. Motors larger than fractional horsepower are typically started and stopped using a motor starter. As illustrated in Figure 13-2, a typical motor starter includes a main disconnect switch, fuses or a circuit breaker, motor protection by temperature monitors on each of the phases, and provisions for remote operation.

Reduced-Voltage Controllers

When the starting current of a motor is so high that it may damage the electrical system or deprive other operating motors of sufficient current, a reduced-voltage controller is used. The controller supplies reduced voltage to start the motor, then applies full voltage when it is about up to speed.

Motor Control Systems

Remote and automatic controls eliminate the need for an operator to be near a pump to operate it. Manual remote controls for pumps are now usually located in a central control room.

Figure 13-3 illustrates how controls can start and stop pumps on several levels to automatically meet customer demands. The system is also furnished with both high- and low-level alarms that will alert the operator of trouble if any of the preset limits are exceeded.

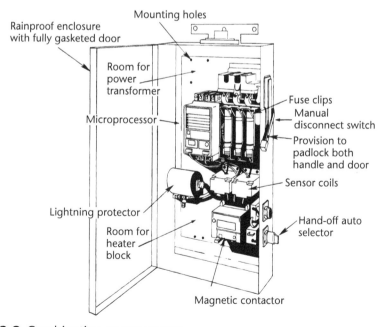

Figure 13-2 Combination motor starter

> **starter**
> A motor-control device that uses a small push-button switch to activate a control relay, which sends electrical current to the motor.
>
> **reduced-voltage controller**
> An electric controller that uses less than the line voltage to start the motor. Used when full line voltage may overload or damage the electrical system.

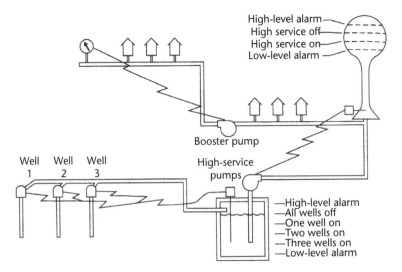

Figure 13-3 Typical pump installation with a number of controls

Internal-Combustion Engines

Internal-combustion engines are used by water utilities to power pumps during emergencies and for portable applications. Also, emergency electric generators are often powered by gas (natural gas, propane, or methane), gasoline, or diesel engines. Remote locations may use internal-combustion engines to provide power for the entire utility operation. Some utilities use internal-combustion engines as part of an energy (cost) efficiency strategy to avoid peak-demand electrical charges.

Gasoline Engines

Most gasoline engines are used for standby or emergency purposes. The initial cost of the engines is low, but fuel and maintenance costs can be relatively high. These engines are available in sizes ranging from 1 hp (0.7 kW) to hundreds of horsepower.

Diesel Engines

Diesel engines are reliable and often used as emergency generator drives that supply power to pump stations and water treatment plants. Initial cost can be higher than similarly sized gasoline engines, but operating costs are usually lower.

Gas Engines

Gas (natural gas, propane, or methane) engines can be converted from gasoline engines or original equipment designed for these fuel sources. Usually, methane fuel is available only from wastewater treatment facilities, where it can be produced. Gas engines are selected where an inexpensive source of fuel is available or where there may be concerns about air emissions.

Steam Engines

Although historically significant, these are not often used in water utilities today. Where they are in use, the steam turbine engine is most often employed to power very large pumps.

Operation and Maintenance

Start-up and operating procedures and maintenance practices for engines are supplied by the equipment supplier. These procedures should be posted near the engine location. The following general procedures are recommended for most engines.

- *Service prior to operation.* Inspect the engine and components before starting. Check fluid levels, belts, and hoses. Make sure cooling water is flowing and the clutch is disengaged.
- *Initial operation service.* Soon after starting the engine, check the idle speed, oil pressure, water temperature, and all other operating indicators. After the recommended warmup period, check all operating indicators again.
- *Service during operation.* Check all gauges and other operating indicators periodically during operation. Fuel levels should be closely monitored for diesel engines.
- *Service after operation.* Any issues observed during operation should be corrected. Lubricant should be examined.
- *Routine preventive maintenance.* Engines that are used to provide emergency service, including power backup, should be exercised under load according to a strict schedule. Many systems operate emergency power engines at least 15 minutes a week. Routine inspection and preventive maintenance should be performed according to the manufacturer's recommendations. All maintenance procedures should be recorded for future reference. Diesel fuel filtering that may be required for injection-type systems must be performed by qualified personnel who have been specially trained.

Pump, Motor, and Engine Records

Detailed records should be maintained on all equipment for use in scheduling maintenance, evaluating operation, and performing repairs. Most water systems maintain a quick-reference data card, notebook sheet, or computer file for each piece of equipment, or use an asset management system. Typical information that should be recorded for each piece of equipment includes the following:

- Make, model, capacity, type, serial number, and warranty information
- Date and location of installation and name of installer
- Part numbers of special components likely to require replacement
- Results of initial tests and of all subsequent tests of the equipment
- Manufacturer's suggested inspection and maintenance schedules
- Names, addresses, and phone numbers for the manufacturer and local representative

A file folder or electronic archive should also be maintained for each piece of equipment, containing the original manufacturer's literature, operating and repair manuals, and copies of correspondence, purchase orders, and other pertinent information.

Study Questions

1. What is the simplest of all AC motors?
   a. Synchronous motor
   b. Squirrel-cage induction motor
   c. Wound-rotor induction motor
   d. Split-phase motor

2. Starters are typically used on motors
   a. larger than fractional horsepower.
   b. of fractional horsepower.
   c. of all horsepower ratings.
   d. when gasoline is unavailable.

3. In what type of motor is power applied to the windings in such a way that a revolving magnetic field is established?
   a. Synchronous motor
   b. Squirrel-cage induction motor
   c. Wound-rotor induction motor
   d. Split-phase motor

4. Operators should check the idle speed, oil pressure, water temperature, and all other operating indicators
   a. before starting the engine.
   b. exactly one hour after starting the engine.
   c. soon after starting the engine.
   d. on an as-needed basis.

5. What is the maximum ambient temperature at which motors are designed to function?

6. What type of motor is used for most large pumps?

7. What is the primary use of diesel engines?

8. How should systems ensure that engines used for emergency service will function when needed?

9. What device is used when the starting current of a motor is so high that it may damage the electrical system or deprive other operating motors of sufficient current?

Chapter 14
Pumps and Pumping Stations

Types of Pumps

Two basic categories of pumps are used in water supply operations: velocity pumps and positive-displacement pumps. Velocity pumps, which include centrifugal and vertical turbine pumps, are used for most distribution system applications. Positive-displacement pumps are most commonly used in water treatment plants for chemical metering.

Velocity Pumps

Velocity pumps use a spinning impeller, or propeller, to accelerate water to high velocity within the pump casing. The high-velocity, low-pressure water leaving the impeller can then be converted to high-pressure, low-velocity water if the casing is shaped so that water moves through an area of increasing cross section. This increasing cross-sectional area may be achieved in two ways:

1. The volute (expanding spiral) casing shape, as in the common centrifugal pump (Figure 14-1A), may be used.
2. Specially shaped diffuser vanes or channels may be used, such as those built into the bowls of vertical turbine pumps (Figure 14-1B).

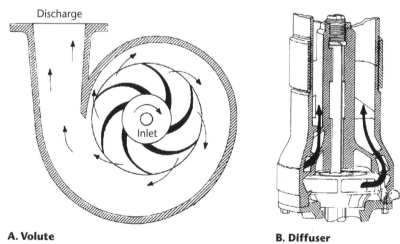

Figure 14-1 Centrifugal pump casings

This material is used with permission of John Wiley & Sons, Inc., from *Centrifugal Pump Design*, John Tuzson, ©2000 by John Wiley & Sons, Inc.

velocity pump
The general class of pumps that use a rapidly turning impeller to impart kinetic energy or velocity to fluids. The pump casing then converts this velocity head, in part, to pressure head.

impeller
The rotating set of vanes that forces water through a pump.

Velocity Pump Design Characteristics

Two designs of velocity pumps are widely used in water systems: centrifugal pumps (volute pumps) and turbine pumps. A feature distinguishing velocity pumps from positive-displacement pumps is that velocity pumps will continue to operate undamaged, at least for a short period, when the discharge is blocked. When this happens, a head builds up that is typically greater than the pressure generated during pumping, and water recirculates within the pump impeller and casing. (Recall that *head* refers to pressure measured in terms of the height of water, in meters or feet.) This flow condition is referred to as slip.

Depending on the casing shape, impeller design, and direction of flow within the pump, velocity pumps can be manufactured with a variety of operating characteristics.

Radial-Flow Designs In the radial-flow (or centrifugal) pump, shown in Figure 14-1, water is thrown outward from the center of the impeller into the volute or diffusers that convert the velocity to pressure. The type of centrifugal pump commonly used in water supply practice is a radial-flow, volute-case type. A cutaway of a typical single-stage pump is shown in Figure 14-2. Centrifugal pumps of this type generally develop very high heads and have correspondingly low-flow capacities.

In general, any centrifugal pump can be designed with a multistage configuration. Each stage requires an additional impeller and casing chamber in order to develop increased pressure, which adds to the pressure developed in the preceding stage.

Although the pressure increases with each stage, the flow capacity of the pump does not increase beyond that of the first stage. There is no theoretical limit to the number of stages that are possible. However, mechanical considerations such as casing strength, packing leakage, and input power requirements do impose practical limitations.

Axial-Flow Designs The axial-flow pump, shown in Figure 14-3, is often referred to as a propeller pump. It has neither a volute nor diffuser vanes. A propeller-shaped impeller adds head by the lifting action of the vanes on the water. As a result, the water moves parallel to the axis of the pump rather than

centrifugal pump
A pump consisting of an impeller on a rotating shaft enclosed by a casing that has suction and discharge connections. The spinning impeller throws water outward at high velocity, and the casing shape converts this high velocity to a high pressure.

turbine pump
(1) A centrifugal pump in which fixed guide vanes (diffusers) partially convert the velocity energy of the water into pressure head as the water leaves the impeller. (2) A regenerative turbine pump.

slip
(1) In a pump, the percentage of water taken into the suction end that is not discharged because of clearances in the moving unit. (2) In a motor, the difference between the speed of the rotating magnetic field produced by the stator and the speed of the rotor.

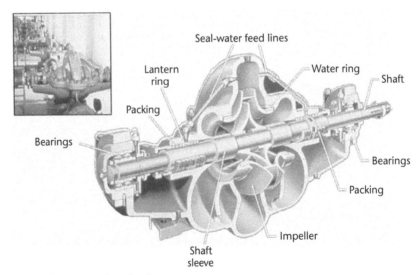

Figure 14-2 Cutaway of a single-stage pump

Pumps and Pumping Stations 263

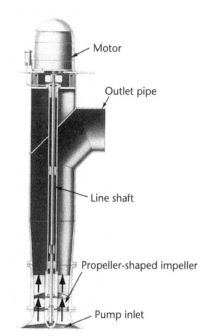

Figure 14-3 Axial-flow pump
Courtesy of Ingersoll-Dresser Pump Company.

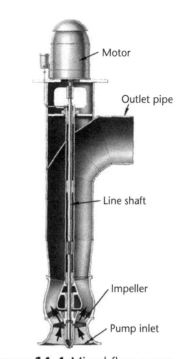

Figure 14-4 Mixed-flow pump
Courtesy of Ingersoll-Dresser Pump Company.

being thrown outward as with a radial-flow pump. Axial-flow pumps handle very high volume but add limited head. Pumps of this design must have the impeller submerged at all times because they are not self-priming.

Mixed-Flow Designs The mixed-flow pump, illustrated in Figure 14-4, is a compromise in features between radial-flow and axial-flow pumps. The impeller is shaped so that centrifugal force will impart some radial component to the flow. This type of pump is useful for moving water that contains solids, as in raw-water intakes.

Centrifugal Pumps

The volute-casing type of centrifugal pump, shown in Figure 14-5, is used in most water utility installations. A wide range of flows and pressures can be achieved by

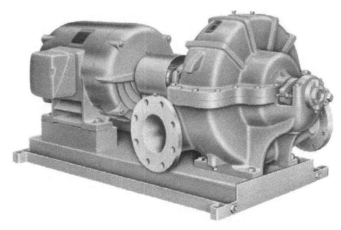

Figure 14-5 Volute-casing type of centrifugal pump

varying the width, shape, and size of the impeller, as well as by varying the clearance between the impeller and casing. The pumps can develop a head up to 250 ft (76 m) per stage and efficiencies up to 75 or 85 percent.

Initial cost is relatively low for a given pump size, and relatively little maintenance is required. However, periodic checks are advised to monitor impeller wear and packing condition.

Advantages and disadvantages vary with the type of centrifugal pump used. Advantages include the following:

- Wide range of capacities (Available capacities range from a few gpm to 50,000 gpm [190,000 L/min]. Heads of 5–700 ft [1.5–210 m] are generally available.)
- Uniform flow at constant speed and head
- Simple construction
- Small amounts of suspended matter, which helps prevent jamming of the pump
- Low to moderate initial cost for a given size
- Ability to adapt to several drive types—motor, engine, or turbine
- Moderate to high efficiency at optimal operation
- No need for internal lubrication
- Little space required for a given capacity
- Relatively low noise level
- Ability to operate against a closed discharge valve for short periods without damage

Disadvantages include the following:

- An efficiency that is at best limited to a narrow range of discharge flows and heads
- Flow capacity that is greatly dependent on discharge pressure
- Generally no self-priming ability
- Potential for running backward if stopped with the discharge valve open
- Potential for impeller to be damaged by abrasive matter in water or become clogged by large quantities of particulate matter

Vertical Turbine Pumps

Vertical turbine pumps have an impeller rotating in a channel of constant cross-sectional area, which imparts mixed or radial flow to the water. As liquid leaves the impeller (Figure 14-6), velocity head is converted to pressure head by diffuser guide vanes. The guide vanes form channels that direct the flow either into the discharge or through diffuser bowls into succeeding stage inlets.

Turbine pumps are manufactured in a wide range of sizes and designs, combining efficiency with high speeds to create the highest heads obtainable from velocity pumps. The clearance between the diffuser and the impeller is usually very small, limiting or preventing internal backflow and improving efficiency. Efficiencies in the range of 90–95 percent are possible for large units. However, the closely fitting impeller prohibits pumping of any solid sediment, such as sand, fine grit, or silt. Turbine pumps have a higher initial cost and are more expensive to maintain than centrifugal volute pumps of the same capacity.

vertical turbine pump
A centrifugal pump, commonly of the multistage diffuser type, in which the pump shaft is mounted vertically.

Figure 14-6 Turbine impeller
Courtesy of Ingersoll-Dresser Pump Company.

Turbine pumps present the following advantages:

- Uniform flow at constant speed and head
- Simple construction
- Individual stages capable of being connected in series, thereby offering multiple head capacities for a single pump model
- Adaptability to several drive types—motor, engine, or turbine
- Moderate to high efficiency under the proper head conditions
- Little space occupied for a given capacity
- Low noise level

The main disadvantages of turbine pumps include the following:

- High initial cost
- High repair costs
- The need to lubricate support bearings located within the casing
- Inability to pump water containing any suspended matter
- An efficiency that is at best limited to a very narrow range of discharge flow and head conditions

Deep-Well Pumps For deep-well service, a shaft-type vertical turbine pump requires a lengthy pipe column housing, a drive unit, a driveshaft, and multiple pump stages. In this type of pump, a drive unit is located at the surface, with the lower shaft, impeller, and diffuser bowls submerged (Figure 14-7). This type of pump requires careful installation to ensure proper alignment of all shafting and impeller stages throughout its length. Deep-well turbines have been installed in wells with lifts of over 2,000 ft (610 m).

Submersible Pumps Multistage mixed-flow centrifugal pumps or turbine pumps with an integral or close-connected motor may be designed for operation while completely submerged, in which case they are termed submersible pumps. As shown in Figure 14-8, the entire pump-and-motor unit is placed below the water level in a well.

Booster Pumps Vertical turbine pumps are often used for in-line booster service to increase pressure in a distribution system. The unit is actually a turbine pump that has the motor and pumps mounted close together and is installed in

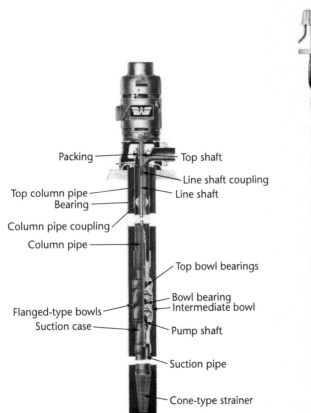

Figure 14-7 Deep-well pump
Courtesy of Ingersoll-Dresser Pump Company.

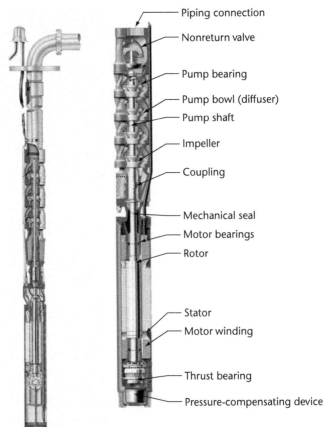

Figure 14-8 Vertical turbine pump driven by a submersible motor (left) and a cross-sectional view of a submersible pump (right)
Images provided courtesy of Flowserve Corporation.

a sump. As shown in Figure 14-9, this type of unit is commonly called a "can" pump. The sump receives fluid and maintains an adequate level above the turbine pump suction.

Centrifugal–Jet Pump Combination

Figure 14-10 illustrates a centrifugal–jet pump combination at the ground surface that generates high-velocity water that is directed down the well to an ejector. Jet pumps are widely used for small, private wells because of their low initial cost and low maintenance. They are rarely used for public water systems because of their relatively low efficiency.

 WATCH THE VIDEO
Pumps—Types (www.awwa.org/wsovideoclips)

Positive-Displacement Pumps

Early water systems used reciprocating positive-displacement pumps powered by steam engines to obtain the pressure needed to supply water to customers. These pumps have essentially all been replaced with centrifugal pumps, which are much more efficient. The only types of positive-displacement pumps used in current water systems are some types of portable pumps used to dewater excavations, as well as chemical feed pumps.

Pumps and Pumping Stations **267**

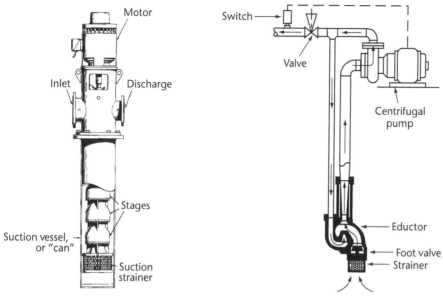

Figure 14-9 Turbine booster pump

Figure 14-10 Centrifugal–jet pump combination
Reproduced with permission of McGraw-Hill Companies from *Pump Handbook* by Karassik et al., 2001. Published by McGraw-Hill.

Reciprocating Pumps

As illustrated in Figure 14-11, reciprocating pumps have a piston that moves back and forth in a cylinder. The liquid is admitted and discharged through check valves. Flow from reciprocating pumps generally pulsates, but this can be minimized by the use of multiple cylinders or pulsation dampeners. Reciprocating pumps are particularly suited for applications where very high pressures are required or where abrasive or viscous liquids must be pumped.

Rotary Pumps

Rotary pumps use closely meshed gears, vanes, or lobes rotating within a close-fitting chamber. The two most common types, which use gears or lobes, are shown in Figure 14-12.

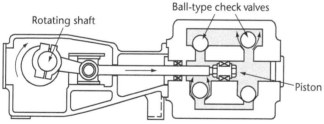

Figure 14-11 Double-acting reciprocating pump
Courtesy of the Hydraulic Institute.

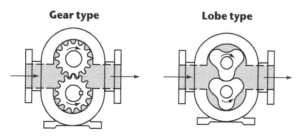

Figure 14-12 Rotary pumps
Courtesy of the Hydraulic Institute.

reciprocating pump
A type of positive-displacement pump consisting of a closed cylinder containing a piston or plunger to draw liquid into the cylinder through an inlet valve and force it out through an outlet valve. When the piston acts on the liquid in one end of the cylinder, the pump is termed single-action; when the piston acts in both ends, the pump is termed double-action.

rotary pump
A type of positive-displacement pump consisting of elements resembling gears that rotate in a close-fitting pump case. The rotation of these elements alternately draws in and discharges the water being pumped. Such pumps act with neither suction nor discharge valves, operate at almost any speed, and do not depend on centrifugal forces to lift the water.

Operation of Centrifugal Pumps

The procedures for centrifugal pump operation vary somewhat from one brand of pump to another. The manufacturer's specific recommendations should be consulted before operating any unit. The procedures described in this chapter are typical and will serve as a guide if the manufacturer's instructions are not available.

Pump Starting and Stopping

A major consideration in starting and stopping large pumps is the prevention of excessive surges and water hammer in the distribution system. Large pump-and-motor units have precisely controlled automatic operating sequences to ensure that the flow of water starts and stops smoothly. General procedures for starting pumps are as follows:

- Check pump lubrication.
- Prime the pump:
    - Where head exists on the suction side of the pump, open valve on the suction line and allow any air to escape from the pump casing through the air cocks.
    - Where no head exists on the suction side and a foot valve is provided on the suction line, fill the case and suction pipe with water from any source, usually the discharge line.
    - Where no head exists on the suction side and no foot valve is provided, the pump must be primed by a vacuum pump or ejector operated with steam, air, or water.
- After priming, start the pump with the discharge valve closed.
- When the motor reaches full speed, open the discharge valve slowly to obtain the required flow. To avoid water hammer, do not open the valve suddenly.
- Avoid throttling the discharge valve; doing so wastes energy.
- Before shutting down the pump, close the discharge valve slowly to prevent water hammer in the system.

Pump Starting

Centrifugal pumps do not generate any suction when dry, so the impeller must be submerged in water for the pump to start operating. If a pump is located above water level, a foot valve is often provided on the suction piping to hold the pump's prime (i.e., to keep some startup water in the pump). The foot valve, which is a type of check valve, prevents water from draining out of the pump when the pump is shut down.

Pump prime can also be maintained by placing a vacuum connection connected to both the pump suction and the high point of the pump, as illustrated in Figure 14-13. The priming valve automatically removes any air that accumulates and keeps the pump completely full of water at all times. (See the details of an air-and-vacuum relief valve in Chapter 9, Figure 9-15.)

Controlling water hammer is important when a pump is being started. Large pumps are furnished with a valve on the discharge that is opened slowly after the pump gets up to speed. As a result, the surge of water does not produce a serious shock in the distribution system.

---

foot valve
A check valve placed in the bottom of the suction pipe of a pump, which opens to allow water to enter the suction pipe but closes to prevent water from passing out of it at the bottom end.

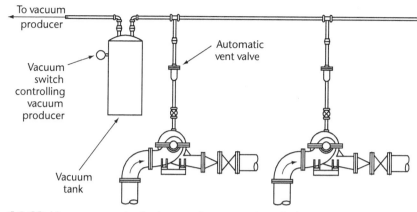

Figure 14-13 Vacuum-controlled central automatic priming

Reproduced with permission of McGraw-Hill Companies from *Pump Handbook* by Karassik et al., 2001. Published by McGraw-Hill.

Pump Stopping

A check valve is usually installed in the discharge piping of small pumps to stop flow immediately after the pump stops. This will prevent reverse flow through the pump. However, the sudden shutdown of a pump may cause water hammer. Relief valves or surge chambers may be installed to absorb the pressure shock.

On large pumps, smooth shutdown is ensured by closing the discharge valve slowly while the pump is still running and then shutting off the pump just as the valve finally closes. In this manner, the pumping unit is eased off the system. Some form of power-activated valve is necessary to obtain slow valve closure. Figure 14-14 shows power-operated discharge valves installed on large pumps. The valve operators are equipped with handwheels for manual operation in the event of a power failure.

Whenever a power failure occurs, the motor will stop while the discharge valve is still open, so there must be a way for the valve to close very rapidly before the pump reverses itself and begins to run backward. Battery power or an emergency hydraulic system is usually provided to operate the valves in an emergency.

Whenever a pump must be shut down for more than a short period in freezing weather, the pump and exposed suction and discharge piping must be drained of water to prevent freezing. If the pump will be out of service for an extended

Figure 14-14 Pump discharge valves

Courtesy of Henry Pratt Company.

time, the pump and motor bearings should be flushed and regreased, and packing should be removed from the stuffing box. The units should also be covered to prevent moisture damage to the motor windings and bearings.

Flow Control

Pumps are usually operated at constant speed. System pressure is controlled by having pumps of various sizes start or stop as necessary. Throttling the discharge valve or using variable-speed motors or pump drives are other ways to control the flow rate.

A major disadvantage of cycling pumps on and off as a means of controlling output is excessive motor wear. Medium-size motors should not be cycled (i.e., started and stopped) more frequently than every 15 minutes, and larger motors should be cycled even less frequently. Frequent starting also increases power costs. Frequent cycling of pumps is an indication that the system probably does not have adequate distribution system storage.

Throttling the discharge valve in an attempt to approximate the required system flow should be done only when elevated storage is not available or when other, smaller pumping units are out of operation. In general, throttling should be avoided because it wastes energy. It is also necessary to make sure the valves used for throttling are appropriate for this purpose. Gate valves should not be used for throttling because the gate is loose in its guides and will vibrate when it is not fully open or shut. The best valves for throttling are plug, ball, self-actuating, or altitude valves. Butterfly valves can be used for throttling for short periods, but extended use may damage them.

If the system design requires continually varying pump discharge rates, variable-speed drives should be provided. Numerous variable-speed package drives are on the market, including continuously variable and stepped-speed motors, as well as constant-speed motors driving variable-speed electrical, hydraulic, and mechanical speed reducers coupled to the pump. Pumps can also be driven by variable-speed motors, which have either variable-voltage or variable-frequency controls.

Monitoring Operational Variables

A primary requirement at every pumping station is to measure the amount of water pumped and provide a record of water delivered to the system. It is also usually necessary to monitor pressure in the system and elevated tank levels as a way to control pump operation. Pumping station production records also provide the basis for the plant maintenance schedule. Past records are usually reviewed to determine the need for equipment replacement.

Suction and Discharge Heads

Pressure gauges should be connected to both the suction and discharge sides of a pump at the pressure taps supplied on the pump. The gauges should be mounted in a convenient location so the operator can frequently check pump performance. The pressure readings can also be electronically transmitted to a control room.

Bearing and Motor Temperature

The most common way to check bearing and motor temperatures in a small- to medium-size plant is by feel. Experienced operators check pump operation by putting a hand on the motor and the pump bearing surfaces. They know how warm the surfaces should be. If a surface is substantially hotter than normal, the unit should be shut down and the cause of excessive heat investigated.

Special thermometers or temperature indicators are also available for monitoring the temperature at critical points in the pump and motor. It is particularly wise to have these monitors installed on equipment at unattended pump stations. The monitors will sound an alarm and automatically shut down the unit if the temperature gets too high.

Vibration

As with temperature, experienced operators get to know the normal feel and sound of each pump unit. They should investigate any change they notice. Vibration detectors are sometimes used on large pump-and-motor installations to sense equipment malfunctions, such as misalignment and bearing failure, that will cause excessive vibration. The detectors can also be used to shut down the unit if vibration increases beyond a preset level.

Speed

Monitoring the pump speed of variable-speed pumps is important because these pumps may experience cavitation (the creation of vapor bubbles) at low speeds. Centrifugal-speed switches can be installed on the pumping unit, or contacts can be provided on a mechanical speed-indicating instrument to sound alarms or shut off the system if the speed goes too high or too low. Other systems use a tachometer generator that generates a voltage in proportion to speed. This voltage is used to drive a standard indicator near the pump or at a remote location. Underspeed and overspeed alarms can be activated by the speed-sensing device.

General Observations

An operator should also monitor surge-tank air levels, recording meters, and intake-pipe screens. Pumps with packing seals should be adjusted so that there is always a small drip of water leaking around the pump shaft. Idle pumps should be started and run weekly. All operations and maintenance should be recorded in log books.

Finally, operators must remain attentive to the general condition of the pump on a day-to-day basis. Unusual noises, vibrations, excessive seal leakage, hot bearings or packing, or overloaded electric motors are all readily apparent to the alert operator who is familiar with the normal sound, smell, sight, and feel of the pump station. Reporting and acting on such problems immediately can prevent major damage that might occur if the problem were allowed to remain until the next scheduled maintenance check.

Mechanical Details of Centrifugal Pumps

Size and construction may vary greatly from one volute-type centrifugal pump to another, depending on the operating head and discharge conditions for which the pumps are designed. However, the basic operating principle is the same. Water enters the impeller eye from the pump suction inlet. There it is picked up by curved vanes, which change the flow direction from axial to radial. Both pressure and velocity increase as the water is impelled outward and discharged into the pump casing. The major components of a typical volute-type centrifugal pump are described in the following paragraphs.

cavitation
A condition that can occur when pumps are run too fast or water is forced to change direction quickly. During cavitation, a partial vacuum forms near the pipe wall or impeller blade, causing potentially rapid pitting of the metal.

tachometer generator
A sensor for measuring the rotational speed of a shaft.

Casing

Water leaving the pump impeller travels at high velocity in both radial and circular directions. To minimize energy losses due to turbulence and friction, the casing is designed to convert the velocity energy to additional pressure energy as smoothly as possible. In most water utility pumps, the casing is cast in the form of a smooth volute, or spiral, around the impeller. Casings are usually made of cast iron, but ductile iron, bronze, and steel are usually available on special order.

Single-Suction Pumps

Single-suction pumps are designed with the water inlet opening at one end of the pump and the discharge opening placed at right angles on one side of the casing. Single-suction pumps, also called end-suction pumps, are used in smaller water systems that do not have a high volume requirement. These pumps are capable of delivering up to 200 psi (1,400 kPa) pressure if necessary, but for most applications they are usually sized to produce 100 psi (700 kPa) or less.

The impeller on some single-suction pump units is mounted on the shaft of the motor that drives the pump, with the motor bearings supporting the impeller (Figure 14-15A). This arrangement is called the close-coupled design. Single-suction pumps are also available with the impeller mounted on a separate shaft, which is connected to the motor with a coupling (Figure 14-15B). In this design, known as the frame-mounted design, the impeller shaft is supported by bearings placed in a separate housing, independent of the pump housing.

The casing for a single-suction pump is manufactured in two or three sections or pieces. All housings are made with a removable inlet-side plate or cover, held in place by a row of bolts located near the outer edge of the volute. Removing the side plate provides access to the impeller. The pump does not have to be removed from its base for the side plate to be removed. However, all suction piping must be removed to provide sufficient access.

Some manufacturers cast the volute and the back of the pump as a single unit. Other manufacturers cast them as two separate pieces, which are connected by a row of bolts, similar to the inlet side plate. In units with separate backs, the impeller and drive unit can be removed from the pump without having to disturb any piping connections.

Double-Suction Pumps

Water enters the impeller of a double-suction pump from two sides and discharges outward from the middle of the pump. Although water enters the impeller from each side, it enters the housing at one location (usually on the opposite side of the discharge opening). Internal passages in the pump guide the water to the impeller suction and control the discharge water flow.

single-suction pump
A centrifugal pump in which the water enters from only one side of the impeller.

double-suction pump
A centrifugal pump in which the water enters from both sides of the impeller.

A. Close-coupled pump

B. Frame-mounted pump

Figure 14-15 Single-suction pumps

The double-suction pump is easily identified because of its casing shape (Figure 14-16). The motor is connected to the pump through a coupling, and the pump shaft is supported by ball or roller bearings mounted external to the pump casing.

The double-suction pump is usually referred to as a horizontal split-case pump. The term *horizontal* does not indicate the position of the pump. It refers to the fact that the housing is split into two halves (top and bottom) along the center line of the pump shaft, which is normally set in the horizontal position. However, some horizontal split-case pumps are designed to be mounted with the driveshaft in a vertical position, with the drive motor placed on top. Double-suction pumps can pump over 10,000 gpm (38,000 L/min), with heads up to 350 ft (100 m). They are widely used in large systems.

Removing the bolts that hold the two halves of the double-suction casing together makes it possible to remove the casing's top half. Most manufacturers place two dowel pins in the bottom half of the casing to ensure proper alignment between the halves when they are reassembled. It is important that the machined surfaces not be damaged when the halves are separated.

Impeller

Most pump impellers for water utility use are made of bronze, although a number of manufacturers offer cast iron or stainless steel as alternative materials. The overall impeller diameter, width, inlet area, vane curvature, and operating speed affect impeller performance and are modified by the manufacturer to attain the required operating characteristics. Impellers for single-suction pumps may be of the open, semiopen, or closed design, as shown in Figure 14-17. Most single-suction pumps

Figure 14-16 Double-suction pump casing shape
Courtesy of Ingersoll-Dresser Pump Company.

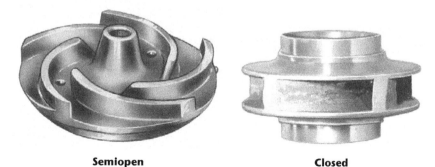

Semiopen   **Closed**

Figure 14-17 Types of impellers
Courtesy of Goulds Pumps, ITT Industries.

in the water industry use impellers of the closed design, although a few have semi-open impellers. Double-suction pumps use only closed-design impellers.

Wear Rings

In all centrifugal pumps, a flow restriction must exist between the impeller discharge and suction areas to prevent excessive circulation of water between the two. This restriction is made using wear rings. In some pumps, only one wear ring is used, mounted in the case. In others, two wear rings are used, one mounted in the case and the other on the impeller. The wear rings are identified in Figure 14-18.

The rotating impeller wear ring (or the impeller itself) and the stationary case wear ring (or the case itself) are machined so that the running clearance between the two effectively restricts leakage from the impeller discharge to the pump suction. The clearance is usually 0.010–0.020 in. (0.25–0.50 mm). Rings are normally machined from bronze or cast iron, but stainless-steel rings are available. The machined surfaces will eventually wear to the point that leakage occurs, decreasing pump efficiency. At this point, the rings need to be replaced or the wearing surfaces of the case and impeller need to be remachined.

Shaft

The impeller is rotated by a pump shaft, usually machined of steel or stainless steel. The impeller can be secured to the shaft on double-suction pumps using a key and a very tight fit (also called a shrink fit). Because of the tight fit, an arbor press or gear puller is required to remove an impeller from the shaft.

In end-suction pumps, the impeller is mounted on the end of the shaft and held in place by a key nut. The end of the shaft may be machined straight or with a slight taper. However, removing the impeller usually will not require a press. Several other methods are also used for mounting impellers.

wear rings
Rings made of brass or bronze placed on the impeller and/or casing of a centrifugal pump to control the amount of water that is allowed to leak from the discharge to the suction side of the pump.

shaft
(1) The bearing-supported rod in a pump, turned by the motor, on which the impeller is mounted. (2) The portion of a butterfly valve attached to the disc and a valve actuator. The shaft opens and closes the disc as the actuator is operated.

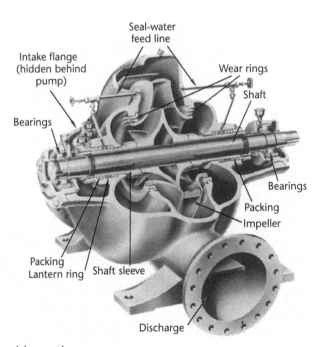

Figure 14-18 Double-suction pump
Courtesy of Ingersoll-Dresser Pump Company.

Shaft Sleeves

Most manufacturers provide pump shafts with replaceable sleeves for the packing rings to bear against. If sleeves are not used, the continual rubbing of the packing can eventually wear out the shaft, which would require replacement. A shaft could be ruined almost immediately if the packing gland were too tight. Where shaft sleeves are used, operators can repair a damaged surface by replacing the sleeve, a procedure considerably less costly than replacing the entire shaft. The sleeves are usually made of bronze alloy, which is much more resistant than steel to the corrosive effects of water. Stainless-steel sleeves are usually available for use where the water contains abrasive elements.

Packing Rings

To prevent leakage at the point where the shaft protrudes through the case, either packing rings or mechanical seals are used to seal the space between the shaft and the case. Packing consists of one or more (usually no more than six) separate rings of graphite-impregnated cotton, flax, or synthetic materials placed on the shaft or shaft sleeves (Figure 14-19). Asbestos material, once common for packing, is no longer used on potable water systems. The section of the case in which the packing is mounted is called the stuffing box. The adjustable packing gland maintains the packing under slight pressure against the shaft, stopping air from leaking in or water from leaking out.

To reduce the friction of the packing rings against the pump shaft, the packing material is impregnated with graphite or polytetrafluoroethylene to provide a small measure of lubrication. It is important that packing be installed and adjusted properly.

Lantern Rings

When a pump operates under suction lift, the impeller inlet is actually operating in a vacuum. Air will enter the water stream along the shaft if the packing does not provide an effective seal. It may be impossible to tighten the packing sufficiently to prevent air from entering without causing excessive heat and wear on the packing and shaft or shaft sleeve. To solve this problem, a lantern ring (Figure 14-20) is placed in the stuffing box. Pump discharge water is fed into the ring and flows out through a series of holes leading to the shaft side of the packing.

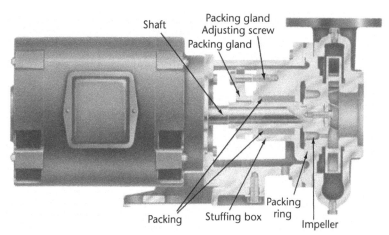

Figure 14-19 Pump packing locations
Courtesy of Aurora Pump.

packing
Rings of graphite-impregnated cotton, flax, or synthetic materials, used to control leakage along a valve stem or a pump shaft.

lantern ring
A perforated ring placed around the pump shaft in the stuffing box. Water from the pump discharge is piped to the lantern ring so that it will form a liquid seal around the shaft and lubricate the packing.

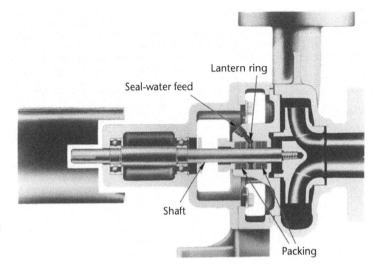

Figure 14-20 Lantern ring placed in the stuffing box
Courtesy of Aurora Pump.

From there, water flows both toward the pump suction and away from the packing gland. This water acts as a seal, preventing air from entering the water stream. It also provides lubrication for the packing.

Mechanical Seals

If the pump must operate under a high suction head (60 psig [400 kPa (gauge)] or more), the suction pressure itself will compress the packing rings, regardless of operator intervention. Packing will then require frequent replacement. Most manufacturers recommend using a mechanical seal under these conditions, and many manufacturers use mechanical seals for low-suction-head conditions as well. The mechanical seal (Figure 14-21) is provided by two machined and polished surfaces; one is attached to and rotates with the shaft, and the other is attached to the case. Contact between the seal surfaces is maintained by spring pressure.

The mechanical seal is designed so that it can be hydraulically balanced. The result is that the wearing force between the machined surfaces does not vary regardless of the suction head. Most seals have an operating life of 5,000–20,000 hours. In addition, there is little or no leakage from a mechanical seal; a leaky mechanical seal indicates problems that should be investigated and repaired. A major

> **mechanical seal**
> A seal placed on the pump shaft to prevent water from leaking from the pump along the shaft. Also prevents air from entering the pump. Mechanical seals are an alternative to packing rings.

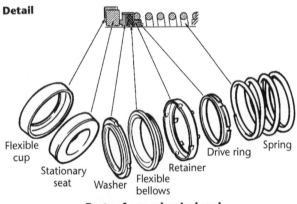
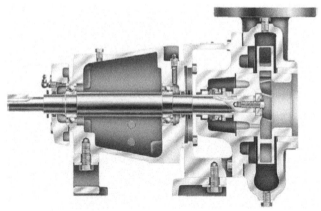

Figure 14-21 Mechanical seal parts and placement
Courtesy of Aurora Pump.

advantage of mechanical seals is that there is no wear or chance of damage to shaft sleeves.

A major disadvantage of mechanical seals is that they are more difficult to replace than packing rings. Replacing the mechanical seal often requires removing the shaft and impeller from the case. Another disadvantage is that failure of a mechanical seal is usually sudden and accompanied by excessive leakage. Packing rings, by contrast, normally wear gradually, and the wear can usually be detected long before leakage becomes a problem. Mechanical seals are also more expensive than packing.

Bearings

Most modern pumps are equipped with ball-type radial and thrust bearings. These bearings are available with either grease or oil lubrication and provide good service in most water utility applications. They are reasonably easy to maintain when manufacturer's recommendations are followed, and new parts are readily available if replacement is required. Ball bearings will usually start to get noisy when they begin to fail, enabling operators to plan a shutdown for replacement.

Couplings

Frame-mounted pumps have separate shafts connected by a coupling. The primary function of couplings is to transmit the rotary motion of the motor to the pump shaft. Couplings are also designed to allow slight misalignment between the pump and motor and to absorb the startup shock when the pump motor is switched on. Although the coupling is designed to accept a little misalignment, the more accurately the two shafts are aligned, the longer the coupling life will be and the more efficiently the unit will operate (Figure 14-22).

Various coupling designs are supplied by pump manufacturers. Couplings may be installed dry or lubricated. Most couplings are of the lubricated style and require periodic maintenance, usually lubrication at 6-month or annual intervals. Dry couplings using rubber or elastomeric membranes do not require any maintenance, except for periodic visual inspection to make sure they are not cracking or wearing out. The rubber or elastomer used for the membrane must be carefully selected for the pump, because the corrosive chemicals used in water treatment plants could affect the life and operation of the coupling.

WATCH THE VIDEO
Pumps—Centrifugal (www.awwa.org/wsovideoclips)

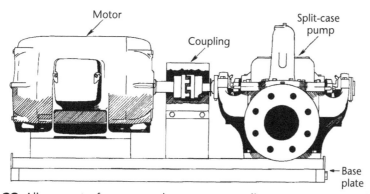

Figure 14-22 Alignment of motor and pump at coupling

bearing
Antifriction device used to support and guide pump and motor shafts.

coupling
A device that connects the pump shaft to the motor shaft.

Centrifugal Pump Maintenance

A regular inspection and maintenance program is important in maintaining the condition and reliability of centrifugal pumps. Bearings, seals, and other parts all require regular adjustment or replacement because of normal wear. General housekeeping is also important in prolonging equipment life.

Study Questions

1. The "heart" of a pump is called the
    a. volute case.
    b. impeller.
    c. motor.
    d. pump.

2. Which device serves the same function as the packing?
    a. Inline suction gland
    b. Packing gland
    c. Mechanical seal
    d. Lantern seal

3. Which of the following is used to stop air leakage into the casing around a pump shaft?
    a. Packing gland
    b. Lantern ring
    c. Seals
    d. Shaft sleeves

4. Which of the following is at the top of a stuffing box?
    a. Packing gland
    b. Lantern ring
    c. Mechanical seal
    d. Seal cage

5. Which assembly holds the lantern ring and packing?
    a. Shaft assembly
    b. Casing ring assembly
    c. Packing gland casing
    d. Stuffing box

6. Which type of valve is used to isolate a pump on the suction side?
    a. Butterfly valve
    b. Globe valve
    c. Gate valve
    d. Ball valve

7. Packing is designed to
   a. add lubricant to the shaft.
   b. expand and deteriorate with normal use.
   c. protect the shaft.
   d. wear and deteriorate with normal use.

8. Why is it so important to monitor the speed of a variable-speed pump?
   a. To prevent excessive temperatures from developing
   b. To prevent vibration from developing
   c. To prevent speed oscillation from occurring
   d. To prevent cavitation from occurring

9. List three disadvantages of turbine pumps.

10. What type of pump has an impeller rotating in a channel of constant cross-sectional area, which imparts mixed or radial flow to the water?

11. What type of pump uses closely meshed gears, vanes, or lobes rotating within a close-fitting chamber?

12. To prevent excessive circulation of water between the impeller discharge and suction areas, what components are used to create a flow restriction?

13. To prevent leakage at the point where the shaft protrudes through the case, what components may be used to seal the space between the shaft and the case?

Chapter 15
Meters

Water meters are used to measure and record the volume of water flowing through a line. The primary functions of metering are to help a water utility account for water pumped to the system and to equitably charge customers for the water they use.

The meter types most commonly used on water systems include the following:

- Positive-displacement
- Compound
- Current
- Detector-check
- Proportional
- Venturi
- Orifice
- Pitometer
- Magnetic
- Sonic

Various types of water meters are covered by various industry standards (e.g., AWWA Standards C700–C704, C706–C708, C710, and C712, and ANSI/NSF Standard 61).

Each type of meter has certain advantages and disadvantages and has been found to work best in specific applications. Desirable meter characteristics include the following:

- Accuracy within the range of anticipated flows
- Minimal head loss
- Durability
- Ease of repair
- Availability of spare parts
- Quiet operation
- Reasonable cost

Customer Water Meters

Most public water systems meter the water used by each service connection or customer. The principal reason is to determine billing charges. A secondary reason is to track water use to ensure that there is no undue waste or leakage in the

> **water meter**
> A device installed in a pipe under pressure for measuring and registering the quantity of water passing through.

distribution system. In addition, when customers are billed for the exact amount of water used, they have an incentive to use water wisely.

> **WATCH THE VIDEO**
> Water Meters—Types (www.awwa.org/wsovideoclips)

Positive-Displacement Meters

The most common type of meter for measuring water use through customer services is the positive-displacement meter. This type of meter consists of a measuring chamber of known size that measures the volume of water flowing through it by means of a moving piston or disk. The movement of each oscillation of the piston or disk is then transmitted to the register to record the amount of water. There are two types of positive-displacement meters: the piston type and the nutating-disk type.

Piston meters (Figure 15-1) utilize a piston that moves back and forth as water flows through the meter. A known volume is measured for each rotation, and the motion is transmitted to a register through a magnetic drive connection and series of gears.

Nutating-disk meters (Figure 15-2) use a measuring chamber containing a flat disk. When water flows through the chamber, the disk nutates (i.e., wobbles and rotates) and "sweeps out" a specific volume of water on each cycle. The rotary motion of the disk is then transmitted to a register that records the volume of water flowing through the meter.

Positive-displacement meters are generally used for residences and small commercial services in sizes from 0.6 in. to 2 in. (16 mm to 51 mm) because of their excellent sensitivity to low flow rates and their high accuracy over a wide range of flow rates.

Positive-displacement meters underregister when they are excessively worn. To avoid excessive wear, they should not be operated in excess of the flow rates listed in Table 15-1. Continuous operation of a meter at maximum flow will quickly destroy it. A meter is generally sized so that its expected maximum rate will be one half of its safe maximum operating capacity.

positive-displacement meter
A type of meter consisting of a measuring chamber of known size that measures the volume of water flowing through it by means of a moving piston or disk.

nutating-disk meter
A type of positive-displacement meter that uses a hard rubber disc that wobbles (rotates) in proportion to the volume of water flowing through the meter.

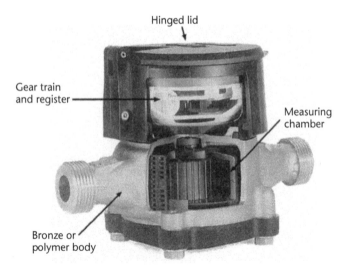

Figure 15-1 Piston meter
Courtesy of AMCO Water Metering Systems Inc.

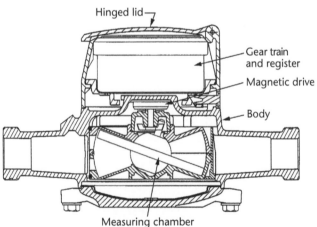

Figure 15-2 Nutating-disk meter with a plastic housing
Courtesy of Neptune Technology Group Inc.

Table 15-1 Maximum flow rates for positive-displacement meters

Meter Size, in.	(mm)	Safe Maximum Operating Capacity, gpm	(L/sec)	Recommended Maximum Rate for Continuous Operations, gpm	(L/sec)
0.5	(15)	15	(0.95)	7.5	(0.47)
0.5 × 0.75	(15 × 20)	15	(0.95)	7.5	(0.47)
0.6	(17)	20	(1.3)	10	(0.63)
0.625 × 0.75	(17 × 20)	20	(1.3)	10	(0.63)
0.75	(19)	30	(1.9)	15	(0.95)
1	(25)	50	(3.2)	25	(1.6)
1.5	(40)	100	(6.3)	50	(3.2)
2	(50)	160	(10)	80	(5.0)

Large-Customer Meters

Examples of customers that use large quantities of water are hospitals, golf courses, large public buildings, apartment houses, and industries. Industries that always use a large amount of water are those that must do a great deal of cleaning and those that incorporate water into their manufactured products. The types of meters most often used for these customers are compound meters, current meters, and detector-check meters.

Compound Meters

Compound meters are usually used for customers that have wide variations in water use. There may be some times of the day when their water demand is very high and other times when there is little or no use. The meters furnished for these customers must be relatively accurate at both low and high flow rates.

A standard compound meter consists of three parts: a turbine meter, a positive-displacement meter, and an automatic valve arrangement, all incorporated into one body (Figure 15-3). The automatic valve opens when high flows are sensed, enabling the water to flow with little restriction through the turbine side of the meter. Under low flows, the valve shuts and directs water through a small displacement meter for measurement. The unit therefore combines the favorable characteristics of both turbine and displacement meters into one unit. Compound meters may have separate registers for each meter or their output can be combined to indicate total use on a single register.

Another type of compound meter utilizes two standard meters connected together, as shown in Figure 15-4.

Current Meters

Current meters are sometimes used to meter water to large industrial customers. However, they are appropriate only when the minimum use by the customer is within the lower limit of the meter's accuracy. It is recommended that a strainer be installed ahead of a current meter to protect the meter from damage by sediment or other objects in the water.

compound meter
A water meter consisting of two single meters of different capacities and a regulating valve that automatically diverts all or part of the flow from one meter to the other. The valve senses flow rate and shifts the flow to the meter that can most accurately measure it.

current meter
A device for determining flow rate by measuring the velocity of moving water. Turbine meters, propeller meters, and multijet meters are common types. Compare with positive-displacement meter.

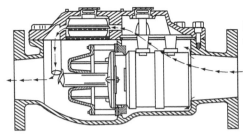

Low Flow
All of the water passes through the nutating-disk measuring element.

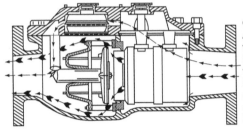

Crossover
As the control valve opens under higher flow rates, water passes through both measuring elements while the disk-side throttling begins.

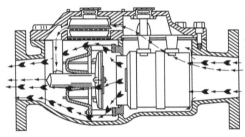

Full Flow
At high flow rates the control valve is fully open. The bulk of the water passes through the turbine measuring element, and the disk side is throttled to a minimal amount.

Figure 15-3 Compound meter
Courtesy of Neptune Technology Group Inc.

Figure 15-4 Compound meter arrangement that uses two standard meters
Courtesy of AMCO Water Metering Systems Inc.

Detector-Check Meters

Detector-check meters are designed for service where daily use is relatively low but where very high flow rates may be required in an emergency. The prime example is a building with a fire sprinkler system. The meter consists of a weight-loaded check valve in the mainline that remains closed under normal usage. A bypass around the check valve has a displacement-type meter to measure domestic use. When the sprinkler system calls for water, the loaded valve detects the decrease in line pressure in the building and swings completely open to allow full flow through the line.

> **detector-check meter**
> A meter that measures daily flow but allows emergency flow to bypass the meter. Consists of a weight-loaded check valve in the main line that remains closed under normal usage and a bypass around the valve containing a positive-displacement meter.

Mainline Metering

Larger water meters are used at various points in water treatment and distribution systems, including the following locations:

- Well discharges (to record the amount of water being supplied to the distribution system)
- The intake of a surface source (for determining chemical feed and control treatment plant operation)
- Intermediate points in the treatment process (for process control)
- The treatment plant discharge (for pump control and for comparison with customer metering to determine unaccounted-for water)

In addition, if water is purchased from another system, meters at the purchase point determine the payment that must be made. Meters installed on the various zones in the distribution system are used to provide pumpage and pressure control. If water is blended from multiple sources, multiple meters are used to help maintain a uniform blend.

System metering also helps in administering water rights and in checking the capacity of pumps and pipelines. Meter records are also often used in the engineering design of water system improvements.

Types of Mainline Meters

The following discussion briefly describes the more commonly used mainline meters.

Current Meters

Current meters are also commonly called velocity meters. They are principally used to measure flow in lines that are 3 in. (76 mm) and larger. The principal types include turbine, multijet, and propeller meters.

Turbine Meters A turbine meter (Figure 15-5) has a measuring chamber with a rotor that is turned by the flow of water. The volume of water recorded on the meter's register is almost in direct proportion to the number of revolutions made by the rotor. Turbine meters have little friction loss, but water must be moving at a sufficient speed before the rotor will start to rotate.

Turbine meters can underregister if the blades of the wheel become partially clogged or coated with sediment. This condition can generally be kept under control if a strainer is installed ahead of the meter. Periodic inspection and testing are also necessary. Current models of horizontal turbine meters have improved low-flow accuracy and higher maximum flows than previous instruments.

Multijet Meters A multijet meter has a multiblade rotor mounted on a vertical spindle within a cylindrical measuring chamber. Water enters the measuring chamber through several tangential orifices around the circumference and leaves the measuring chamber through another set of tangential orifices placed at a different level in the measuring chamber.

If the jets in a multijet meter become clogged, the meter will overregister. If the jet orifices become worn, the meter will underregister.

Propeller Meters A propeller meter (Figure 15-6) has a propeller that is turned by the flow of the water. This movement is transmitted to a register.

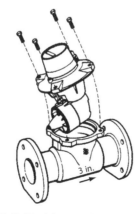

Figure 15-5 Turbine meter
Courtesy of Badger Meter, Inc., Milwaukee, Wisconsin.

---

turbine meter
A meter that measures flow rates by measuring the speed at which a turbine spins in water, indicating the velocity at which the water is moving through a conduit of known cross-sectional area.

multijet meter
A type of current meter in which a vertically mounted turbine wheel is spun by jets of water from several ports around the wheel.

propeller meter
A meter that measures flow rate by measuring the speed at which a propeller spins as an indication of the velocity at which the water is moving through a conduit of known cross-sectional area.

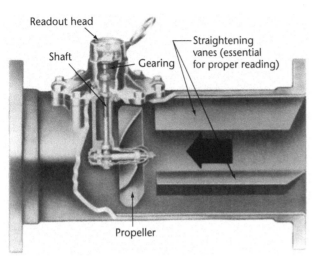

Figure 15-6 Propeller flowmeter

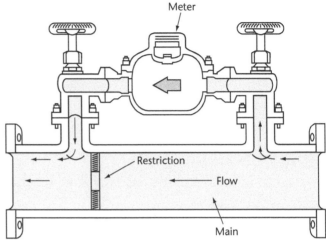

Figure 15-7 Proportional flowmeter

proportional meter
Any flowmeter that diverts a small portion of the main flow and measures the flow rate of that portion as an indication of the rate of the main flow. The rate of the diverted flow is proportional to the rate of the main flow.

Venturi meter
A pressure-differential meter used for measuring flow of water or other fluids through closed conduits or pipes, consisting of a Venturi tube and a flow-registering device. The difference in velocity head between the entrance and the contracted throat of the tube is an indication of the rate of flow.

orifice meter
A type of flowmeter consisting of a section of pipe blocked by a disc pierced with a small hole or orifice. The entire flow passes through the orifice, creating a pressure drop proportional to the flow rate.

The propeller may be small in diameter in relation to the internal diameter of the pipe, especially in larger sizes. Propeller meters are primarily used for mainline measurement where flow rates do not change abruptly, since the propeller has a slight lag in starting and stopping. Propeller meters can be built within a section of pipe or they can be saddle mounted.

Proportional Meters

A proportional meter has a restriction in the line to divert a portion of water into a loop that holds a turbine or displacement meter (Figure 15-7). The diverted flow in the loop is proportional to the flow in the mainline. A multiplying factor can be applied to the measurement of the diverted flow to record the flow in the pipeline. Proportional meters are relatively accurate but are difficult to maintain. However, they have little friction loss and offer little obstruction to the flow of water.

Venturi Meters

Venturi meters consist of a carefully sized constriction in the pipeline, called a venturi tube (Figure 15-8). The increased velocity of the water through the throat section causes the pressure at that point to be higher than before the throat. The change in pressure is proportional to the square of the velocity. The amount of water passing through the meter is therefore determined by a comparison of the pressure at the throat and at a point upstream from the throat.

Electronic or mechanical instruments are used to compare the pressures, determine the flow rate, and maintain a total of the flow (Figure 15-9). Venturi meters are accurate for a certain range of flows, have little friction loss, require almost no maintenance, and have long been used for measuring flows in larger pipelines.

Orifice Meters

As illustrated in Figure 15-10, orifice meters consist of a thin plate with a circular hole in it. This plate is installed in a pipeline between a set of flanges. As is the case with the venturi meter, the flow is determined by comparing the upstream line pressure with the reduced pressure at the restriction of the orifice. Although an orifice meter is not as expensive as most other large meters and occupies very little space, it has the disadvantage of creating considerably more head loss than is created by other meters.

Meters **287**

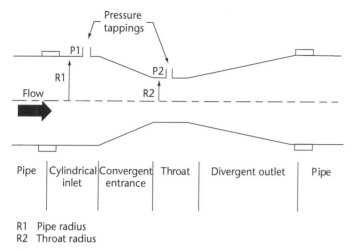

Figure 15-8 Venturi meter

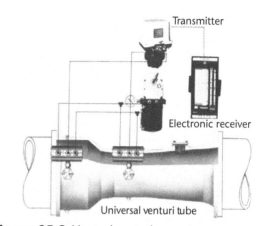

Figure 15-9 Venturi metering system
Courtesy of Honeywell, Industrial Measurement and Control.

Magnetic Meters

Magnetic meters, commonly called "mag meters," measure flow by means of a magnetic field generated around an insulated section of pipe (Figure 15-11). Water passing through the magnetic field induces a small flow of electrical current, proportional to the water flow, between electrical contacts set into the pipe section. The electrical current is measured and converted into a measure of water flow. Mag meters are particularly useful for measuring the flow of dirty or corrosive liquids that would damage a meter with moving parts or would plug the pressure taps of a venturi tube.

Ultrasonic Meters

Ultrasonic meters utilize sound-generating and -receiving sensors (transducers) attached to the sides of the pipe. Sound pulses are alternately sent in opposite

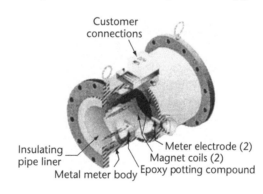

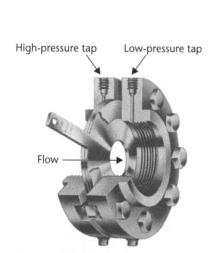

Figure 15-10 Orifice meter
Courtesy of Bristol Babcock Inc., Div. FKI Energy Technology.

Figure 15-11 Magnetic flowmeters
Courtesy of ABB Inc.

magnetic meter
A flow-measuring device in which the movement of water induces an electrical current proportional to the rate of flow.

ultrasonic meter
A meter that utilizes sound-generating and -receiving sensors (transducers) attached to the sides of the pipe.

diagonal directions across the pipe. Because of a phenomenon called the Doppler effect, the frequency of the sound changes with the velocity of the water. The difference between the frequency of the sound signal traveling with the flow of water and the signal traveling against the flow is an accurate indication of the quantity of water passing the meter.

Study Questions

1. What type of meter is most common for individual service lines?

2. What are the advantages of a venturi meter?

3. When are magnetic meters used?

4. How many meters can be tested on most test benches?

5. Which of the following is *not* a meter type commonly used on water systems?
    a. Proportional
    b. Current
    c. Magnetic
    d. Inverse

6. Positive-displacement meters ____ when they are excessively worn.
    a. do not register
    b. underregister
    c. overregister
    d. register intermittently

7. When using turbine meters, clogging of the wheel's blades can be prevented by installing a(n) ____ ahead of the meter.
    a. venturi device
    b. impeller
    c. strainer
    d. restrictor

8. What type of meter is used for service where daily use is relatively low but where very high flow rates may be required in an emergency?
    a. Magnetic meter
    b. Detector-check meter
    c. Venturi meter
    d. Compound meter

9. What type of meter is used for customers that have wide variations in water use?
    a. Magnetic meter
    b. Detector-check meter
    c. Venturi meter
    d. Compound meter

Chapter 16
Basic Chlorination

Basics of Chemical Disinfection

There are some fundamental concepts that operators need to know about chemical disinfection. When an oxidant such as chlorine is applied to water, it reacts with many of the impurities in the water to form compounds, some of which do not have the power to disinfect. Because the chlorine is used up in these reactions, it is said that the water has a "demand" for chlorine. Of course, some of the chlorine is used up destroying the pathogenic organisms, which also contributes to the demand for chlorine. Other impurities that use up chlorine include ammonia, iron and manganese, hydrogen sulfide, and organic material, including organic nitrogen. As long as these impurities are found in the source water, some of the disinfectant will be used up in the treatment process. Some reactions, like that of chlorine and ammonia, take place quickly. Others, like the reactions between chlorine and organic material, may take longer.

The oxidation of soluble inorganic contaminants must take place before filtration because the oxidized form of some constituents will stain fixtures if allowed to enter the distribution system. Additionally, all inorganics will create an oxidant demand. Table 16-1 shows the amount of chlorine needed to oxidize some of these inorganic compounds. It should be noted that the process also consumes alkalinity.

Most disinfection regulations require that an excess of disinfectant remain in the water as it travels through the distribution system. This excess, or residual, is desirable because it helps to continue the disinfection process, as well as provides added protection from unforeseen contamination or slime growth in the distribution system. When all of the demand for disinfectant is met, any added amount of chemical will produce this residual.

Table 16-1 Chlorine requirements and alkalinity consumption with inorganics

Inorganic Reactant	Part of Chlorine Required per Part of Inorganic	Alkalinity Consumed per Part of Chlorine Added
Iron	0.6	0.9
Manganese	1.3	1.5
Nitrite	1.5	1.8
Sulfide to sulfur	2.1	2.6
Sulfur to sulfate	6.2	7.4

Source: Connell 1996.

disinfection
The water treatment process that kills disease-causing organisms in water, usually by the addition of chlorine.

Effective disinfection takes place when the needed amount of disinfectant comes in contact with the microorganisms in the water and has sufficient time to disrupt the normal life processes of the organism. Pathogens such as viruses and *Giardia lamblia* have minimum requirements for disinfection concentration and contact time, and these requirements are temperature and pH dependent. The concept of disinfection sufficiency is represented by the equation $C \times T$, where C represents a known concentration of disinfectant, in milligrams per liter, that is multiplied by T, the time in minutes. Achieving the appropriate $C \times T$ value, in mg/L-min, is a requirement of the Surface Water Treatment Rule (SWTR) and is explained later in this chapter.

Practical Aspects of Chlorination

Disinfection using chlorine, or chlorination, is usually accomplished by adding gaseous chlorine, liquid sodium hypochlorite, or solid calcium hypochlorite (Figure 16-1). The application of any of these chemicals results in "available chlorine," or hypochlorite, and is represented by the formula OCl^-. Gaseous chlorine is considered pure, and so 100 percent of it goes into producing hypochlorite. Liquid sodium hypochlorite is purchased or generated at the site of application. It has a working percentage of 1–16 percent available hypochlorite, which must be factored by using mathematical formulas such as those found in Appendix D. Granular calcium hypochlorite also is not pure, available chlorine; generally, it is found to be 65–70 percent calcium hypochlorite. Operators should note the effect that chlorination has on pH. Gaseous chlorine tends to lower pH, while the hypochlorite compounds tend to raise it. The degree to which the pH is affected is dependent on the water's buffering capacity. Low-alkalinity waters can exhibit very noticeable swings in pH when chlorinated.

Properties of Chlorine

Chlorine is a greenish-yellow gas with a penetrating and distinctive odor, and it is very harmful to humans, even at low concentrations in air. The gas is 2.5 times heavier than air. For this reason, it is necessary to move to high ground when a chlorine leak occurs (Figure 16-2). Chlorine gas has a high coefficient of expansion, meaning that it expands easily and with great volume when heated. This

$C \times T$ value
The product of the residual disinfectant concentration C, in milligrams per liter, and the corresponding disinfectant contact time T in minutes. Minimum $C \times T$ values are specified by the Surface Water Treatment Rule as a means of ensuring adequate kill or inactivation of pathogenic microorganisms in water.

chlorination
The process of disinfecting water through the controlled use of chlorine; usually accomplished by adding gaseous chlorine, liquid sodium hypochlorite, or solid calcium hypochlorite.

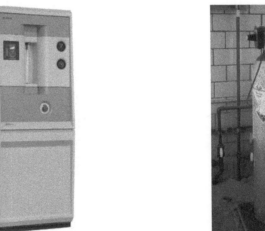

Figure 16-1 Gas chlorinator
Source: Siemens Industry Inc.

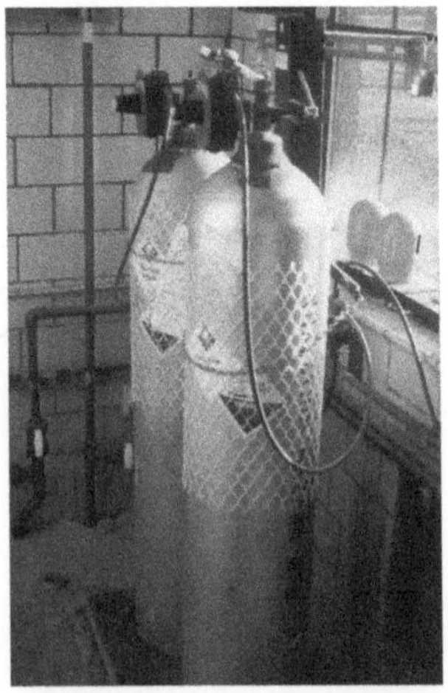

Figure 16-2 Installation of 150-lb chlorine cylinders. Tandem unit allows for dual withdrawal.
Source: Conneaut, Ohio, Water Department.

latter property makes it dangerous because it could easily burst its container if heat is applied. Chlorine will not burn but it will support combustion. Also, water makes chlorine gas corrosive to steel. For these reasons, operators should follow these commonsense rules when using chlorine:

- Always use a self-contained, positive-pressure breathing apparatus when working on chlorine cylinders.
- Never work on cylinders alone.
- Never apply heat to a chlorine cylinder.
- Never apply water to a leaking or burning cylinder, because the leak will grow worse.

Chlorine Reactions

When chlorine is added to water, several chemical reactions take place. Some involve the water molecules and some involve organic and inorganic substances in the water. To study these reactions, it is necessary to become familiar with some terms that are commonly used in the practice of chlorination. Small amounts of chlorine added to water will combine with organic and inorganic materials to form chlorine compounds. As more chlorine is added, a point is reached at which the reaction with these organic and inorganic materials stops. At this point, the chlorine demand has been satisfied.

The chemical reactions that took place in satisfying the chlorine demand created compounds, some of which have disinfecting properties and some of which do not. In the same fashion, chlorine will react with water molecules to produce compounds that have disinfecting properties. The total of all the compounds with disinfecting properties, plus any remaining uncombined chlorine, is known as chlorine residual. The presence of chlorine residual is measurable and is an indication that chemical reactions have taken place and that there is still some *free* or *combined available residual* to kill the microorganisms that may be introduced to the water.

The amount of chlorine needed should be added together to satisfy the chlorine demand and the amount of chlorine residual needed for disinfection to arrive at the *chlorine dosage*. This is the amount of chlorine that must be added to the water to disinfect it.

Chlorine Dosage Equation

Chlorine Dosage, mg/L = Chlorine Demand, mg/L + Total Chlorine Residual, mg/L

Chlorine Residual Equation

Total Chlorine Residual, mg/L = Combined Chlorine, mg/L + Free Chlorine, mg/L

Reaction With Water

Free chlorine combines with water to form hypochlorous acid (HOCl) and hydrochloric acid (HCl):

$$Cl_2 + H_2O \leftrightarrow HOCl + HCl$$

Depending on the pH of the water, hypochlorous acid may be present in the water as the hydrogen ion and hypochlorite ion dissociate. Hypochlorous acid dissociation is represented by the following equation:

$$HOCl \leftrightarrow H^+ + OCl^-$$

chlorine demand
The amount of chlorine that will combine with organic and inorganic materials to form chlorine compounds when added to water; once the demand is satisfied, additional chlorine will not combine with the organic and inorganic materials.

chlorine residual
The total of all the compounds with disinfecting properties, plus any remaining uncombined chlorine.

In dilute solutions of chlorine in water (such as those found in a water treatment plant), the formation of HOCl is complete and leaves little free chlorine (Cl_2). Hypochlorous acid dissociates into hypochlorite ion, with a reaction that is pH-dependent. Below pH 7.5, the dominant species is hypochlorous acid. Above pH 7.5, the dominant species is hypochlorite ion. Understanding this reaction is extremely important because HOCl and OCl$^-$ differ in disinfection strength. HOCl has a much greater disinfection potential than does OCl$^-$. Normally, 50 percent of the chlorine present in water with a pH of 7.3 will be in the form of HOCl, and 50 percent will be in the form of OCl$^-$. The higher the pH level, the higher the percentage of OCl$^-$ and therefore the less disinfecting ability.

Chlorination With Hypochlorite

The use of hypochlorite compounds to treat water achieves the same results as does the use of chlorine gas, in that each produces HOCl in water as a disinfectant. Hypochlorite may be applied in the form of calcium hypochlorite, $Ca(OCl)_2$ (commonly known as HTH), or sodium hypochlorite, NaOCl (referred to as bleach in liquid form). The chemical reactions of hypochlorite in water are similar to those of chlorine gas.

Calcium Hypochlorite

$$Ca(OCl)_2 + 2H_2O \longrightarrow 2HOCl + Ca(OH)_2$$

calcium hypochlorite + water → hypochlorous acid + calcium hydroxide

Sodium Hypochlorite

$$NaOCl + H_2O \longrightarrow HOCl + NaOH$$

sodium hypochlorite + water → hypochlorous acid + sodium hydroxide

When chlorine gas is used for disinfection, 100 percent of the gas is available as chlorine. Therefore, if the chlorine demand and the desired residual require 50 lb of chlorine to be fed, the chlorinator would be set at exactly 50 lb/day. However, when hypochlorites are used, the compounds contain only a percentage of the available chlorine that the gas does. HTH is about 65 percent available chlorine, and sodium hypochlorite is usually found to be 5–15 percent available chlorine. Because of this, more pounds per day must be fed to obtain the same results that would be expected from chlorine gas. To calculate the pounds-per-day requirement of hypochlorite, first determine the pounds per day of chlorine gas needed, then divide that number by the percentage of available chlorine in the hypochlorite. The calculation is as follows:

lb/day hypochlorite = (million gallons per day)(8.34)(mg/L gas)/ percent available chlorine in hypochlorite

> **Example 1**
>
> A total chlorine dose of 12 mg/L is needed to treat a source water. If the flow is 1.2 mgd and the hypochlorite has 65 percent available chlorine, how many pounds per day should be fed?
>
> *Answer:* 12 mg/L × 1.2 mgd × 8.34/0.65 = 185 lb/d HTH

Example 2

A flow of 850,000 gpd requires a dose of 25 mg/L chlorine. If sodium hypochlorite is 15 percent available chlorine, how many pounds per day are needed?

Answer: 25 mg/L × 0.85 mgd × 8.34/0.15 = 1,181 lb/d sodium hypochlorite.

Normally, these compounds are fed as solutions, and so it is necessary to determine the gallons-per-day feed necessary to deliver the required pounds per day. If the specific gravity is known, multiply it by 8.34 lb/gal and use that number as a divisor of the pounds per day to calculate gallons per day. If the specific gravity is unknown, assume it to be 1, and divide the pounds per day by 8.34 to calculate gallons per day.

Breakpoint Chlorination

Many water systems attempt to produce "free available" chlorine residual in their plant effluent. This process of adding chlorine to water until the chlorine demand has been satisfied is known as breakpoint chlorination. Chlorine in this form has its highest disinfectant ability. Further addition of chlorine will produce a chlorine residual that is directly proportional to the amount of chlorine added beyond the breakpoint. Public water supplies are normally chlorinated past the breakpoint unless they practice chloramination.

Figure 16-3 is a graph showing the breakpoint chlorination curve. As an example, assume that the water being treated contains some manganese, iron, nitrite, organic matter, and ammonia. When a small amount of chlorine is added, it reacts with (oxidizes) the manganese, iron, and nitrite. And that is all that happens—no disinfection takes place and no chlorine residual forms (point 1 to point 2 on the graph). If a little more chlorine is added, enough to react with the organics and

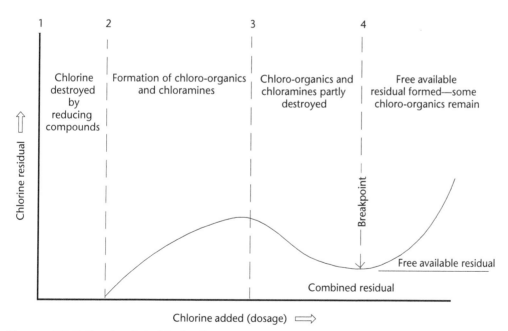

Figure 16-3 Breakpoint chlorination curve

breakpoint
The point at which the chlorine dosage has satisfied the chlorine demand.

the ammonia, chloro-organics and chloramines will form. The chloramines produce combined chlorine residual—a residual that has combined with other components and has lost some of its disinfecting power. These components may cause tastes and odors (from points 1 to 3).

If just a little more chlorine is added, the chloramines are destroyed (points 3 to 4) and breakpoint, the point at which a little more chlorine will begin to produce a free available chlorine residual (beyond point 4), has been reached. This chlorine is free in the sense that it has not reacted with anything, but it is free to do so and it will react if anything gets into the water that needs to be disinfected or oxidized further.

Chlorine Feed Equipment

Chlorine can be fed during pretreatment or post-treatment and often is fed during both to accomplish several tasks (Figure 16-4). Coagulation is usually enhanced in the presence of a strong oxidant, and chlorine is routinely used for this purpose in the pretreatment scheme. Use of pretreatment chlorine is sometimes minimized if disinfection by-product (DBP) formation is an issue.

Chlorine gas is fed using a pressure-feeder or vacuum-feeder system. Most often, the vacuum type is used. This equipment operates by sending a vacuum signal to the gas container. A mechanism draws the gas to a stream of water and produces a solution, which is fed at the application point(s). These units are safer than pressure feeders and generally provide more even dosage control.

Gas Chlorination Facilities

Chlorine gas, Cl_2, is about 2.5 times as dense as air. It has a pungent, noxious odor and a greenish-yellow color, although it is visible only at a very high concentration. The gas is very irritating to the eyes, nasal passages, and respiratory tract,

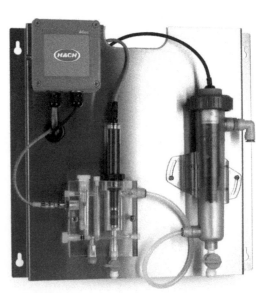

Figure 16-4 Chlorine residual analyzer
Source: Hach Company.

and it can kill a person in a few breaths at concentrations as low as 0.1 percent (1,000 ppm) by volume. Its odor can be detected at concentrations above 0.3 ppm.

Chlorine liquid is created by compressing chlorine gas. The liquid, which is about 99.5 percent pure chlorine, is amber in color and about 1.5 times as dense as water. It can be purchased in cylinders, containers, tank trucks, and railroad cars (Figures 16-5 through 16-8).

Liquid chlorine changes easily to a gas at room temperatures and pressures. One volume of liquid chlorine will expand to about 460 volumes of gas. Dry chlorine gas will not corrode steel or other metals, but it is extremely corrosive to most metals in the presence of moisture.

Chlorine will not burn. But, like oxygen, *it will support combustion*; that is, it takes the place of oxygen in the burning of combustible materials. Chlorine is not explosive, but it will react violently with greases, turpentine, ammonia, hydrocarbons, metal filings, and other flammable materials. Chlorine will not conduct electricity, but the gas can be very corrosive to exposed electrical equipment. Because of the inherent hazards involved, chlorine requires special care in storage and handling.

Handling and Storing Chlorine Gas

Safe handling and storage of chlorine are vital to the operator and to the communities immediately surrounding a treatment plant. An error or accident in chlorine handling can cause serious injuries or even fatalities.

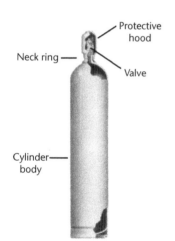

Figure 16-5 Chlorine cylinder
Courtesy of the Chlorine Institute.

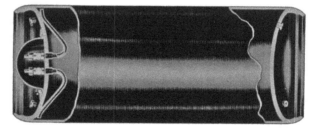

Figure 16-6 Chlorine ton container
Courtesy of the Chlorine Institute.

Figure 16-7 Chlorine ton container truck
Courtesy of PPG Industries, Inc.

Figure 16-8 Chlorine tank car
Courtesy of the Chlorine Institute.

The containers commonly used to supply chlorine in smaller water treatment plants are 150-lb (68-kg) cylinders. Larger plants find it more economical to use ton containers. Some very large plants are equipped to draw chlorine directly from tank cars.

The decision of whether to use cylinders or ton containers should be based on cost and capacity. The cost per pound (kilogram) of chlorine in cylinders is usually substantially more than that of chlorine in ton containers. If a plant's needs for chlorine are lower than 50 lb/d (23 kg/d), cylinders should usually be selected. For systems that use large amounts, ton containers will probably be more economical.

Cylinders

Chlorine cylinders hold 150 lb (68 kg) of chlorine and have a total filled weight of 250–285 lb (110–130 kg). They are about 10.5 in. (270 mm) in diameter and 56 in. (1.42 m) high. As illustrated in Figure 16-5, each cylinder is equipped with a hood that protects the cylinder valve from damage during shipping and handling. The hood should be properly screwed in place whenever a cylinder is handled and should be removed only during use.

Cylinders are usually delivered by truck. Each cylinder should be unloaded to a dock at truck-bed height if possible. If a hydraulic tailgate is used, the cylinders should be secure to keep them from falling. Cylinders must never be dropped, including "empty" cylinders, which actually still contain some chlorine.

The easiest and safest way to move cylinders in the plant is with a hand truck. As shown in Figure 16-9, the hand truck should be equipped with a restraining chain that fastens snugly around the cylinder about two-thirds of the way up. Slings should never be used to lift cylinders, and a cylinder should never be lifted by the protective hood because the hood is not designed to support the weight of the cylinder. Cylinders should not be rolled to move them about a plant. Tipping the cylinders over and standing them up can lead to employee injury. In addition, the rolled cylinders might strike something that could break off the valve.

Cylinders can be stored indoors or outdoors. If cylinders are stored indoors, the building should be fire resistant, have multiple exits with outward-opening doors, and be adequately ventilated. Outdoor storage areas must be fenced and protected from direct sunlight, and they should be protected from vehicles or

chlorine cylinder
A container that holds 150 lb (68 kg) of chlorine and has a total filled weight of 250–285 lb (110–130 kg).

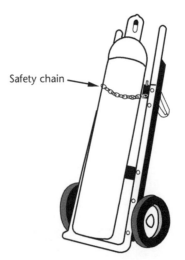

Figure 16-9 Hand truck for moving chlorine cylinders
Courtesy of the Chlorine Institute.

falling objects that might strike the cylinders. If standing water accumulates in an outdoor storage area, the cylinders should be stored on elevated racks. Avoiding contact with water will help minimize cylinder corrosion.

Some operators find it convenient to hang "full" or "empty" identification tags on cylinders in storage, so that the status of the chlorine inventory can be quickly determined. Other plants maintain separate storage areas for full and empty cylinders. Regardless of the system in place, all cylinders, full or empty, should receive the same high level of care. In addition, protective hoods should be placed on empty and full cylinders in storage. Even when a cylinder no longer has sufficient chlorine for plant use, a small amount of gas remains and could escape if the cylinder or valve were damaged. Both full and empty cylinders should always be stored upright and secured with a chain to prevent them from tipping over.

Ton Containers

The ton container is a reusable, welded tank that holds 2,000 lb (910 kg) of chlorine. Containers weigh about 3,700 lb (1,700 kg) when full and are generally 30 in. (0.76 m) in diameter and 80 in. (2.03 m) long. As shown in Figure 16-6, the ends are concave. The containers are crimped around the perimeter of the ends, forming good gripping edges for the hoists used to lift and move them. The ton container is designed to rest horizontally both in shipping and in use. It is equipped with two valves that provide the option of withdrawing either liquid or gaseous chlorine. The upper valve will draw gas, and the lower valve will draw liquid.

Handling the heavy containers is, by necessity, far more mechanized than handling cylinders. Containers are loaded or unloaded by a lifting beam in combination with a manual or motor-operated hoist mounted on a monorail that has a capacity of at least 2 tons (1,815 kg) (Figure 16-10). To prevent accidental rolling, containers are stored on trunnions, as illustrated in Figure 16-11. The trunnions allow the container to be rotated so that it can be positioned correctly for connection to the chlorine supply line.

Ton containers can be stored indoors or outdoors and require the same precautions as chlorine cylinders. The bowl-shaped hood that covers the two valve

> **ton container**
> A reusable, welded tank that holds 2,000 lb (910 kg) of chlorine. Containers weigh about 3,700 lb (1,700 kg) when full and are generally 30 in. (0.76 m) in diameter and 80 in. (2.03 m) long.

Figure 16-10 Lifting beam with motorized hoist for ton containers

Figure 16-11 Ton containers stored on trunnions

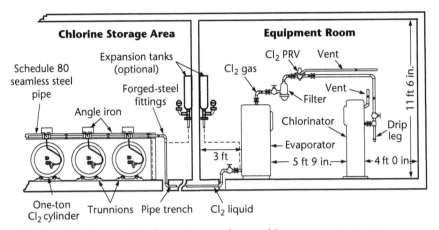

Figure 16-12 Chlorination feed equipment located in a separate room
Source: *Handbook of Chlorination and Alternative Disinfectants.* 4th ed. by Geo. Clifford White, copyright © 1998. Reprinted by permission of John Wiley & Sons, Inc.

assemblies when the tank is delivered should be replaced each time the container is handled, as well as right after it has been emptied.

The chlorine storage area should provide space for a 30- to 60-day supply of chlorine. Some systems feed chlorine directly from this storage area. When ton containers are used, the chlorination feed equipment is usually housed in a separate room (Figure 16-12).

 WATCH THE VIDEO
Chlorine—Transportation and Storage (www.awwa.org/wsovideoclips)

Feeding Chlorine Gas

Chlorine feeding begins where the cylinder or ton container connects to the manifold that leads to the chlorinator. The feed system ends at the point where the chlorine solution mixes into the water being disinfected. The system is composed of the following main components:

- Weighing scale
- Valves and piping
- Chlorinator
- Injector or diffuser

Weighing Scales

It is important that an accurate record be kept of the amount of chlorine used and the amount of chlorine remaining in a cylinder or container. A simple way to do this is to place the cylinders or containers on weigh scales. The scales can be calibrated to display either the amount used or the amount remaining. By recording weight readings at regular intervals, the operator can develop a record of chlorine-use rates. Figure 16-13 shows a common type of two-cylinder scale. Figure 16-14 shows a portable beam scale. Figure 16-15 shows a combination trunnion and scale for a ton container; this scale operates hydraulically and has a dial readout.

Basic Chlorination 299

Figure 16-13 Two-cylinder scale
Courtesy of US Filter/Wallace & Tiernan.

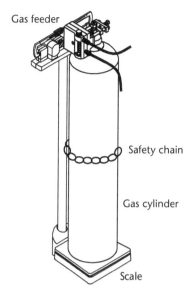

Figure 16-14 Portable beam scale
Courtesy of De Nora Water Technologies.

Figure 16-15 Combination trunnion and scale for a ton container
Courtesy of Force Flow.

Valves and Piping

Chlorine cylinders and ton containers are equipped with valves as shown in Figures 16-16 and 16-17. The valves must comply with standards set by the Chlorine Institute.

It is standard practice for an auxiliary tank valve to be connected directly to the cylinder or container valve, as illustrated in Figure 16-18. The connection is made with either a union-type or a yoke-type connector. The auxiliary valve can be used to close off all downstream piping, thus minimizing gas leakage during container changes. The auxiliary tank valve will also serve as an emergency

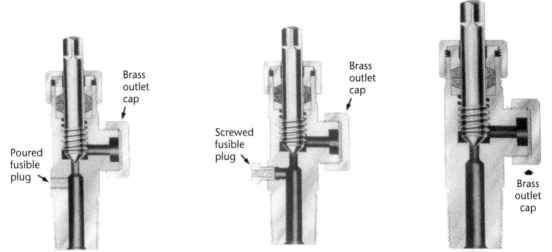

Figure 16-16 Standard cylinder valves: poured-type fusible plug (left) and screw-type fusible plug (right)
Courtesy of the Chlorine Institute.

Figure 16-17 Standard ton container valve
Courtesy of the Chlorine Institute.

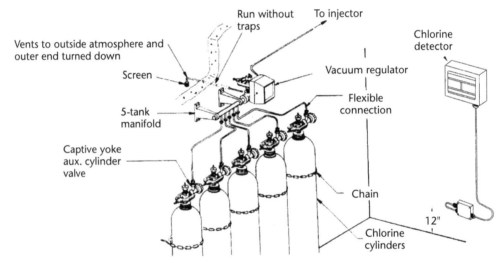

Figure 16-18 Auxiliary tank valve connected directly to container valve
Courtesy of US Filter/Wallace & Tiernan.

shutoff if the container valve fails. If a direct-mounted chlorinator is used, an auxiliary tank valve is not required (Figure 16-19).

The diagram in Figure 16-18 is of a typical valve assembly. The figure shows that the assembly is connected to the chlorine-supply piping by flexible tubing, which is usually ⅜-in. (10-mm) copper rated at 500 psig (3,500 kPa).

When more than one container is connected, a manifold must be used, as shown in Figure 16-18. The manifold channels the flow of chlorine from two or more containers into the chlorine-supply piping. The manifold and supply piping must meet the specifications of the Chlorine Institute. Manifolds may have from 2 to 10 connecting points. Each point is a union nut suitable for receiving flexible connections. Notice in Figure 16-18 that the header valve is connected at the manifold discharge end, providing another shutoff point. Additional valves are used along the chlorine supply line for shutoff and isolation in the event of a leak.

Chlorinators

The chlorinator can be a simple direct-mounted unit on a cylinder or ton container, as shown in Figure 16-19. This type of chlorinator feeds chlorine gas directly to the water being treated. A free-standing cabinet-type chlorinator is illustrated in Figure 16-20. Cabinet-type chlorinators, which operate on the same principle as cylinder-mounted units, have a sturdier mounting, and are capable of higher feed rates. Schematic diagrams of two typical chlorinators are shown in Figures 16-21 and 16-22.

The purpose of the chlorinator is to meter chlorine gas safely and accurately from the cylinder or container and then accurately deliver the set dosage. To do this, a chlorinator is equipped with pressure and vacuum regulators that are actuated by diaphragms and orifices for reducing the gas pressure. The reduced pressure allows a uniform gas flow, accurately metered by the rotameter (feed rate indicator). In addition, a vacuum is maintained in the line to the injector for safety purposes.

Figure 16-19 Direct-mounted chlorinator
Courtesy of De Nora Water Technologies.

Figure 16-20 Free-standing chlorinator cabinet
Courtesy of De Nora Water Technologies.

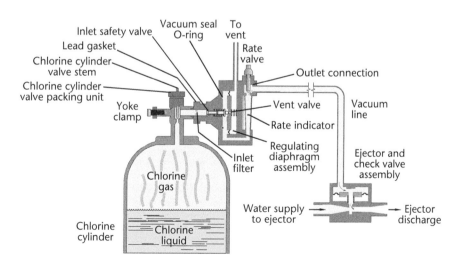

Figure 16-21 Schematic of direct-mounted gas chlorinator
Courtesy of De Nora Water Technologies.

chlorinator
Any device that is used to add chlorine to water.

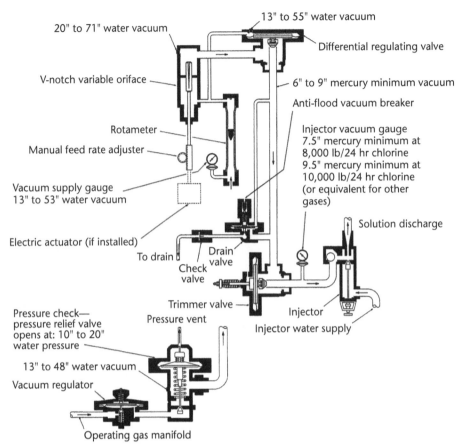

Figure 16-22 Schematic of cabinet-style chlorinator
Courtesy of US Filter/Wallace & Tiernan.

If a leak develops in the vacuum line, air will enter the atmospheric vent, causing the vacuum relief valve to close and stopping the flow of chlorine gas. To vary the chlorine dosage, the operator manually adjusts the setting of the rotameter.

It is normally required that each treatment plant have at least one standby chlorinator ready for immediate use in the event that the primary chlorinator should fail. Automatic switchover equipment is also strongly recommended.

Injectors

An injector (or ejector) is located within or downstream of the chlorinator, as illustrated in Figure 16-21. It is a venturi device that pulls chlorine gas into a passing stream of dilution water, forming a strong solution of chlorine and water. The injector also creates the vacuum needed to operate the chlorinator. The highly corrosive chlorine solution (pH of about 2–4) is carried to the point of application in a corrosion-resistant pipeline. The type of pipe typically used is polyvinyl chloride (PVC), fiberglass, or steel pipe lined with PVC or rubber. A strainer should be installed on the water line upstream of the injector. This strainer prevents any grit, rust, or other material from entering and blocking the injector or causing wear of the injector throat.

Diffusers

A diffuser is one or more short lengths of pipe, usually perforated, that quickly and uniformly disperses the chlorine solution into the main flow of water. There are two types of diffusers: those used in pipelines and those used in open channels

---

injector
The portion of a chlorination system that feeds the chlorine solution into a pipe under pressure.

diffuser
(1) A section of a perforated pipe or porous plates used to inject a gas, such as carbon dioxide or air, under pressure into water. (2) A type of pump.

and tanks. A properly designed and operated diffuser is necessary for the complete mixing needed for effective disinfection.

The diffuser used in pipelines less than 3 ft (0.9 m) in diameter is simply a pipe protruding into the center of the pipeline. Figure 16-23 shows a diffuser made from Schedule 80 PVC, and Figure 16-24 shows how the turbulence of the flowing water completely mixes the chlorine solution throughout the water. Complete mixing should occur downstream at a distance of 10 pipe diameters.

Figure 16-25 shows a perforated diffuser for use in larger pipelines. A similar design is used to introduce chlorine solution into a tank or open channel, as shown in Figure 16-26. (During normal operations, the diffuser would be completely submerged, but in the figure, the water level has been dropped, for illustrative purposes only, to show the chlorine solution passing out of each perforation.)

Gas Chlorination Auxiliary Equipment

A variety of auxiliary equipment is used for chlorination. The following discussion describes the functions of the more commonly used items.

Booster Pumps

A booster pump (Figure 16-27) is usually needed to provide the water pressure necessary to make the injector operate properly. The booster pump is usually a low-head, high-capacity centrifugal type. It must be sized to overcome the pressure in the line that carries the main flow of water being treated, and it must be rugged enough to withstand continuous use.

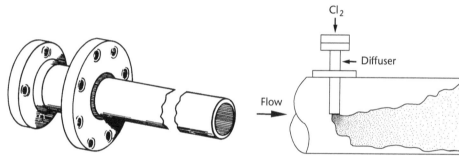

Figure 16-23 Diffuser made from Schedule 80 PVC

Figure 16-24 Chlorine solution mixing in a large-diameter pipeline

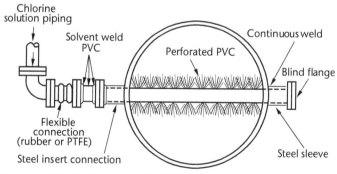

Figure 16-25 Perforated diffuser for pipelines larger than 3 ft (0.9 m) in diameter

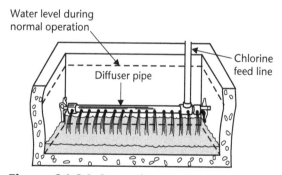

Figure 16-26 Open-channel diffuser

Source: *Handbook of Chlorination and Alternative Disinfectants.* 4th ed. by Geo. Clifford White, copyright © 1998. Reprinted by permission of John Wiley & Sons, Inc.

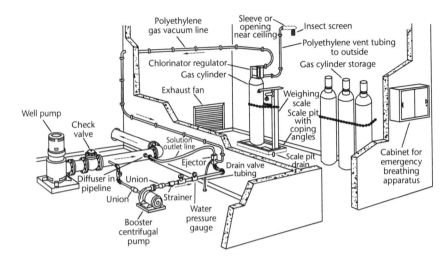

Figure 16-27 Typical chlorinator deep-well installation showing booster pump

Courtesy of De Nora Water Technologies.

Automatic Controls

If a chlorination system is to be manually operated, adjustments must be made each time the flow rate or the chlorine demand changes. For constant or near-constant flow rate situations, a manual system is suitable.

However, when flow rate or chlorine demand is continually changing, the operator is required to change the rotameter settings frequently. In these situations, automatic controls are valuable. Although many automatic control arrangements are possible, there are two common types: flow proportional control and residual flow control.

Flow Proportional Control If chlorine demand rarely changes and it is necessary to compensate only for changes in the pumping rate, flow proportional control works well. It will automatically increase or decrease the chlorine feed rate as the water flow rate increases or decreases. The required equipment includes a flowmeter for the treated water, a transmitter to sense the flow rate and send a signal to the chlorinator, and a receiver at the chlorinator. The receiver responds to the transmitted signal by opening or closing the chlorine flow rate valve.

Residual Flow Control If the chlorine demand of the water changes periodically, it is necessary to make corresponding changes in the rate of feed to provide adequate disinfection. Residual flow control, also called *compound loop control*, automatically maintains a constant chlorine residual, regardless of chlorine demand or flow rate changes. The system uses an automatic chlorine residual analyzer (Figure 16-28) in addition to the signal from a meter measuring the flow rate. The analyzer uses an electrode to determine the chlorine residual in the treated water. Signals from the residual analyzer and flow element are sent to a receiver in the chlorinator, where they are combined to adjust the chlorine feed rate to maintain a constant residual in the treated water.

Evaporators A chlorine evaporator is a heating device used to convert liquid chlorine to chlorine gas. Ton containers are equipped with valves that will draw either liquid or gas. At 70°F (21°C), the maximum gas withdrawal rate from a ton container is 400 lb/d (180 kg/d). If higher withdrawal is required, the liquid feed connection is used and connected to an evaporator. The evaporator accelerates the evaporation of liquid chlorine to gas, so that withdrawal rates up to 9,600 lb/d (4,400 kg/d) can be obtained.

chlorine evaporator
A heating device used to convert liquid chlorine to chlorine gas.

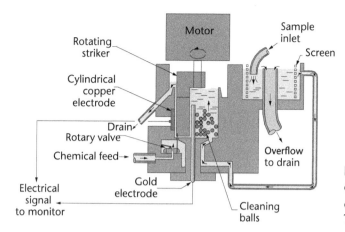

Figure 16-28 Automatic chlorine residual analyzer
Courtesy of De Nora Water Technologies.

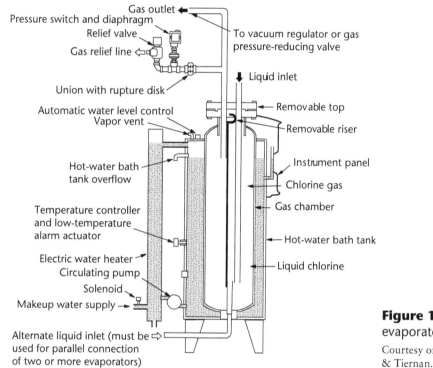

Figure 16-29 Chlorine evaporator
Courtesy of US Filter/Wallace & Tiernan.

An evaporator (Figure 16-29) is a water bath heated by electric immersion heaters to a temperature of 170–180°F (77–82°C). The pipes carrying the liquid chlorine pass through the water bath, and liquid chlorine is converted to gas by the heat.

Automatic Switchover Systems

For many small water systems, it is either impossible or uneconomical to have an operator available to monitor operation of the chlorination system at all times. An automatic switchover system provides switchover to a new chlorine supply when the online supply runs out. The switchover is either pressure or vacuum activated. The vacuum type of installation is shown in Figure 16-30. The automatic changeover mechanism has two inlets and one outlet. As the online supply is exhausted, the vacuum increases, causing the changeover mechanism to close on the exhausted supply and open the new chlorine supply. The unit can also send

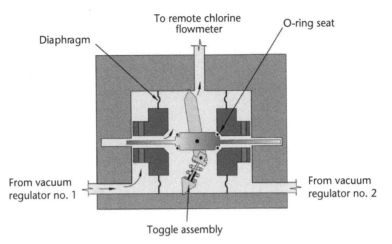

Figure 16-30 Automatic switchover unit
Courtesy of De Nora Water Technologies.

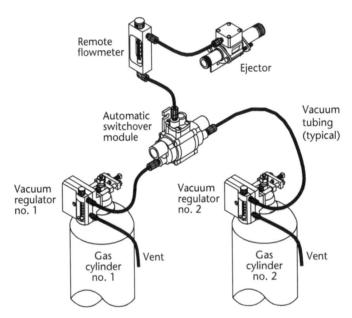

Figure 16-31 Typical installation of switchover system
Courtesy of De Nora Water Technologies.

a signal to notify operating personnel that the one tank is empty and should be replaced. Figure 16-31 shows a typical installation. This system is ideal for remote locations to ensure uninterrupted chlorine feeding.

Chlorine Alarms

Chlorinators are often equipped with a vacuum switch that triggers an alarm when it senses an abnormally low or high vacuum. A low-vacuum condition can mean an injector failure, vacuum line break, or booster pump failure. A high-vacuum condition can be caused by a plugged chlorine supply line or by empty chlorine tanks.

Safety Equipment

Safety in and around the gas chlorination process is important to prevent serious accidents and equipment damage. Certain items of equipment, such as the following, are essential for the safe operation of a chlorination facility:

- Chlorine detectors
- Self-contained breathing apparatus
- Emergency repair kits

Hypochlorination Facilities

Hypochlorination is a chlorination method increasingly used by water treatment plants because of its relative safety (as compared to gaseous chlorine) and its ease of use. Sodium hypochlorite is fed as a liquid, and many operators prefer to feed liquids rather than gases.

Hypochlorite Compounds

The two most commonly used compounds are calcium hypochlorite and sodium hypochlorite. Table 16-2 lists the properties of both compounds.

Calcium Hypochlorite

Calcium hypochlorite, $Ca(OCl)_2$, is a dry, white or yellow-white, granular material. It is also available in compressed tablets. It normally contains 65 percent available chlorine by weight. This means that when 1 lb (0.5 kg) of the powder is added to water, only 0.65 lb (0.3 kg) of pure chlorine is being added. Conversely, if 1 lb (0.5 kg) of chlorine is added, 1.5 lb (0.7 kg) of calcium hypochlorite must be added (Table 16-3).

Calcium hypochlorite requires special storage to avoid contact with organic material. Its reaction with any organic substances can generate enough heat and oxygen to start and support a fire. When calcium hypochlorite is mixed with water, heat is given off. To provide adequate dissipation of the heat, the dry chemical should be added to the water; the water should *not* be added to the chemical.

Calcium hypochlorite is used mostly for disinfection of new and repaired water mains, water storage tanks, and small water volumes, such as swimming pools.

Table 16-2 Properties of hypochlorites

Property	Sodium Hypochlorite	Calcium Hypochlorite
Symbol	NaOCl	$Ca(OCl)_2$
Form	Liquid	Dry granules, powder, or tablets
Strength	Up to 15% available chlorine	65–70% available chlorine, depending on form

Table 16-3 Chlorine content of common disinfectants

Compound	Chlorine Percentage	Amount of Compound Needed to Yield 1 lb of Pure Cl
Chlorine gas or liquid (Cl_2)	100	1 lb (0.454 kg)
Sodium hypochlorite (NaOCl)*	15	0.8 gal (3 L)
	12.5	1.0 gal (3.8 L)
	5	2.4 gal (9.1 L)
	1	12.0 gal (45.4 L)
Calcium hypochlorite [$Ca(OCl)_2$]	65	1.54 lb (0.7 kg)

*Sodium hypochlorite is available in four standard concentrations of available chlorine. Ordinary household bleach contains 5% chlorine.

hypochlorination
Chlorination using solutions of calcium hypochlorite or sodium hypochlorite.

Sodium Hypochlorite

Sodium hypochlorite, NaOCl, is a clear, light-yellow liquid commonly used for bleach. Ordinary household bleach contains 5–6 percent available chlorine. Industrial bleaches are stronger, containing from 9 to 15 percent.

The sodium hypochlorite solution is alkaline, with a pH of 9–11, depending on the available chlorine content. For common strengths, Table 16-3 shows the amount of solution needed to supply 1 lb (0.5 kg) of pure chlorine. Large systems can purchase the liquid chemical in carboys, drums, and railroad tank cars. Very small water systems often purchase it in 1-gal (3.8-L) plastic jugs.

There is no fire hazard in storing sodium hypochlorite, but the chemical is quite corrosive and should be kept away from equipment susceptible to corrosion damage. At its maximum strength of 12–15 percent, sodium hypochlorite solution can lose 2–4 percent of its available chlorine content per month at room temperature. It is therefore recommended that it not be stored for more than 15–25 days. Instability of the chemical increases with increasing temperature, solution strength, and exposure to sunlight. Sodium hypochlorite solutions of 6 percent are more stable. Water plant design specifications usually call for the dilution of the 12 percent stock chemical to 6 percent, unless special storage facilities are built that keep out sunlight and heat.

Common Equipment

Disinfecting facilities using calcium hypochlorite should be equipped with a cool, dry storage area to stockpile the compound in the shipping containers. A variable-speed chemical feed pump (hypochlorinator), such as a diaphragm pump, is all that is required for feeding the chemical to the water. A mix tank and a day tank (Figure 16-32) are also required. After calcium hypochlorite is mixed with water, impurities and undissolved chemicals settle to the bottom of the mix tank. The clear solution is then transferred to the day tank for feeding. This prevents any of the solids from reaching and plugging the hypochlorinator or rupturing the diaphragm.

Because sodium hypochlorite is a liquid, it is simpler to use than calcium hypochlorite. It is fed neat (as the 12 percent stock chemical) or at the 6 percent strength, usually with peristaltic pump equipment that uses a quality-grade tubing. Redundant pumps are needed and operators must get used to changing the tubing on a frequent basis to prevent failure.

Off-gassing is a major issue with the equipment used for storing and feeding sodium hypochlorites. Equipment failure and damage are common occurrences when hypochlorite feed systems are poorly designed or poorly maintained.

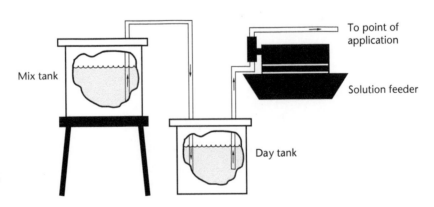

Figure 16-32 Mix tank and day tank

C × T Values

The adequacy of disinfection is determined by the product of the final residual concentration (C) of the disinfectant and the contact time (T) for that disinfectant. The contact time is evaluated as that which is exceeded by 90 percent of the water; that is, 10 percent of the water has less contact time than the T value. These T values are commonly determined by performing tracer studies or by computations based on the ratio of the length and width of the reaction chambers (clearwells or contactors), including any baffling that may be present. Baffling of clearwells, which involves adding separation walls that force snakelike flow patterns, provides greater length-to-width ratios and therefore greater contact times for the disinfectant.

The US Environmental Protection Agency has published tables that show the C × T values needed for basic disinfectants such as chlorine, chloramine, ozone, and chlorine dioxide. C × T values will vary with temperature, pH, and concentration. Surface water treatment plants must provide the minimum C × T value in order to be in compliance with the SWTR. The C × T calculation, which is performed every 24 hours, is derived at the highest hourly flow rate and lowest contactor volume for that period.

Disinfection By-products

There are many DBPs, the chief of which are total trihalomethanes (TTHMs) and haloacetic acids (HAAs). These by-products are formed when chlorine is added to water that contains the necessary types and amounts of natural organic matter (NOM).

Chapter 1 discusses the regulatory concerns for DBPs, but operators should understand the kinetics of DBP formation and know which mechanisms are under their control. In general, the following may be considered true for DBP formation:

- DBP formation is affected by contact time between the disinfectant and the NOM in water. Therefore, minimizing that contact time, either by adding disinfectant later in the treatment process or by "turning over" the water in distribution storage tanks more frequently, may decrease DBP formation.
- The *water age*—the amount of time that water resides in the distribution system or storage tanks before being consumed—is of critical importance. Longer water age usually leads to increased formation of DBPs.
  - Water age can be estimated by simply performing tracer studies of the distribution system. Operators choose representative sample locations in the system and begin sampling them each day for a tracer such as fluoride. The fluoride feed at the plant is turned off (with permission of the regulatory authority), and the time it takes for the fluoride at each sample location to reach background levels will give a good snapshot of relative water age at these locations. Operators can then make improvements by changing valve positions, booster settings, and so forth, which may decrease this age.
  - Another way to estimate water age is to use a *calibrated* system hydraulic model. Many hydraulic modeling computer programs can generate calculated water age values for locations throughout the distribution system.

> **tracer study**
> A study using a substance that can readily be identified in water (such as a dye) to determine the distribution and rate of flow in a basin, pipe, or channel.

- Increasing disinfectant dose and residual increases DBP formation. **Caution:** Operators must avoid the risk of pathogen passage that can accompany disinfectant losses. A balance must be achieved to simultaneously remove microbiological contaminants while controlling DBP formation.
- As temperature increases, DBP formation may increase. Warmer water favors the formation of DBPs, as does summer heat, which bakes the system storage tanks.
- The water pH can control the type of formation. Lower pH favors HAA formation, while higher pH favors TTHM formation.
- The nature and concentration of the precursor material will affect the types and amounts of DBP formation. Bromide in the water, or as a contaminant in the chemical feed system, will influence formation of brominated DBP species.

Booster Disinfection

Most systems perform disinfection only at the water treatment plant or wellhead. The dosage applied is adequate to provide a measurable residual throughout the distribution system. In some cases, this is not possible without adding a prohibitive amount at the source. Booster disinfection (redisinfection) is needed in this instance. Some utilities have employed booster disinfection as a strategy to maintain a more uniform disinfection residual at a lower level, reduce DBPs, and lower total disinfection cost (in some cases).

Where chloramines are used to provide the disinfectant residual, booster disinfection requires special consideration. If free chlorine is applied for booster disinfection, it is possible for free chlorine from this process to blend with chloramine residual in the system in an uncontrolled manner. The potential problems of blending are related to the breakpoint curve (Figure 16-3) for free and combined chlorine. Various resultant concentrations of mono-, di-, and free chlorine are possible. This may lead to undesirable tastes and odors or possible reduction of chlorine residual in some areas. Some utilities have dealt with this situation by measuring the ammonia residual remaining in the water at the point of free chlorine addition. They have then added the correct ratio of free chlorine to reform predominantly monochloramine. In this case, blending with residual chloramine is not an issue. Other utilities have found that free chlorine booster disinfection in a chloraminated supply does not cause problems. The best strategy for each utility depends on many factors, some of which are unique to the site. Therefore, each situation should be fully evaluated before implementing any disinfection strategy.

Study Questions

1. One of chlorine's advantages is that it
   a. is not influenced much by pH changes.
   b. does not produce chlorinated by-products.
   c. has a persistent residual.
   d. does not cause taste and odor problems.

2. Chlorine gas is _____ times heavier than air.
   a. 1.5
   b. 2.5
   c. 3.5
   d. 4.5

3. After a water storage tank has been chlorinated, which bacteriological test must prove negative before the tank is put back into service?
   a. Gram-negative test
   b. HPC test
   c. Coliform test
   d. Chloramine test

4. Sodium hypochlorite (NaOCl) solution is available with _____ available chlorine.
   a. 2–5%
   b. 5–20%
   c. 25–50%
   d. 50–70%

5. Booster chlorination is chlorine added
   a. in the coagulation mixing chamber.
   b. before the filters.
   c. at the clearwell.
   d. somewhere in the distribution system.

6. Chlorine cylinders hold _____ of chlorine.
   a. 100 lb (45 kg)
   b. 150 lb (68 kg)
   c. 200 lb (97 kg)
   d. 350 lb (159 kg)

7. Why would a system need a booster chlorination station?

8. In the $C \times T$ equation, what does C represent?

9. What is the term for the total of all the compounds with disinfecting properties, plus any remaining uncombined chlorine?

10. What is the term for a venturi device that pulls chlorine gas into a passing stream of dilution water, forming a strong solution of chlorine and water?

11. What is the term for a chlorination method increasingly used by water treatment plants because of its relative safety (as compared to gaseous chlorine) and its ease of use?

Chapter 17
System Operations

Operating and maintaining a water distribution system is an involved process. There are two major objectives for drinking water distribution system operational policies: (1) maintain water quality from the point of entry into the distribution system to the point of use and (2) maintain adequate pressure and deliver adequate flow to satisfy customer demands and protect from fire losses. This chapter discusses operational practices and procedures that are designed to address these two objectives.

Maintaining Water Quality

System operators have traditionally focused their attention on achieving adequate pressure and flow. Customers (and regulators) are now demanding an increasing emphasis on the water quality.

The distribution and storage system is one component of a multiple-barrier approach to preventing contamination (Table 17-1). Each component must be optimized to provide the maximum level of protection from contamination as water is delivered to consumers. Several groups (departments) within a utility have key roles in maintaining and operating distribution systems. To make sure all groups can provide their input, standard operating procedures (SOPs) should be jointly developed.

Five steps are suggested to optimize water quality in the distribution system:

1. Gather information to understand the system and define the cause of any problems.
2. Set water quality goals that go beyond regulatory requirements and establish preliminary performance objectives.
3. Evaluate alternative solutions to address problems and satisfy the preliminary performance objectives.
4. Implement good management practices and monitor effectiveness of these practices in maintaining water quality.
5. Prepare final performance objectives and SOPs with input from all affected operating groups (or departments).

These steps should result in a distribution system that is optimized to provide water of the highest quality possible.

Table 17-1 Multiple protection barriers

1. Source protection and management
2. Treatment (may include a number of internal barriers)
3. Disinfection
4. Distribution system operation and maintenance
5. Monitoring and response

WATCH THE VIDEO
Maintaining Water Quality in the Distribution System
(www.awwa.org/wsovideoclips)

Water Quality Monitoring

A comprehensive, well-designed water quality monitoring program is necessary to optimize distribution system operation to deliver high-quality water to consumers. The goal of a well-managed distribution system should be to provide water to customers' taps that has not changed from the point of entry into the system. Attaining this goal will result in increased customer satisfaction and compliance with all regulatory requirements.

Routine Monitoring

The analysis of water samples collected from the distribution system provides evidence that the water being delivered to customers is safe to drink and desirable to use. The tests ensure that water treatment processes are functioning properly and that the water has not objectionably degraded while in the distribution system. Testing also indicates whether there is corrosion or scale accumulation in the distribution piping.

Development of a Sampling Plan

The purpose of a routine sampling plan is to examine characteristics of water quality in the distribution system as the water moves from source to tap. The plan must include all samples to meet regulatory requirements. Of particular interest to distribution system operators is the US Environmental Protection Agency (USEPA) Lead and Copper Rule. This regulation requires special testing of samples collected from residences and in the distribution system. The sampling plan should also include sites, parameters, and frequencies necessary to fully describe the distribution system. The following lists identify key components of a water quality sampling plan.

Site Selection Criteria
- Age of the water
- Locations where multiple sources mix
- Storage facilities
- A selection of main materials and conditions
- Locations of booster disinfection stations
- Critical facilities (hospitals, high-usage customers)

Test Parameter Selection
- All regulated parameters
- Disinfectant residual
- Color, turbidity, pH
- Heterotrophic plate count bacteria
- Total coliform bacteria
- Other targeted parameters (corrosion inhibitors, coagulant residuals, nitrite, ammonia, etc.)

Sampling Frequency Criteria
- Satisfy regulatory requirements
- Continuous (may be useful at some strategic locations—e.g., storage facilities)
- Daily (disinfectant residuals, bacteria, pH, temperature, at some locations)
- Weekly (storage facility grab samples, dead ends, targeted parameters)
- Monthly (long-term trends or special surveillance samples)

Nonroutine Monitoring

Special sampling studies are needed at times to help manage the system, investigate water quality problems, respond to emergencies, and evaluate alternatives for system improvements. Operators frequently establish monitoring programs that go beyond regulatory requirements. Special sampling is often necessary to identify areas where water quality is deteriorating. Some examples of nonroutine monitoring studies for specific purposes are given below.

Finished Water Storage Facilities Many distribution system water quality problems can be traced to finished water storage facilities. Water quality monitoring at these sites can be used as a quality control check. In addition to more routine parameters, additional parameters such as *Cryptosporidium*, nitrite, ammonia, odor, total trihalomethanes, and iron may be examined.

Customer Complaint Investigations Water quality testing is an important part of an onsite investigation. Coliform bacteria, temperature, pH, and disinfectant residual should normally be tested. In addition, other tests may be necessary depending on the complaint (odor, color, iron, particle identification). The results of water quality tests should be communicated to the customer as the inquiry is resolved.

Construction Activities Main rehabilitation and replacement activities usually require disinfection. The mains must be tested and found acceptable prior to a return to service. Test parameters may include coliform bacteria, pH, turbidity, and disinfectant residual. Other possible tests are odor, color, and volatile organic compounds.

Emergency Monitoring Main breaks, treatment upsets, and backflow events are only a few possible emergency situations that may occur in distribution systems. Water quality monitoring is important to help determine the extent of the problem and verify when corrective measures have been successful. Generally, quick field tests are needed to provide information in a timely manner. It may be useful to monitor the following parameters in an emergency: bacteria, disinfectant residual, turbidity, color, pH, conductivity, alkalinity, and fluoride.

System Design for Water Quality Enhancement

Drinking water distribution systems have, historically, been designed to satisfy potability and fire flow demands. Little consideration was given to the impact of distribution system design on water quality. Many potential water quality problems can be reduced or eliminated by including design features that are specific to water quality.

Planning Considerations

As utilities consider distribution system improvements, it is critical that they include factors such as water quality regulations, customer expectations, customer growth, budget limitations, and financial impacts. Long-range master plans that include a capital improvement program must incorporate water quality changes as well as growth demands.

Hydraulic and Water Quality Modeling

Water distribution system models simulate the behavior of physical facilities and water use patterns within the system. These computer models (Figure 17-1) are

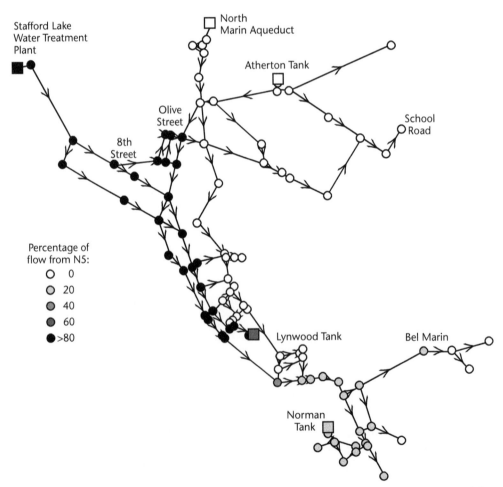

Figure 17-1 North Marin hydraulic calibrations for 6:00 a.m. showing direction of flow and percentage of flow from N5 (Stafford Lake Water Treatment Plant)

essential tools for designing distribution system improvements or extensions. The model can simulate actual or proposed operating practices such as operation of pumps, opening or throttling valves, fire flow events, and major main breaks.

Water quality models are extensions of hydraulic models and can estimate water quality changes in a distribution system. An accurate, calibrated, hydraulic model is a necessary building block for a water quality model since the flows, pressures, and volumes in storage are major contributing factors that influence water quality. These computer simulations of a distribution system can be used to illustrate the effect of blending between sources for parameters such as hardness, total dissolved solids, or conductivity. Water age, pH, chlorine residual, trihalomethane formation, or the spread of a waterborne contaminant can also be estimated using sophisticated calculations to account for changing characteristics of the parameter. Information from water quality models can provide valuable information that can be used to establish water quality monitoring sites based on system characteristics, thus providing an opportunity to better diagnose the cause of changes in the system.

Pipeline Design

Water quality considerations need to be included in the design of pipeline networks. By integrating these concepts with volume and pressure requirements, many future water quality problems can be avoided. Special attention should be given to issues affecting pipeline water main sizing. Fire flow requirements generally determine the need for larger mains. This requirement may lead to stagnation, bacterial regrowth, and depletion of disinfectant residual. When designing a pipeline network, a balance between flow requirements and water quality consequences must be considered. The pipeline network design should strive to eliminate (or at least reduce) the number of dead ends, which are notorious for producing water quality problems. The network design should avoid low- or negative-pressure areas, prevent uncontrolled pipe scouring, and reduce the collection of suspended sediments during routine operations.

By selecting the best pipeline construction materials for the given water quality, future problems can be avoided. Generally, lined ductile-iron, concrete pressure, and polyvinyl chloride pipe are ranked by utilities as the most favorable materials for maintaining water quality. Selection of the most suitable material depends on many factors, including water quality, pressure, size of main, structural loads, soil conditions, and cost.

Pressure Zone Adjustments

In some systems it is necessary to divide the service area into defined pressure service zones. These zones may be connected in series (water must flow through one zone to reach the next). This practice may increase the age of the water and thus incur the water quality problems associated with this issue. Pressure zone boundaries may also form barriers, in effect establishing artificial dead ends in the system. The dead ends are often the cause of multiple water quality problems: tastes and odors, decay of disinfectant residual, bacterial regrowth, increased corrosion, changes in pH, and collection of sediment.

Finished Water Storage Facilities

Monitoring, inspection, maintenance, and mixing to enhance water quality (Table 17-2) must be considered when designing storage facilities. Many water quality problems are directly associated with water retention in storage facilities.

Table 17-2 Examples of water quality design considerations for storage facilities

1. Conform to latest AWWA/NSF standards.
2. Provide for isolation.
3. Provide for bypass.
4. Minimize dead storage.
5. Protect drain from cross-connections.
6. Provide redundant units for maintenance.
7. Isolate altitude valve.
8. Ensure roof is watertight.
9. Ensure vents and overflow are tamper resistant, lockable, and screened.
10. Be sure that overflow pipe is sized equal to or greater than maximum inflow rate.
11. Be sure that inlet and outlet pipe are sized and positioned to promote mixing.
12. Ensure access hatches are lockable, hinged, and constructed to avoid stormwater inflow.
13. Check that safety ladders conform to regulations.
14. Install lockable security fence if not already in place.
15. Provide cathodic protection if needed.
16. Provide monitoring instruments for level, flow, and water quality parameters.
17. Identify sampling locations for inflow and outflow.
18. Provide security alarms.

The size and type of facility is normally heavily influenced by hydraulic considerations such as maximum hour demand, fire flow and emergency supply considerations, and future growth demands. Major factors in storage facility design that can mitigate water quality problems are selecting and maintaining the correct materials of construction, providing adequate and appropriate cathodic protection to prevent or control corrosion, and preventing dead zones (Figure 17-2), thus reducing any problems that are associated with stagnant water.

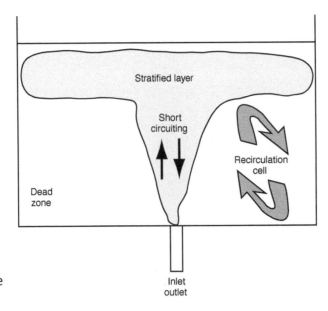

Figure 17-2 Nonideal flow patterns in a storage tank

Study Questions

1. Precipitative softening water treatment plants try to end up with distribution system water that is
   a. slightly scale forming.
   b. moderately scale forming.
   c. neutral.
   d. in equilibrium.

2. There are two major objectives for drinking water distribution system operational policies: (1) maintain water quality from the point of entry into the distribution system to the point of use and (2)
   a. keep operating expenses under budget.
   b. make sure customers do not waste water.
   c. educate customers to about water safety.
   d. maintain adequate pressure and deliver adequate flow.

3. The goal of a well-managed distribution system should be to provide water to customers' taps that has _____ from the point of entry into the system.
   a. not changed
   b. gained valuable nutrients
   c. decreased dramatically in pH
   d. become better tasting

4. What factor generally determines the need for larger mains?
   a. Fire flow requirements
   b. pH concerns
   c. Potability issues
   d. Number of customers served

5. Which of the following is *not* a major factor in storage facility design?
   a. Preventing dead zones
   b. Selecting and maintaining the correct materials of construction
   c. Staying under budget to allow for design modifications
   d. Providing adequate and appropriate cathodic protection to prevent or control corrosion

6. Where do many distribution system water quality problems develop?

7. What is a hydraulic system model?

8. What may occur at pressure zone boundaries?

9. What should be considered when designing storage facilities?

10. In assessing water quality, what rule is of particular interest to distribution system operators?

Chapter 18
Water Quality Testing

Sampling

Importance of Sampling

Sampling is a vital part of monitoring the quality of water in a water treatment process, distribution system, and supply source. However, errors occur easily when recording water quality information. Every precaution must be taken to ensure that the sample collected is as representative as is feasible of the water source or process being examined.

Water treatment decisions based on incorrect data may be made if sampling is not correctly performed. Representative analytical results depend on the water treatment plant operator ensuring the following:

- The sample is representative of the water source under consideration
- The proper sampling techniques are used
- The samples are protected and preserved until they are analyzed
- The proper sample containers are used

Types of Samples

Waterworks operators collect grab samples and composite samples, depending on the requirements of the operation or on regulations.

Grab Samples

A grab sample is a single water sample collected at any time. Grab samples show the water characteristics at the time the sample was taken. A grab sample may be preferred over a composite sample in the following situations:

- The water to be sampled does not flow on a continuous basis.
- The water's characteristics are relatively constant.
- The water is to be analyzed for water quality indicators that may change with time, such as dissolved gases, coliform bacteria, residual chlorine, disinfection by-products, temperature, volatile organics, certain radiological parameters, and pH.

Figures 18-1 and 18-2 illustrate this point. Figure 18-1 shows the changes in surface water dissolved oxygen (DO) over a 24-hour period. A grab sample represents the DO level only at the time the sample was taken. DO can change rapidly—for example, because of the growth of algae or plants in the water (diurnal effect). Online process instruments are good examples of instruments that perform grab sample analyses; they analyze a continuous string of grab samples

> **grab sample**
> A single water sample collected at one time from a single point.
>
> **dissolved oxygen (DO)**
> The oxygen dissolved in water, wastewater, or other liquid, usually expressed in milligrams per liter, parts per million, or percent of saturation.

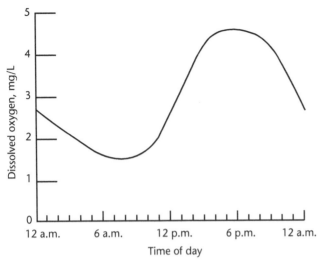

Figure 18-1 Example of hourly changes in dissolved oxygen for a surface water source

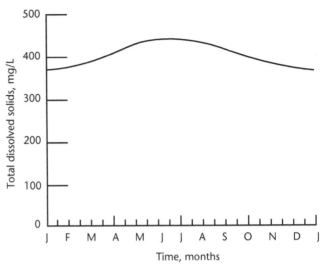

Figure 18-2 Example of monthly changes in total dissolved solids for the surface water source shown in Figure 18-1

and produce a series of individual analyses that, when plotted, illustrate trends such as those in the figures.

Figure 18-2 shows that levels of total dissolved solids (TDS) in the same water change very little. A grab sample can be representative of the water quality in a stable supply such as a deep well for perhaps a month. TDS levels are a function of the minerals dissolved from rocks and soil as the water passes over or through them and may change only in relation to seasonal runoff patterns. TDS levels in groundwater (e.g., wells) may also change if certain water-bearing zones in the well become plugged, changing the dilution or zones from which the water is being drawn.

Composite Samples

In many processes, water quality changes with time. A continuous sampler–analyzer provides the most accurate results in these cases. Often the operator is the sampler–analyzer, and continuous analysis could prove costly. Except for tests that cannot wait because of rapid physical, chemical, or biological changes of the sample (such as tests for DO, pH, and temperature), a fair compromise may be reached by taking samples throughout the day at hourly or 2-hour intervals. Each sample should be refrigerated immediately after collection. At the end of 24 hours, each sample is vigorously mixed and a portion of each sample is then withdrawn and mixed with the other samples. The size of the portion is in direct proportion to the flow when the sample was collected (aliquot) and the total size of sample needed for testing. For example, if hourly samples are collected when the flow is 1.2 mgd, use a 12-mL portion of the sample, and when the flow is 1.5 mgd, use a 15-mL portion of the sample. The resulting mixture of portions of samples is a composite sample. *In no instance should a composite sample be collected for bacteriological examination.*

When the samples are taken, they can either be set aside or combined as they are collected. In both cases, they should be stored at a temperature of less than 40°F (4°C) but above freezing until they are analyzed.

> **composite sample**
> A series of individual or grab samples taken at different times from the same sampling point and mixed together.

Continuous Sample

Continuous sampling is used in online or process control sampling devices/instruments. Some of the new regulations call for this type of sampling for the larger systems for chlorine residual under the new Ground Water Rule (GWR) and for surface water filtration or groundwater under the direct influence of surface water (GWUDI) filtration systems. Continuous sampling is also being used in certain circumstances to monitor distribution systems for chlorine levels and for other parameters associated with security monitoring. It is also used by larger systems on the incoming surface water for turbidity, pH, and streaming current measurements for treatment control. As technology becomes more sophisticated and affordable, this type of monitoring will become more prevalent in the industry. Some systems use this technology to monitor levels of nitrate and other specific ions during the treatment process and fluoride levels of water leaving the treatment facility. Examples of the online instruments are shown in Figures 18-3 and 18-4.

Sampling Point Selection

Careful selection of representative sample points is an important step in developing a sampling procedure that will accurately reflect water quality. The criteria used to select a sample point depend on the type of water sampled and the purpose of the testing. Check with primacy regulations as to *compliance samples* versus *process samples*. Any sample taken from a *compliance sample tap* may have to be reported as a *performance sample* even if it is just being collected for process control. Samples are generally collected from three broad types of areas:

- Raw-water supply
- Treatment plant
- Distribution system

Figure 18-3 Online chlorine residual analyzer
Courtesy of HACH Inc.

Figure 18-4 Online particle counter
Courtesy of HACH Inc.

> **representative sample**
> A collected sample that accurately reflects the composition of water to be tested.

Because this text is focused on distribution systems, only distribution system sampling will be discussed here.

Distribution System Sample Points

Representative sampling in the distribution system is an indication of system water quality. Results of sampling should show if there are quality changes in the entire system or parts of it, and they may point to the source of a problem (such as tastes and/or odors). Sampling points should be selected, in part, to trace the course from finished-water source (at the well or plant) through the transmission mains, and then through the major and minor piping of the system. A sampling point on a major transmission main, or on an active main directly connected to it, would be representative of the plant effluent water quality.

Sample points in the distribution system are used to determine the quality of water delivered to consumers. In some cases, the distribution system samples may be of significantly different quality than samples of finished water at the point of entry to the distribution system. For example, corrosion in distribution system pipelines can cause increases in water color, turbidity, taste and odor, and physical constituents such as lead and copper. Microbiological growth may also be taking place in the water mains, which degrades water quality. In addition, a cross-connection between the distribution system and a source of contamination can result in chemical or microbiological contamination of the water in the system.

Most of the samples collected from the distribution system will be used to test for coliform bacteria and chlorine residual. Others may be used to determine water quality changes. Still others will be used to test for maximum contaminant levels (MCLs) of inorganic and organic contaminants and for compliance with the Lead and Copper Rule, as required by the applicable drinking water standards. Distribution system sampling should always be performed at locations representative of conditions within the system. Radiological samples must be from the source water supply under current regulation.

There are two major considerations in determining the number and location of sampling points, other than those required by regulation:

- They must be representative of each different source of water entering the system.
- They must be representative of conditions within the system, such as dead ends, loops, storage facilities, and pressure zones.

The precise location of sampling points depends on the configuration of the distribution system. The following examples provide some general guidance for sample point selection.

Example 1

Figure 18-5 provides an example of how sample points may be selected for a small surface water distribution system serving a population of 4,000. This is a typical small branch system having one main water line and several branch or dead-end water lines. For this system, a single point, *A*, is sufficient for turbidity monitoring. This point is representative of all treated water entering the distribution system.

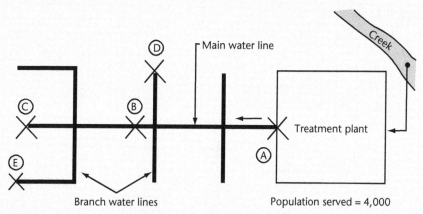

Figure 18-5 Sampling points (indicated by x) in a typical small-branch distribution system

For a community of 4,000, the NPDWRs require a minimum of four bacteriological samples per month to be taken at four different points in the system. Point *B* represents water in the mainline, and point *C* represents water quality in the main-line dead end. Points *D* and *E* were selected to produce samples representative of a branch line and a branch-line dead end, respectively.

Consideration of how often and at what times these points are sampled is also necessary to ensure that the samples accurately represent conditions in the distribution system. Although the minimum requirement of four samples per month could be met by collecting samples from all points on one day, this sampling frequency would not produce samples that represented bacteriological conditions within the system throughout the month. A better program would be to sample points *B* and *E* at the beginning of the month and points *C* and *D* at mid-month. Sampling should be representative both in location and in time.

Although this type of program is adequate to meet the minimum monitoring requirements of the NPDWRs, good operating practices would include periodic sampling at each dead end and several additional sampling points within the distribution system, with samples taken each week. The exact number and location of these operational sampling points depend on the characteristics of the specific system and on state requirements.

Chlorine residual samples should be taken from each sample point when bacteriological samples are collected and should be analyzed within 15 minutes of sampling, preferably at the sample location. Sampling for routine water chemistry, along with the required sampling for inorganic and organic chemicals, also can be conducted at one of the coliform sampling points.

Sampling for a similar system using a groundwater source would be the same, except that turbidity sampling generally is not required and samples for organic chemical analysis must be collected at each well.

Example 2

Figure 18-6 illustrates a typical small-loop distribution system having one main loop and several branch loops, serving a population of 4,000. One turbidity sample point, A, is sufficient because that point is representative of all treated water entering the distribution system.

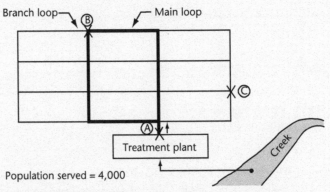

Figure 18-6 Sampling points (indicated by x) in a typical small-loop distribution system

For bacteriological sampling, two sampling points, B and C, are adequate. Point B is representative of water in the main-line loop, and point C is representative of water in one of the branch-line loops. To produce the required minimum of four samples per month, points B and C can be sampled on alternate weeks, or additional similar sampling points can be selected. However, good operating practice would include two to three times this number of samples, depending on the characteristics of the particular system. As with the system in the previous example, chlorine residual samples should be taken whenever bacteriological sampling is performed.

Example 3

Figure 18-7 illustrates a system serving a population of 17,440 that obtains water from both a creek and a well. The distribution system has the features of both the branch and the loop systems shown in Figures 18-5 and 18-6.

To determine sample-point locations, the following four questions should be considered:

1. What tests must be run?
2. From what locations will the samples be collected?
3. How often must the samples be taken?
4. How many sampling points will be needed?

The answers to the first and third questions—what tests must be run and how often—may vary from state to state, and they are also likely to change periodically in response to changes in federal requirements.

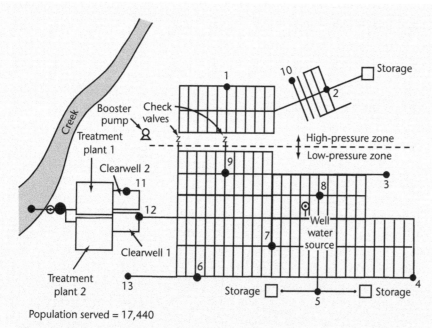

Figure 18-7 Sampling points (indicated by solid dots) in a medium-size system with surface and groundwater sources

Additional samples may also be required for the system's own quality control program. Examples include taste and odor, color, pH, TDS, iron, manganese, and heterotrophic plate count.

Once the tests and test frequencies have been determined, the number and specific locations of sampling points must be selected. The National Primary Drinking Water Regulations (NPDWRs) require a turbidity sample to be taken at each point representative of the filtered surface water that enters the distribution system. Because waters from parallel treatment plants enter two separate clearwells in Figure 18-7, two turbidity sampling points are required (points 11 and 12). The well will not have to be sampled for turbidity, but periodic sampling directly from the well for chemical quality analysis will be required as directed by the state.

In the selection of sample points that will be representative for coliform analysis, a variety of factors must be considered, including the following:

- Uniform distribution of the sample points throughout the system
- Location of sample points in both loops and branches
- Adequate representation of sample points within each pressure zone
- Location of points so that water coming from storage tanks can be sampled
- For systems with more than one water source, location of sample points in relative proportion to the number of people served by each source

On the basis of these fundamental considerations, bacteriological sample points can be selected. A treatment plant serving a community with a population of 17,440 must test 20 coliform bacteria samples per month, according to the NPDWRs. After a careful review of the configuration of the distribution system layout, 10 coliform bacteria sample sites were selected. The reasons for the selection of each point shown in Figure 18-7 are as follows for bacteriological purposes only:

- Point 1 is on the main loop in the high-pressure zone; it should produce representative samples for that part of the system.
- Point 2 is on the branch loop in the high-pressure zone, representative of storage flow to the system.
- Point 3 is on a dead end. Some authorities advise against dead-end sampling points because they do not produce representative samples. However, consumers do take water from branch-line dead ends. In the example, there are seven branch-line dead ends that no doubt serve significant numbers of consumers. It is representative to have one or two sample points on these branch lines at or near the end. If there are indications of chlorine residual decline or bacteriological problems in water sampled at branch-line dead ends, hydrants and blowoff valves should be flushed and branch lines resampled immediately to determine if the problem has been corrected. If the problem persists, additional investigation is needed to locate the condition contributing to the problem.
- Point 4 is located on the main loop of the low-pressure zone and represents water from treatment plant 2, the well water source, the storage tanks, or any combination of these (depending on system demand at sampling time).
- Point 5 allows for sampling of water flowing into the system from storage.
- Points 6 through 9 were selected by uniformly distributing points in the low-pressure zone, the zone that serves the major part of the community.
- Point 10 was selected as representative of a branch-line dead end in the high-pressure zone, just as point 3 was selected in the low-pressure zone.
- Points 11 and 12, as stated previously, are used as turbidity monitoring points.
- Point 13 was added to monitor a dead-end branch that is fairly isolated from other sampling points yet serves a large population.

Sample Faucets

Once representative sample points have been located on the distribution system map, specific sample faucets must be selected. In many cases, suitable faucets can be found inside public buildings such as fire stations or school buildings, inside the homes of water system or municipal employees, or inside the homes of other consumers. The sites selected should have a service line of reasonable size and a good record of water usage. (For example, in some instances in public buildings such as firehouses or schools, a small-diameter service line off a large-diameter fire service line does not provide a representative sample. Because the fire line may never be used, or may be tested only on a scheduled program, water in the line could become stale [i.e., lose its chlorine residual], become oxygen depleted, dissolve some of the main or sediment in the stagnant pipe.)

In smaller water systems, special sample taps are not available. Therefore, customers' faucets must be used to collect samples. Indoor taps are best, if available. Front-yard outside faucets on homes supplied by short service lines (i.e., homes on the same side of the street as the water main) will suffice if there are no other options. Submerging the ends of these faucets in bleach or swabbing the tap with bleach or hydrogen peroxide (again, check with your drinking water primacy agency for acceptable method) first is one way to ensure that the tap will not taint the sample. However, the disinfectant must be flushed from the tap so it does not enter the sample and produce possible false negative results.

Contact the person in the home and obtain permission to collect the sample. If no one is to be home, disconnect the hose from the faucet if one is attached, and do not forget to reconnect the hose after you have collected the sample. Open the faucet to a convenient flow for sampling (usually about half a gallon [2 L] per minute). Allow the water to flow until the water in the service line has been replaced twice. Because 50 ft (15 m) of 0.75-in. (19-mm) pipe contains more than 1 gal (3.8 L), 4 or 5 minutes will be required to replace the water in the line twice. You can check the water temperature and/or chlorine residual to determine if water coming from the tap matches the quality of the water in the area. Collect the sample, being sure the sample container does not touch the faucet.

Do not try to save time by turning the faucet handle to wide open to flush the service line. This high pressure will disturb sediment and incrustations in the line that must be flushed out before the sample can be collected.

For sampling, it is also best to try to find a faucet that does not have an aerator. If a faucet with an aerator must be used, follow the state primacy agency's recommendation on whether the aerator should be removed for sample collection.

Once a representative sample point has been selected, it should be described on the sample record form and placed in the appropriate sample plan such as the one required for the Total Coliform Rule or the Lead and Copper Rule so it can be easily located for future sample collection.

Collection of Samples

The steps described in the following sections are general sample collection procedures that should be followed regardless of the constituent tested. Special collection procedures required for certain tests are described later in this chapter.

Only containers designed for water sampling and provided by the laboratory should be used. Mason jars and other recycled containers cannot be trusted to function properly no matter how well they are cleaned, and they are generally not accepted by a laboratory for water analysis. Some laboratories reuse sample containers by washing them under carefully controlled conditions and sterilizing them prior to reuse. In other cases, it has been found more economical to dispose of used bottles and provide only new ones for collection.

When a container with a screw-on lid is used, the lid should be removed and held threads-down while the sample is collected in the container. The lid can easily be contaminated if the inside is touched or if it is set facedown or placed in a pocket. A contaminated lid can contaminate the sample, which will necessitate resampling, costing a great deal of time and money unnecessarily.

Raw-Water Sample Collection

If no raw-water sample tap is available and the sample must be taken from an open body of water, the following procedures should be used.

On a well supply, if no raw-water sample tap is available, the well should be put to waste with any treatment shut off, the samples collected, the treatment restarted, and the well placed back in service. A clean, wide-mouth sampling bottle should be used for raw-water sampling. The bottle should not be rinsed; this is especially important if the bottle has been pretreated or contains a preservative.

The open bottle should be held near its base and plunged neck downward below the surface of the water. The bottle should then be turned until the neck points slightly upward for sampling, with the mouth directed toward any current present. Care must be taken to avoid floating debris and sediment. In a water body with no current, the bottle can be scooped forward to fill the bottle. Once the bottle has been filled, it is retrieved, capped, and labeled.

If the sampler is wading, the sample bottle should be submerged upstream from that person. If a boat is being used for stream sampling, the sample should be taken on the upstream side.

When samples are being taken from a large boat or a bridge, the sample bottle should be placed in a weighted frame that holds the container securely. The opened bottle and holder are then slowly lowered toward the water with a rope or with the handle that comes with certain devices available through water supply equipment catalogs, DO sampling cans, "swing samplers" on poles, long-handled dippers, or weighted bailers. When the bottle or sample device approaches the surface, the unit is dropped quickly into the water. Slack should not be allowed in the rope because the bottle could hit bottom and break, or it could pick up mud and silt. After the bottle is filled, it is pulled in, capped, and labeled. There are also specialized sampling devices to be used as required for specific samples. For example, for DO, the device with the sample bottles is lowered into the water and then the stopper is remotely removed, the sample container is filled, and the stopper is replaced before the unit is removed from the water. This type of device can also be used when sampling at a certain depth to ensure the water is from the zone desired.

Treatment Plant Sample Collection

The procedure used to collect samples from an open tank or basin or in an open channel of moving water is essentially the same as for raw-water sampling. Treatment plants should be equipped with sample taps. These faucets provide a continuous flow of water from various locations in the treatment plant, including raw-water sources. In some plants, these taps do not run continuously because of operational constraints, so the operator may have to turn the taps on and run the water for a specific amount of time to obtain a representative sample. To collect a sample, the operator or laboratory technician draws the required volume from the sample tap. Figure 18-8 shows a typical bank of sample faucets in a laboratory.

Distribution System Sample Collection

Once the distribution system sample locations have been selected, sample collection consists of a few simple, carefully performed steps. First, the faucet is turned on and set to produce a steady, moderate flow of water (Figure 18-9). If a steady flow cannot be obtained, the tap should not be used. The water is allowed to run long enough to flush any stagnant water from the service line. (The important exception to this procedure is with samples collected for lead and copper analyses; these must be first-draw samples collected immediately after the faucet has been opened.) Depending on the length of the service line, as mentioned before, this process can take from 2 to 5 minutes or longer. The line is usually flushed when the water temperature changes (depending on climate and source, the temperature may increase or decrease) and stabilizes. The sample is then collected without the flow changing. The sample bottle lid should be held threads-down during sample collection and replaced on the bottle immediately. If the lid must be set down during the sampling process, it should be placed threads-up and protected from splatter or falling matter (rain, for example) that could contaminate the sample. The final step after sampling is to label the bottle.

Figure 18-8 Sample faucets in a laboratory

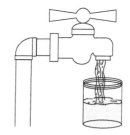

Figure 18-9 Sample faucet should be set to produce a steady, moderate flow

A once common practice was to flame the outside of a faucet. This procedure is no longer recommended. Experience showed that the flame could not be held on the faucet long enough to kill all the bacteria on the outside of the faucet without potential damage to the faucet. Many faucets are now made partly or entirely of plastic, which will quickly melt if high heat is applied. The current common practice is to dip the faucet into a small container of bleach or swab it with bleach or other approved disinfectant.

Samples should not be collected from sill cocks or other faucets with hose threads unless local regulations require it and the threads can be thoroughly cleaned. Because of the way they are constructed, these faucets do not usually throttle to a smooth flow. Also, if any water splashes up onto the threads and then drains into the sample bottle, it will bring with it contaminants from the outside of the faucet.

Special-Purpose Samples

Occasionally a water utility may need to collect samples for special testing purposes. Procedures in such cases depend on the reason for the sampling.

For example, a consumer may have complained about taste, odor, or color in the water. In such a case, samples are collected from the consumer's faucet to determine the source of the problem. The faucet is opened and a sample taken immediately. This sample represents the quality of water standing in the service line. The water is then allowed to run for 2–5 minutes or until the water temperature changes, so that the standing water in the service line is completely flushed out; then a second sample is taken. The second sample is fresh from the distribution system. Comparing test results from the two samples often helps to identify the origin of the problem causing the consumer complaint.

Customer complaints of taste, odor, or color are often caused because the consumer's water heater, water softener, or home water-treatment device is not maintained or operating properly. If the hot-water supply is suspected, the first sample should be collected from the hot-water tap. The tap is turned on and allowed to run until the water is hot before the sample is collected. A second sample representing the water in the service line should be taken from the cold-water tap as previously described. Comparing the test results from the two samples will help identify the origin of the problem unless a whole-house filter or treatment device is in use. In that case, it may be necessary to collect a sample from an outside untreated faucet, the meter connection, or a neighbor's faucet for comparison with an untreated sample.

There are many other reasons for taking special-purpose samples. The previous example emphasizes the importance of knowing what the sample test results will be used for so that the sample collected will be representative of the conditions tested.

Monitoring for Chemical Contaminants

Drinking water may contain contaminants considered a threat to the public. The contaminants of concern may occur naturally in the water, be human-made, or be formed during the water treatment process. The chemicals are broken into four general classes for regulation:

1. Inorganic chemicals (IOCs)
2. Synthetic organic chemicals (SOCs)
3. Volatile organic chemicals (VOCs)
4. Radionuclides

The need to establish regulations for new chemical contaminants has presented the US Environmental Protection Agency (USEPA) with the problem of creating, adapting, proving and promulgating *analytical techniques*. The method must be for a specific chemical and not for a "contaminant" defined by a physical description such as a boiling fraction. An example of this distinction can be found in kerosene, which is a member of the petroleum distillates; these chemicals boil between 150°C and 275°C (300–530°F), which results in a mixture of organic chains of 6 to 16 carbon atoms in each compound. Before a requirement to monitor for a contaminant can be imposed, the testing methods must be developed to ensure that an adequate number of laboratories will be available to perform the tests and that they will get consistent, reliable results. Many of the chemicals now being added to the list of regulated contaminants must be analyzed at the parts-per-billion level or in even smaller concentrations.

Faucets selected should be on the lines connected directly to the main. Only cold-water faucets should be used for sample collection. A sampling faucet must not be located too close to a sink bottom. Contaminated water or soil may be present on the exteriors of such faucets, and it is difficult to place a collection bottle beneath them without touching the neck's interior against the faucet's outside surface. In most instances, samples should not be taken from the following types of faucets (Figure 18-10):

- Leaking faucets, which allow water to flow out around the stem of the valve and down the outside of the faucet
- Faucets with threads
- Faucets connected to home water-treatment units, including water softeners and hot-water tanks
- Faucets that swivel, since the swivel joint may act as a siphon and introduce contamination
- Faucets with single-lever handles that do not guarantee only the cold-water sample is being selected

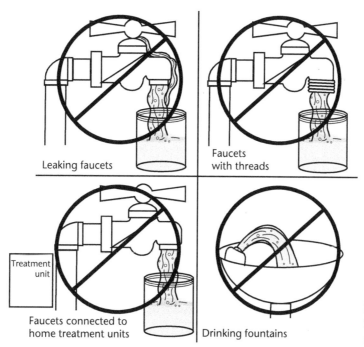

Figure 18-10 Types of faucets that should not be used for sampling

Laboratory Certification

Each of the approximately 155,000 public water systems affected by the Safe Drinking Water Act (SDWA) must routinely monitor water quality to determine if the water is adequately protected from regulated microbiological, chemical, and radiological contaminants. It is imperative that the analyses for all of this monitoring be performed by standard methods approved for compliance testing so that the results are comparable for all systems. Consequently, states are required by federal regulations to consider analytical results from water systems only if samples have been analyzed by a certified laboratory. Some exceptions are measurements for turbidity, chlorine residual, temperature, and pH, which may be performed by a person acceptable to the state using approved equipment and methods.

Federal regulations require each state with primary enforcement responsibility to have available laboratory facilities that have been certified by USEPA, with capacities sufficient to process samples for water systems throughout the state. Certified laboratories fall into the following general classes:

- State-operated laboratories
- Water-system laboratories
- Commercial laboratories

In most states, the necessary capacity is provided by a combination of all three types of laboratories. Some laboratories may be certified to perform only one type of analysis; for instance, some laboratories are set up to handle only microbiological analyses. Analyses requiring expensive equipment and highly trained technicians, such as for organic chemical and radiological monitoring, are also generally handled by specialized laboratories.

Consistency among laboratories in analytical results is overseen by USEPA and by state primacy programs for each of the types of analyses for which the laboratories are certified. Periodically, an independent vendor contracted by USEPA provides to each laboratory carefully prepared proficiency testing samples containing a known concentration of a contaminant. The values of the samples are unknown to the laboratory, and its staff must be able to determine the contaminant concentration within an appropriate tolerance to maintain the laboratory's certification. The results determined by the laboratory are submitted to the agency having primacy for laboratory certification for the particular state.

Historically, most states have operated their own laboratories to process water system samples. But the number of samples has increased severalfold in recent years, so it is difficult for the states to continue providing laboratory service with state funding only. Some states have instituted charges to water systems to help fund the laboratory services. Other states process only a certain number of samples from any one water system, and if more are required, commercial laboratories must be used.

Record Keeping and Sample Labeling

Records should be kept for every sample that is collected. A sample identification label or tag should be filled out at the time of collection. Each label or tag should include at least the following information:

- Water utility name
- Water system's public water system identification number
- Date on which sample was collected
- Time when sample was collected
- Location where sample was collected
- Type of sample—grab or composite
- Tests to be run
- Name of person sampling
- Preservatives used
- Bottle number

The samples provided to laboratories should always be clearly labeled. The information on the label should also be entered on a record-keeping form that is maintained as a permanent part of the water system's records and placed on the chain-of-custody forms submitted to the laboratory. Each laboratory may have its own forms that request the required information for compliance with regulations.

Sample Preservation, Storage, and Transportation

Samples cannot always be tested immediately after they are taken. Ensuring that the level of the constituent remains unchanged until testing is performed requires careful attention to techniques of sample preservation, storage, and transportation. It is also extremely important that records be kept of the chain of custody of samples collected for SDWA compliance. Refer to *Standard Methods for the Examination of Water and Wastewater* for a comprehensive list of sampling, storage, and preservation techniques for common water quality parameters.

Preservation and Storage

After a sample has been collected, its quality may change because of chemical and/or biological activity in the water. Some characteristics (alkalinity, pH, dissolved gases, disinfectant residuals, temperature, and odor) can change quickly and quite significantly, and so samples to be analyzed for these parameters should not be stored under any conditions. The tests for disinfectant residuals, pH, and temperature must be completed in the field at the time of collection. Other parameters, such as pesticides and radium, change more slowly and much less noticeably, and these samples can usually be stored for considerable lengths of time if necessary.

Sample-Preservation Techniques

To extend the storage time of samples requiring chemical analysis, sample-preservation techniques have been developed that slow the chemical or biological activity in the sample. This allows it to be transported to the laboratory and tested before significant changes occur.

Sample preservation usually involves one or both of the following steps:

- Refrigeration
- Chemical preservatives

For some samples, storage time can be prolonged by keeping samples refrigerated until the analysis is performed. In some cases, it is recommended that samples be transported or shipped to the laboratory in a portable cooler containing an ice pack.

Often the laboratory provides bottles for specific analyses with the preservative already added. It is particularly important not to allow these containers to overflow as they are filled; if overflow occurs, some of the preservative will be lost. These containers must also be kept out of the reach of children, because the preservative material could be harmful to a child who opens a container. If preservatives are to be added by the sampler, specific instructions on the procedures should be obtained from the laboratory that will perform the analyses.

Time of Sampling

Most laboratories do not maintain a full staff on weekends, so they generally request that samples with short holding times, such as bacteriological samples, be collected and shipped early in the week. If a sample arrives on a weekend and cannot be processed, the delay will probably exceed the required holding time and the sample will be rejected. However, most laboratories accept emergency samples on weekends.

Samples that must be submitted within a specified compliance period should generally be collected and sent to the laboratory early in the compliance period. Some of the problems that can require resampling are described in the following list:

- The sample is frozen or broken during shipment.
- The sample is lost or delayed in shipment and arrives at the laboratory after the specified holding time has elapsed.
- The laboratory makes an error in processing the sample.
- The laboratory analysis is inconsistent or shows a result above the MCL, and another sample to confirm the results is required.
- The sample was not properly preserved or was too warm, or no preservative was present.
- The sample container used was not the proper container for the required sample, size, volume, or material.

Sampling early in the compliance period ensures that time is available for one or more resamplings, if necessary, before the end of the period. If resampling has not been completed before the end of a compliance period, a water system is usually deemed out of compliance and will be instructed by the state to provide public notification.

Transportation

If samples arrive at a laboratory past the specified holding time following collection, the laboratory must reject the samples. New sample bottles must then be shipped to the water system, and another set of samples will have to be collected and shipped back.

The mail is usually the best and easiest method of shipment, except for microbiological or certain radiological samples that require delivery within about a 24-hour period. If regular mail service fails to deliver samples reliably within the required time period, overnight shipping services or package delivery services may be tried. In some cases, changing to a laboratory at a different location may improve delivery time. Some water system operators who are located near a laboratory have found it best just to drive the samples directly to the laboratory or arrange with the laboratory for pickup of the samples as part of the analysis price. Depending on distance and availability of personnel, a bonded courier service may be used.

If samples are shipped, it is important to make sure the bottle caps are tight to prevent leakage. Systems that have had bottle caps loosen during shipment have found that wrapping the lids with electrical or packing tape is an easy method of further securing them. Samples must be packed in a sturdy container with enough cushioning material to prevent breakage. The box should be marked to indicate which end is up, that the contents are fragile, that they must not be allowed to freeze, and that priority should be given to the shipment.

Chain of Custody

As more and more parameters are added to the list of regulated and unregulated contaminants, and with the MCLs and MCLGs in the micrograms per liter range, the practice of good quality assurance and quality control (QA/QC) procedures becomes very important. One essential part of QA/QC is maintaining a written record of the sample's history from the time of collection to the time of analysis and subsequent disposal. This record, called the chain of custody, is important if the analyses are ever challenged and need to be defended. Chain-of-custody requirements vary by state, so water system operators should be sure that the requirements for their state are being met.

Field Log Sheet

One method of establishing the chain-of-custody record is to use a daily field log sheet, which should contain the following information:

- Date the samples were collected
- Name of the sampler
- List of all the samples collected by the sampler on this date
- List of all the sample locations for this date
- Time of day each sample was collected
- Comments concerning any unusual situations
- Signature of the individual receiving the samples from the sampler
- Date and time the samples were received by the laboratory
- Location or identification of the laboratory

This log sheet states that the samples were in the custody of the sampler until they were turned over to the shipper. The laboratory record then follows the history of the sample to disposal.

> **chain of custody**
> A written record of the sample's history from the time of collection to the time of analysis and subsequent disposal.

Sampler's Liability

If the results of an analysis of a specific sample are ever questioned, the sampler will be asked to verify that the sample was in his or her custody until it was turned over or sent to the laboratory. The sampler will be asked to verify that the sample was collected, stored, and transported using proper procedures and that no other person could have in any way altered the concentrations of any contaminant(s) present.

Sampler's Responsibility

The sampler has the basic responsibility to ensure that the sample is properly collected, labeled, stored, and transported to the laboratory. The sample collector must be able to testify that the sample was under his or her custody at all times. The sample collector is also responsible for knowing and performing the proper sampling routine for each type of analysis required, including preservation.

Common Water Quality Tests

Chlorine (Free or Total)

For water plants that use chlorine for disinfection or oxidation, this is one of the most important tests performed by operators. After any addition of chlorine, a measurement should be taken routinely. This test will verify the correct dosage and reveal changes that may affect plant performance and the safety of the water supply. Many plants use both free and total (or combined) forms of chlorine in their processes.

The test most often used is the N,N diethyl-1,4 p-phenylenediamine sulfate (DPD) color test. The DPD (either for free or total chlorine testing) is added to a water sample, and the intensity of the color indicates the amount of chlorine present in the sample. Most plants use a digital read-out colorimeter to give an accurate result. Some color comparison portable test devices are used as well; however, they are not always accurate. Another test method is amperometric titration. This test is usually performed in the laboratory. A special meter is used to determine the end points when titrating a measured sample with a standard phenylarsine oxide solution. This method is very accurate and is capable of determining many chlorine species that may be present. There are several other chlorine residual test methods that can be considered.

Both the DPD color test and amperometric titration are used in online instruments for continuous chlorine monitoring. As with any instrument, these devices must be calibrated and checked frequently to ensure accuracy.

Chlorine demand can be determined using the residual test method. This measurement can be used to predict residual chlorine over a specified time. Jar testing apparatus may be used for this test. A sample is taken and the chlorine residual is measured immediately. After a specified time, the residual is measured again and the difference is the demand. Care is needed to duplicate the conditions (e.g., light, temperature, holding time) that are of interest.

Coliform

Coliforms are a group of bacteria that produces gas bubbles in lactose or lauryl tryptose broth at 35.5°C (95.9°F) within 24 to 48 hours. They are considered indicator organisms, as their presence may indicate the presence of other more harmful bacteria and organisms. Because total coliforms are easier to analyze in the average water utility laboratory than the actual diseasecausing micro-organisms, total coliform testing is used in place of more tedious, more expensive testing for these other organisms.

Most laboratories use one of two methods to test for coliforms. One method is the membrane filter technique, in which water samples are passed through a 0.45-μm filter. The filter paper is fine enough to trap bacterial particles as the water passes through. The filter paper is then placed onto a growth medium (such as M-endo) and incubated. Any bacterial colonies that are present will grow in size and will be visible to the naked eye; they can be counted after a period of time (usually 24 hours).

Another method is the MMO-MUG (minimal medium) technique. This method allows for inoculation of water sample bottles with powder. The substance in the bottle will feed any total coliform that may be present and produce a color change during incubation.

Samples for coliform testing are always collected in sterile bottles and in quantities sufficient for testing (Figure 18-11). Most bacteriological samples require a minimum of 100 mL (approximately 4 oz) for analysis. Samples should be analyzed the same day as they are collected but can be refrigerated for 8 hours prior to analysis. Coliform samples are taken in the plant at various stages to test for process efficiency; they are also taken in the distribution system for regulatory compliance. The number of samples that must be taken in the distribution system is a function of the population served. Many water treatment plants also take coliform samples of the raw or source water.

For more information about coliform testing, see Chapter 1.

 WATCH THE VIDEO
Water Sampling—Coliform (www.awwa.org/wsovideoclips)

Nitrate, Nitrite, and Ammonia

These three inorganic nitrogen compounds are often encountered in water supplies. Contamination from agricultural activities is often the source in surface water, but another significant source is wastewater discharges. Nitrate and nitrite

Figure 18-11 Autoclave used for sterilization. Bacteriological equipment must be sterilized in steam under pressure.
Source: Conneaut, Ohio, Water Department.

are regulated contaminants with enforceable MCLs. Ammonia is not regulated in drinking water but is a compound that may encourage the growth of nitrifying bacteria. Ammonia is often associated with the chloramination process. Free ammonia testing may be used to help control the chloramination process, while the presence of nitrite in the distribution system can be an indicator of nitrification. Ammonia and nitrate monitoring in the distribution system are especially important for chloraminated systems that may be at risk for nitrification.

Several test methods are available for these substances. Operational control testing is usually conducted using colorimetric methods. Special precautions are needed when testing for free ammonia in the presence of chloramines in order to ensure accurate results.

pH

This is one of the most common tests performed in water treatment and drinking water monitoring. The pH value is an indicator of the acidity or basicity of the water. There is not an MCL for pH in drinking water. However, there is a secondary standard range of 6.5–8.5.

pH testing uses a scale from 0 to 14, with the midpoint of 7 being neutral (i.e., the acidity and basicity are balanced). Below 7, the acidity of the water predominates; above 7, the basicity of the water predominates. With each unit increase or decrease, the concentration or intensity changes tenfold.

For example, for the pH to change from 5 to 6, the acidity must decrease by a factor of 10. The pH of water is significant because it affects the efficiency of chlorination, coagulation, softening, and corrosion control. Also, pH testing can provide early warning of unit process failure. For example, the addition of alum to the rapid-mix stage should produce a predictable drop in pH. If it does not, a malfunction of coagulant feed could be indicated.

Samples for pH should be collected in glass or plastic containers and analyzed as quickly as possible. Samples should not be agitated because dissolved carbon dioxide could be liberated, which will change the pH.

A pH meter is used for the test in combination with a suitable probe. The meter must be periodically calibrated using known standards called buffers. The pH value for a water sample may change while standing due to a change in temperature or exposure to air. Therefore, measurements are usually taken immediately upon sampling.

Study Questions

1. Samples to be tested for coliforms are collected in plastic bottles that must contain
   a. sodium thiocarbonate.
   b. sodium thiooxalate.
   c. sodium thiosulfate.
   d. sodium thiocyanate.

2. The volume of a sample for coliform compliance is
   a. 100 mL.
   b. 200 mL.
   c. 300 mL.
   d. 0; there is no volume compliance for coliforms.

3. If a water sample is not analyzed immediately for chlorine residual, it is acceptable if it is analyzed within
   a. 10 minutes.
   b. 15 minutes.
   c. 20 minutes.
   d. 30 minutes.

4. The best choice to collect a water sample from a customer's faucet when responding to a complaint would be a
   a. faucet without threads.
   b. faucet that can swivel.
   c. single-lever handle faucet.
   d. faucet with an aerator.

5. When measuring for free chlorine residual, which method is the quickest and simplest?
   a. DPD colorimetric method
   b. Orthotolidine method
   c. Amperometric titration
   d. 1,2 nitrotoluene di-amine method

6. A _____ is a single-volume sample collected at one time from one place.
   a. DPD colorimetric method
   b. grab sample
   c. random sample
   d. continuous sample

7. Under the Ground Water Rule (GWR), what type of sampling is required for larger systems for chlorine residual?

8. What are three broadly classified areas from which samples are generally collected?

9. Consistency among laboratories in analytical results is overseen by state primacy programs and by what organization?

10. If samples arrive at a laboratory past the specified holding time following collection, the laboratory must do what?

Chapter 19
Backflow Prevention and Cross-Connection Control

As potable water is transported from the treatment facility to the user, opportunities exist for unwanted substances to cause contamination. Water in the distribution system can become contaminated by backflow of nonpotable substances through cross-connections. The utility operator is responsible for protecting the public water system from hazards that originate on the customers' premises and from temporary connections that may impair or alter the water in the public water system.

All utility operators need to be concerned with this problem, because cross-connections and backflow can and do occur in systems of all sizes. The consequences can be serious.

Terminology

The following definitions are important because the conditions they describe aid in understanding and eliminating the potential hazards posed by cross-connections. State or local statutes may define these conditions more specifically.

Backflow

Backflow is the flow of any water, foreign liquids, gases, or other substances back into a potable water system. Two conditions that can cause backflow are backpressure and backsiphonage.

Backpressure is a condition in which a substance is forced into a water system because that substance is under a higher pressure than system pressure.

Backsiphonage is a condition in which the pressure in the distribution system is less than atmospheric pressure. In other words, something is "sucked" into the system because the main is under a vacuum.

Cross-Connections

A cross-connection is any connection or structural arrangement between a potable water system and any other water source or system through which backflow can occur. The existence of a cross-connection does not always result in backflow; however, where a cross-connection exists, the potential for backflow is always present if either backpressure or backsiphonage should occur.

---

backflow
A hydraulic condition, caused by a difference in pressures, in which nonpotable water or other fluids flow into a potable water system.

backpressure
A condition in which a pump, boiler, or other equipment produces a pressure greater than the water supply pressure.

backsiphonage
A condition in which the pressure in the distribution system is less than atmospheric pressure, which allows contamination to enter a water system through a cross-connection.

cross-connection
Any arrangement of pipes, fittings, fixtures, or devices that connects a nonpotable system to a potable water system.

Table 19-1 Some cross-connections and potential hazards

Connected System	Hazard Level	Connected System	Hazard Level
Access hole flush	High	Car wash	Moderate to high
Agricultural pesticide mixing tanks	High	Photographic developers	Moderate to high
Aspirators	High	Pump primers	Moderate to high
Boilers	High	Baptismal fonts	Moderate
Chlorinators	High	Dishwashers	Moderate
Cooling towers	High	Swimming pools	Moderate
Flush valve toilets	High	Watering troughs	Moderate
Laboratory glassware or washing equipment	High	Auxiliary water supply	Low to high
Plating vats	High	Garden hose (sill cocks)	Low to high
Sewage pumps	High	Irrigation systems	Low to high
Sinks	High	Solar energy systems	Low to high
Sprinkler systems	High	Water systems	Low to high
Sterilizers	High	Commercial food processors	Low to moderate

Cross-Connections and Locations

Cross-connections can be found in almost any type of facility where water is used, including houses, factories, restaurants, hospitals, laboratories, and water and wastewater treatment plants. Many pieces of equipment within these facilities draw water from the potable water distribution system for use in cooling, lubricating, or washing, or as an additive to a process. Table 19-1 provides a partial list of the types of fixtures that are particularly likely to have cross-connections that could contaminate a potable water supply.

Cross-connections are frequently created by individuals who are not familiar with the hazards, even though they may be otherwise well trained and experienced in plumbing, steam fitting, pipe fitting, or water distribution work. Many such connections are made simply as a matter of convenience, with no regard given for the problems that may result.

Types of Cross-Connections

A cross-connection will appear to be simply a pipe, hose connection, or any water outlet with no specific or outstanding features. The real distinction is where the connection leads. A cross-connection exists if a connection leads from a potable line to anything other than potable service. Cross-connections can be found in many different places, such as homes, farms, laboratories, and factories. Each connection leads to a vessel containing materials that should not be allowed to enter the potable water system. Cross-connections can be categorized as actual or potential.

Actual Cross-Connections

An actual cross-connection is one for which the connection exists at all times. Examples would be solid piping to an auxiliary supply (Figure 19-1) or into a boiler. An auxiliary supply could be a nonpotable source used for emergencies or for reducing the cost of cooling or washing. The situation could also apply to an older home that has retained its private well (which could be contaminated) but has also been connected to a municipal system, with only a valve separating the two water sources.

Potential Cross-Connections

A potential cross-connection is one for which something must be done to complete the connection. In Figures 19-2 and 19-3, the slop sink and water tank are examples of potential cross-connections. In both cases, water has to be added to the vessel before the connection is completed. Although the cross-connection shown in Figure 19-3 appears to pose a very low hazard, the tank could be used for mixing a pesticide, weed killer, or fertilizer, which would greatly increase the danger.

Cross-Connection Examples

Figure 19-4 depicts a very common cross-connection, formed when a chemical dispenser is connected to a garden hose and the hose is attached to a sill cock. If a vacuum should occur in the water system while the dispenser is in use, the chemical solution could be sucked back into the house plumbing.

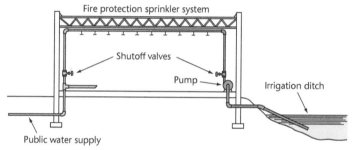

Figure 19-1 Cross-connection between a potable water supply and an auxiliary fire sprinkler system

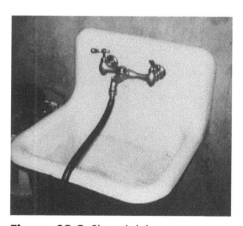

Figure 19-2 Slop-sink hose cross-connection

Figure 19-3 Water tank cross-connection
Courtesy of USEPA, Region VII, Water Supply Division.

actual cross-connection
Any arrangement of pipes, fittings, or devices that connects a potable water supply directly to a nonpotable source at all times.

potential cross-connection
Any arrangement of pipes, fittings, or devices that indirectly connects a potable water supply to a nonpotable source. This connection may not be present at all times but it is always there potentially.

Figure 19-4 Garden-hose cross-connection

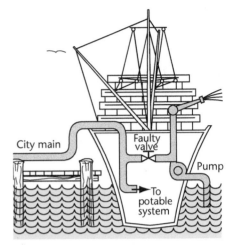

Figure 19-5 Unprotected potable water supply from dock to ship in port

Figure 19-5 shows an unprotected hose from a dock used to fill the water tanks on a ship. Ships in port are possible hazards to the potable water supply because their onboard high-pressure fire systems use seawater.

Cooking vessels in hospitals, restaurants, and canneries use potable water (Figure 19-6). So do chemical reaction tanks. If the water inlet to one of these vessels is below the overflow rim, it can become a cross-connection.

In hard-water areas, many buildings, particularly homes, have individual water softeners. The softeners form a cross-connection that is not normally hazardous. However, when the drain is connected directly to the sanitary sewer, as shown in Figure 19-7, it becomes potentially very hazardous to both the residents and the community water supply. At least 36 persons were struck by a hepatitis epidemic in California as a result of an arrangement similar to that shown in Figure 19-7.

Figure 19-8 illustrates a cross-connection where the water supply has been protected. An atmospheric vacuum breaker is attached between the soap dispenser and the water faucet.

Figure 19-6 Cooking vessel with water inlet beneath overflow rim

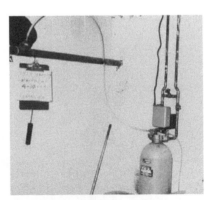

Figure 19-7 Water softener cross-connection

Figure 19-8 Atmospheric vacuum breaker installed between soap dispenser and water faucet

Backflow due to Backpressure

A typical example of backflow due to backpressure is illustrated in Figure 19-9. When pressure inside the boiler exceeds that of the water supply, backflow can result. A similar situation occurs in Figure 19-10, where a chemical storage tank pressurized by an air compressor is connected to a potable water supply line. A third common type of cross-connection involves the recirculation of potable or nonpotable water on a premises for the purpose of meeting fire demand or processing requirements (Figure 19-11). When the auxiliary system pressure exceeds the pressure in the potable supply, any feeder connections become potential sources of backflow.

Backflow due to Backsiphonage

Because backsiphonage is actually caused by atmospheric pressure, the height to which siphoned water can be lifted is limited to 33.9 ft (10.3 m) at sea level. This is the point at which the downward pressure caused by the weight of a column of water equals the pressure of the atmosphere forcing it upward (Figure 19-12). A typical example of backflow due to backsiphonage is shown in Figure 19-13.

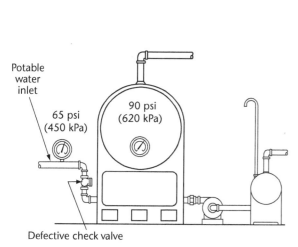

Figure 19-9 Backflow from high-pressure boiler due to backpressure

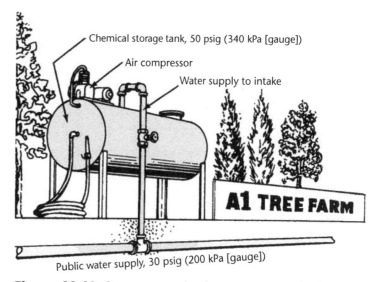

Figure 19-10 Cross-connection between pressurized chemical storage tank and lower-pressure potable system

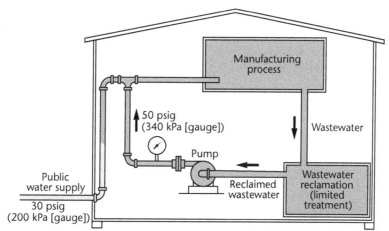

Figure 19-11 Backflow from recirculated system

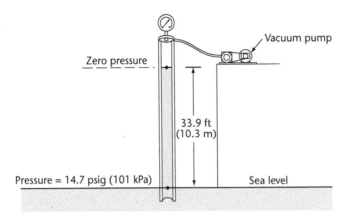

Figure 19-12 Effect of evacuating air from a column

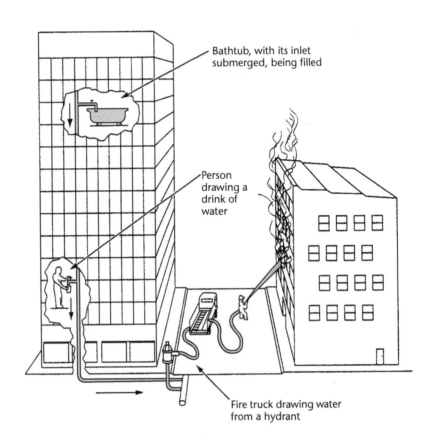

Figure 19-13 Backsiphonage due to pressure loss

In this situation, the partial vacuum in the potable system (created by high water flow out of the hydrant) sucks nonpotable materials into the system.

Another example, illustrated in Figure 19-14, can occur if a hose is being used to fill a sink. If someone opens a large valve on a lower floor, the water from the sink can be backsiphoned into the building piping.

A common cause of the less-than-atmospheric pressure (negative gauge pressure) needed to create a backsiphonage condition is overpumping by a fire or booster pump (Figure 19-15). Undersized distribution piping can also create negative pressures. Undersized piping creates a high velocity of water, which in turn causes severe pressure drops. A water service in an undersized area may have negative gauge pressure when water is flowing through the main supplying the service.

A broken main or fire hydrant, particularly at low elevations, can cause backsiphonage, as illustrated in Figure 19-16. When a break occurs in a main, the entire distribution system could become contaminated between the break and the

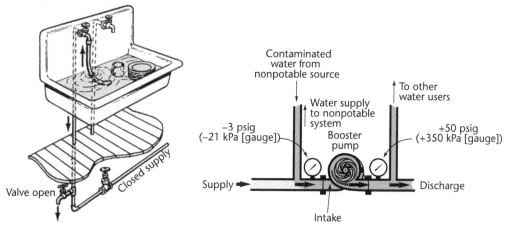

Figure 19-14 Backsiphonage (hose forms cross-connection)

Figure 19-15 Backsiphonage from a booster pump

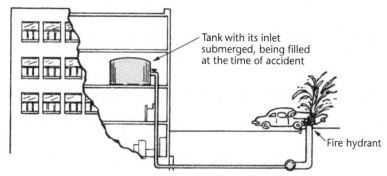

Figure 19-16 Backsiphonage due to a broken fire hydrant

cross-connections. Subsequently reestablishing the pressure could, in turn, contaminate all of the system downstream of the break.

 **WATCH THE VIDEO**
Backflow Prevention (www.awwa.org/wsovideoclips)

Backflow Control Methods and Devices

When a cross-connection is found, one of two actions must be taken. Either the cross-connection must be removed or some method to protect the potable water supply from possible contamination must be devised. Where removal is impractical, a protective device should be installed.

The preventive measure chosen depends on the degree of hazard involved, how accessible the premises are, and the type of water distribution system within the cross-connection location. Some protective methods and backflow-prevention devices, in order of decreasing effectiveness, are as follows:

- Air gaps
- Reduced pressure zone backflow preventers
- Double check valve assemblies
- Vacuum breakers (atmospheric and pressure)
- Barometric loops

Air Gaps

When correctly installed and maintained, an air gap (Figure 19-17) is the best method available for protecting against backflow. It is acceptable in all cross-connection situations and for all degrees of risk. Another advantage is that there are no moving parts to break or wear out. Only surveillance is needed to ensure that no bypasses are added.

The only requirement for installation is that the gap between the supply outlet and the overflow level of the downstream receptacle measure at least two times the inside diameter (ID) of the outlet's tip but no less than 1 in. (25 mm). In situations where an isolated water system is needed to supply nonpotable uses in a factory, a surge tank and booster pump may be installed, as illustrated in Figure 19-18. In this case, the air gap provides positive isolation from the potable water system.

Reduced Pressure Zone Backflow Preventers

The second type of device that can be used in every cross-connection situation and with every degree of risk is the reduced pressure zone backflow preventer (RPZ), also referred to as reduced-pressure backflow preventer (RPBP). This device consists of two spring-loaded check valves with a pressure-regulated relief valve located between them, as shown in Figures 19-19 and 19-20. Flow from the supply at left enters the central chamber against the pressure exerted by check valve 1 (Figure 19-21A). Water loses pressure passing through this valve, so the central chamber is known as the reduced pressure zone. The amount of pressure loss through the check varies with the valve size, flow rate, and valve

air gap
In plumbing, the unobstructed vertical distance through the free atmosphere between (1) the lowest opening from any pipe or outlet supplying water to a tank, plumbing fixture, or other container and (2) the overflow rim of that container.

reduced pressure zone backflow preventer (RPZ)
A mechanical device consisting of two independently operating, spring-loaded check valves with a reduced-pressure zone between the check valves. Designed to protect against both backpressure and backsiphonage.

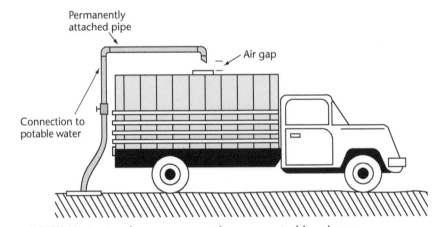

Figure 19-17 Water truck cross-connection prevented by air gap

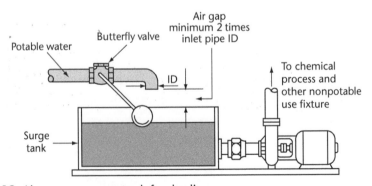

Figure 19-18 Air gap on surge-tank feeder line

Figure 19-19 Reduced pressure zone backflow preventer
Courtesy of Watts Regulator Co.

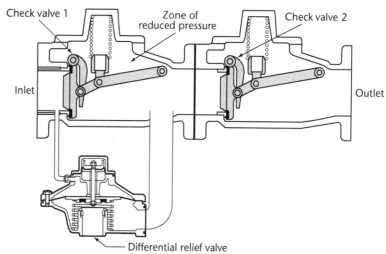

Figure 19-20 Cutaway view of a reduced pressure zone backflow preventer
Courtesy of Cla-Val Co., Backflow Preventer Division.

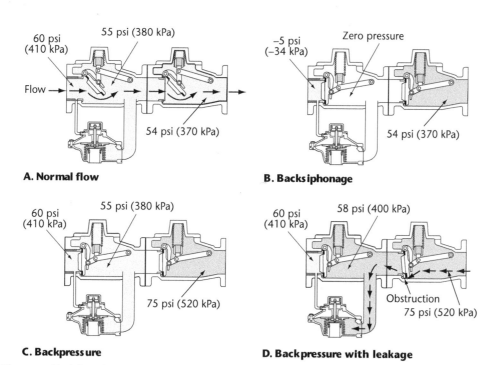

Figure 19-21 Valve position and flow direction in a reduced pressure zone
Courtesy of Cla-Val Co., Backflow Preventer Division.

manufacturer. The second check valve is loaded considerably less (causing about 1 psi [7 kPa] further pressure drop) to keep the total pressure loss within reason.

Connecting two standard check valves together is not considered sufficient protection for most hazardous locations, because all valves can leak from wear or obstruction. For this reason, a relief valve is positioned between the two checks. When the unit is operating correctly and the supply pressure exceeds the downstream pressure, the supply pressure opposes the relief valve's spring tension and keeps the valve closed. If a vacuum should occur in the supply line, as shown in Figure 19-21B, both check valves will close and the relief valve will open to drain water from the reduced pressure zone. Backsiphonage is therefore positively prohibited. If the second check valve should leak at this time, the leakage will be harmlessly discharged through the relief valve.

If backpressure on the building side of an RPZ should exceed the supply pressure, as illustrated in Figure 19-21C, both check valves will close. If the second check valve should fail to seal completely under this condition, the leakage will be discharged through the relief valve, as shown in Figure 19-21D.

A definite sign of an RPZ malfunction is continuous drainage from the relief port. Because of the multiple protective systems, the probability of backflow occurring across an RPZ backflow preventer is very small. A typical installation of an RPZ device is illustrated in Figure 19-22.

Even though the RPZ backflow preventer is highly dependable, it is a mechanical device that requires maintenance. Valve faces, springs, and diaphragms deteriorate with age. Also, solids can lodge in or damage the check valves, causing leakage. For these reasons, each installed unit must be periodically inspected, tested, and maintained. RPZ devices should be tested in accordance with the manufacturer's specifications and regulatory requirements.

The backflow preventer's performance can be affected by how and where it is installed. To achieve continued satisfactory performance, the following installation procedures should be followed:

- Install the unit where it will be accessible. Accessibility will provide for ease of testing and maintenance. Installation at least 12 in. (300 mm) above the floor is one criterion for accessibility. If the device can be easily inspected, it will increase the probability that any malfunction (leakage) will be noticed and promptly repaired.
- Install the unit so that the relief-valve port cannot be submerged. Submergence of the port creates another cross-connection and may prevent the unit from operating properly.
- Protect the device from freezing, which will damage the unit.
- Protect the device from vandals. Accessible units—particularly the associated gate valves—are a temptation to vandals.
- If a drain is needed, an air gap must be provided below the relief-valve port. Most manufacturers have designed attachments to fit the unit.
- Install a screen upstream of the backflow preventer to eliminate the possibility of debris becoming lodged in the unit, which could render it inoperative. Screens are available commercially.
- Most models must be installed horizontally. Check with the manufacturer to determine the allowable positions for each model.

Figure 19-22 Typical installation of reduced pressure zone backflow preventer

Double Check Valve Assemblies

The double check valve backflow preventer is designed basically the same way as the RPZ but without the relief valve (Figure 19-23). The absence of the relief valve significantly reduces the level of protection provided. The unit will not give any indication that it is malfunctioning.

A double check valve assembly is not recommended as protection in situations where a health hazard may result from failure. Before installing such a unit in a potable water line, contact the agency having statutory jurisdiction to determine whether its use is permitted.

The installation and maintenance of a double check valve assembly should follow these procedures:

- The unit must be an approved model. Two single check valves in series will not be satisfactory.
- The installation must be protected from freezing and vandals.
- The unit must be periodically tested and maintained. Frequency of testing and maintenance is governed by state and local codes.
- The unit should be installed in a position that will allow testing and maintenance to be performed.
- The unit should be installed in an area that does not flood. The test cocks can form a cross-connection when submerged.

In recent years, many water systems have become interested in installing check valves on customer water services. This is particularly the case where many homes still have old private wells in use. There is always a possibility that a connection within the home plumbing will allow the well water to flow out into the public water system. The principal problem has been finding a location to install one or preferably two check valves for this purpose.

Several manufacturers have recently developed double check valve assemblies for this use. Various configurations are available, each placing the assembly near the meter, as illustrated in Figure 19-24. Locating the check valves on the discharge side of a meter has the advantage of preventing water from draining out of the residence plumbing while a meter is being replaced.

Vacuum Breakers

Backsiphonage occurs when a partial vacuum pulls nonpotable liquids back into the supply lines. If air enters the line between a cross-connection and the source of the vacuum, the vacuum will be broken and backsiphonage will be prevented. This is the principle behind a vacuum breaker.

double check valve backflow preventer
A mechanical designed basically the same way as the reduced pressure zone backflow preventer but without the relief valve.

vacuum breaker
A mechanical device that allows air into the piping system, thereby preventing backflow that could otherwise be caused by the siphoning action created by a partial vacuum.

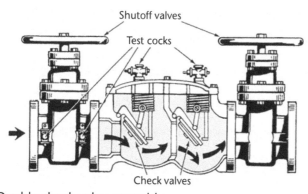

Figure 19-23 Double check valve assembly

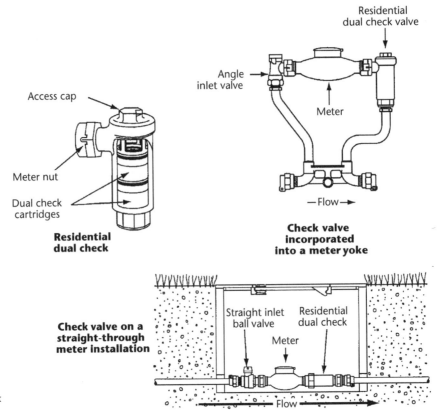

Figure 19-24 Examples of residential dual check valves

Courtesy of the Ford Meter Box Company, Inc.

Figure 19-25 shows an atmospheric vacuum breaker, consisting of a check valve operated by water flow and a vent to the atmosphere. When flow is forward, the valve lifts and shuts off the air vent. When flow stops or reverses, the valve drops to close the water supply entry and open an air vent. A version of the unit designed for a hose bibb is shown in Figure 19-26.

Atmospheric vacuum breakers are not designed to protect against backpressure, nor are they reliable under continuous use or pressure because the gravity-operated valve may stick. Atmospheric vacuum breakers may be used only where there is no possibility of backpressure being applied. These breakers must be installed at least 6 in. (150 mm) above the highest point of the downstream outlet.

Figure 19-27 shows a pressure vacuum breaker. The unit illustrated has a spring-loaded check valve that opens during forward flow and is closed by the spring when flow stops. When pressure in the line drops to a low value, a second valve opens and allows air to enter this breaker. With this arrangement, the breaker can remain under supply pressure for long periods without sticking and can be installed upstream of the last shutoff valve. The placement and use of the pressure vacuum breaker is restricted to situations where no backpressure will occur and where it can be installed 12 in. (300 mm) above the highest point of the downstream outlet.

atmospheric vacuum breaker
A mechanical device consisting of a float check valve and an air-inlet port designed to prevent backsiphonage.

pressure vacuum breaker
A device designed to prevent backsiphonage, consisting of one or two independently operating, spring-loaded check valves and an independently operating, spring-loaded air-inlet valve.

▶ **WATCH THE VIDEO**
Backflow Prevention Devices (www.awwa.org/wsovideoclips)

Barometric Loops

The barometric loop is a simple installation that prevents only backsiphonage. To create a barometric loop, an inverted U is inserted into the supply pipe upstream of the cross-connection. Figure 19-12 illustrates how the height of a column of

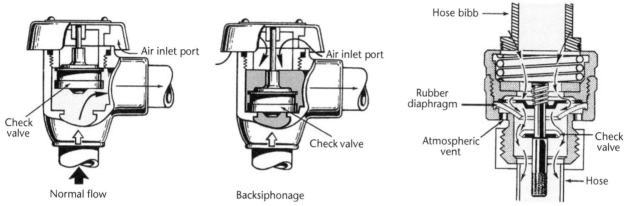

Figure 19-25 Atmospheric vacuum breaker

Figure 19-26 Hose bibb type of atmospheric vacuum breaker

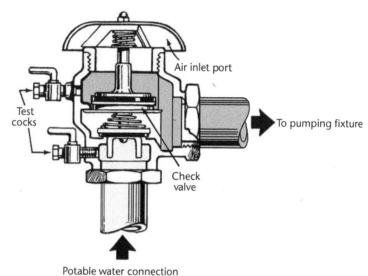

Figure 19-27 Pressure vacuum breaker

water, open to the atmosphere at the bottom, is limited to about 33.9 ft (10.3 m). If the barometric loop is taller than the 33.9-ft (10.3-m) limitation, siphonage or backsiphonage will not occur.

Although the barometric loop is effective against backsiphonage and requires no maintenance or surveillance (other than to ensure that it remains leak-free), the space requirement is a serious disadvantage. The loop must extend about 35 ft (11 m) above the highest liquid level. In addition, the barometric loop is completely ineffective against backflow due to backpressure. For these reasons, barometric loops are no longer installed.

Other Methods and Devices

One of the first methods for cross-connection control was the complete separation of potable and nonpotable piping systems. When water piping systems from two different sources are located in the same building, they are usually identified by color coding. This is an effective approach when adequate surveillance is provided to prevent connections between systems. In some instances, however, separate systems have been interconnected by such devices as spool pieces, flexible temporary connections, and swing connections. None of these interconnections is recommended for use regardless of the degree of risk involved. Each type should be removed from any system. However, these types of connections (as opposed to cross-connections) may be useful when two *potable* systems need to be connected for emergency or other approved service.

Level of Protection

The type of backflow-prevention device that is to be used for each installation depends on the degree of hazard present and whether backflow could result from backpressure or backsiphonage. Table 19-2 provides a listing of the minimum protection suggested for various applications. Local codes or regulations must be followed before a method is selected for protecting an installation. Table 19-3 lists the types of cross-connections discussed in this chapter and suggests the level of protection appropriate for each.

Table 19-2 Recommended minimum protection requirements

Type of Hazard on Premises	Minimum Protection at Meter					Minimum Options to Isolate Area of Plant Affected*					Comments
	AG	RPZ	DCVA	AVB or PVB	None	AG	RPZ	DCVA	AVB or PVB	None	
1. Sewage treatment plant	x	x				See comments					RPZ at meter with air gap in plant also
2. Sewage lift pumps	x	x							x		
3. Domestic water booster pumps			x					x			
4. Equipment or containers manufactured for industrial use without proper backflow protection											
A. Dishwashing								x			Normally machine has built-in AG
B. Clothes washing			x					x			If no health hazard exists, DCVA is acceptable
C. Food processing		†	†			†	†				
D. Pressure vessels			x					x			A vacuum breaker or DCVA may be used if no health hazard exists
E. Tank or vat containing a nonpotable or objectionable solution		x	x			x	x				
F. Sinks with hose threads on inlet									x		

Table 19-2 Recommended minimum protection requirements (continued)

Type of Hazard on Premises	Minimum Protection at Meter					Minimum Options to Isolate Area of Plant Affected*					Comments
	AG	RPZ	DCVA	AVB or PVB	None	AG	RPZ	DCVA	AVB or PVB	None	
G. Any dispenser connected to a potable water supply									x		
H. Aspirator equipment									x		
I. Portable spray equipment								x			
5. Reservoirs, cooling towers, circulating systems		x				x	x				System where no health hazard or potential for a health hazard exists, a DCVA may be used
6. Commercial laundry	x	x							x		
7. Steam-generating facilities and lines			x					x			Must be hot water related
8. Equipment under hydraulic test or hydraulically operated equipment	x	x				x	x				
9. Laboratory equipment											
A. Health hazard		x	x			x					
B. Not health hazard				x					x		
10. Plating facilities	x	x				x					
11. Irrigation systems		x	x	x		x	x	x			
12. Firefighting systems	x	x	x								Chemicals are often used in such systems
13. Dockside facilities	x	x				x	x				DCVA may be used on dockside if outlet is protected
14. Tall buildings		x	x								Two devices should be installed in parallel; if a health hazard exists within the building, two RPZs in parallel may be required

(continued)

Table 19-2 Recommended minimum protection requirements (continued)

Type of Hazard on Premises	Minimum Protection at Meter					Minimum Options to Isolate Area of Plant Affected*					Comments
	AG	RPZ	DCVA	AVB or PVB	None	AG	RPZ	DCVA	AVB or PVB	None	
15. Unapproved auxiliary supply	x	x				x	x				
16. Premises where inspection is restricted	x	x									
17. Hospitals, mortuaries, clinics	x	x				x	x				
18. Laboratories	x	x				x	x				DCVA, if no health hazard exists
19. Chemical plants using a water process	x	x				x	x				
20. Petroleum processing or storage plants	x	x				x	x				
21. Radioactive material processing plants or nuclear reactors	x	x				x	x				
22. Swimming pools	x	x				x	x				

Note: The list is not all-inclusive and does not necessarily conform to state and local codes. These codes must be checked before installation of any protective device.

AG = air gap; RPZ = reduced pressure zone backflow preventer; DCVA = double check valve assembly; AVB = atmospheric vacuum breaker; PVB = pressure vacuum breaker.

*In areas where no health hazard or potential for a health hazard exists, a vacuum breaker properly installed in the line to the problem area may be adequate protection (only if there is no backpressure present.)

Table 19-3 Suggested backflow protection for situations discussed in this chapter

Application	Reference Figure	Minimum Protection		Notes
		At Meter	At Cross-Connection	
Auxiliary water system	19-1	RPZ or air gap	RPZ	
Sink	19-2	RPZ or air gap*	Atmospheric or pressure vacuum breaker	
Water truck	19-3	RPZ or air gap*	Air gap, RPZ	See Figure 19-17
Garden hose	19-4	RPZ or air gap*	Atmospheric or pressure vacuum breaker	A double check-valve assembly may be acceptable if no health hazard exists
Ship	19-5	RPZ or air gap	RPZ	The hose bibb type of breaker is used
Cooking vessel	19-6	RPZ or air gap*	RPZ	If no health hazards exist, double-check valve assembly may be used
Water softener	19-7	RPZ or air gap*	RPZ	

*Protection at the meter may not be necessary if protection at the cross-connection is ensured.

RPZ = reduced pressure zone backflow preventer.

Cross-Connection Control Programs

Passage of the Safe Drinking Water Act in 1974 made each water utility responsible for the quality of water at the consumer's tap. Legal proceedings have also established that the utility is responsible for cross-connection control in some jurisdictions. In addition, many states assign the cross-connection control responsibility directly to the water supplier. The size of the utility has no bearing on the degree of risk posed by cross-connections. Therefore, all water systems, large and small, public or private, should maintain an active cross-connection control program.

Developing a Cross-Connection Control Program

A water utility can develop and operate a cross-connection control program or jointly participate in such a program with other municipal agencies. The specific nature of any program depends on state and local laws and regulations, municipal agencies, and the size of the community.

An effective program deals with the two major sources of cross-connection problems: (1) plumbing within the customer's premises and (2) auxiliary water sources. Plumbing within the customer's premises usually falls under the supervision of state and local health departments or local building–engineering departments. Auxiliary water sources can fall under the supervision of a health department or environmental protection agency or its equivalent. When different agencies have an interest, the program must be administered as a cooperative effort.

A utility is more likely to serve as the single program manager in a smaller community. Many small communities have a public water distribution system but still depend on a higher authority, such as a county or district, for health and sanitary support.

A control program will give the water utility (or responsible agency) the legal authority to do the following:

- Take actions to protect the water supply
- Provide a systematic procedure for locating, removing, or protecting all cross-connections in the distribution system
- Establish the procedures for obtaining the cooperation of customers and the public

Anyone attempting to establish such a program should consult the state water program agency or other authorities for legal direction.

The operator's involvement in a cross-connection control program will vary from almost no involvement, to being an inspector, to providing major assistance in establishing and operating the program. Knowledge of program content and procedures is valuable at every level of responsibility.

Program Content

An effective cross-connection control program has the following elements:

- An adequate plumbing and cross-connection control ordinance
- An organization or agency with overall responsibility and authority for administering the program, with adequate staff
- Systematic inspection of new and existing installations with formal record keeping
- Follow-up procedures to ensure compliance

- Backflow-prevention device standards, as well as standards for inspection and maintenance
- Cross-connection control training
- A public awareness and information program

Procedures

The procedure for initiating a program begins with planning, which is normally a function of management. Assistance from the health, building, or plumbing inspection departments should be obtained at the beginning. Planning should include identifying a tentative organization, the appropriate authority, and internal procedures for executing the program.

The next step is to inform the municipal government (or other applicable government agency) and the public of the nature of cross-connections and backflow, as well as the steps needed to protect the public. This can be done through newspaper announcements, interviews, and public presentations. The government agency uses this background information to understand and enact the authorizing control ordinance, which is the legal basis of a local program. The public should continue to be informed even after the program has been implemented.

The authorizing control ordinance provides authority for establishing and operating the program. The amount of detail included depends on state laws and codes. Where the state has an extensive and detailed code, the ordinance can be very simple. In other locations, it may be necessary to provide details to describe the program completely. These details should include at least the following:

- Authority for establishment
- Responsibilities and organization for operating the program
- Authority for inspections and surveys
- Prohibitions and protective requirements (if the state code is not specific, this description can become very detailed)
- Penalty provisions for violations

After establishing the program, the water utility or other agency designated by the ordinance begins inspecting customers' premises. Those premises that, by nature of their activity, present the greatest risk to public health should be inspected first. The inspector, who must be trained to identify cross-connections and the actions needed to protect the potable water supply, should be accompanied by the owner or the owner's representative so that there is an understanding of the procedures and results.

The complete plumbing system should be inspected. Particular attention should be given to pipelines leading to process areas, laboratories, liquid storage, and similar facilities. When cross-connections are found during the inspection, they should be pointed out to the owner and their significance discussed. The following questions should be answered:

- What is the material that could backflow?
- How hazardous is the cross-connection?
- Can the cross-connection be eliminated or must it be protected?
- How can it be protected?

Generally, the water utility is responsible for the water distribution supply to the owner's premises. Therefore, for the purpose of a cross-connection program, the inspector does not need to require that each cross-connection be protected. Instead, protection needs to be provided only at the point of entry to the premises (this tactic is termed *containment*). However, the inspector should recommend that some action be taken on each connection, both for the owner's protection and to fulfill the requirements of the Occupational Safety and Health Administration and local plumbing or building codes.

In addition, local plumbing or health codes may require that each connection be protected. Only when the codes are specific should the inspector specify the item needed for a certain application. Otherwise, a minimum level of protection should be recommended in accordance with the code and Table 19-2. Upon completing the survey, the inspector should prepare a report and notify the owner of the corrective actions needed, the approvals required, and the time limits.

For new installations, jurisdictions normally require building and plumbing plans to be submitted and approved. The plan review stage presents an opportunity to identify potential problem areas. The review should identify cross-connections, with a view toward their elimination or protection. Only after the plans are changed to ensure adequate protection should they be approved. Most jurisdictions inspect new buildings during construction and after completion. A survey for cross-connections should be included in these inspections.

Repeat inspections are needed both to ensure that all corrective actions required from previous inspections have been completed and to look for new cross-connections. These inspections are conducted in the same manner as the initial inspection. Timing of repeat inspections will vary according to their nature and whether compliance checks are routine or special.

Routine inspection intervals will vary according to the degree of hazard involved and human resources available. However, inspections should be conducted as often as feasible. The regulatory code may recommend appropriate inspection intervals, which could be every 3 to 6 months for high-risk installations and annually for others. Special compliance checks should be made immediately after the utility is notified of a plumbing change or a change in activity at an installation (the protection may need to be upgraded if the risk has increased). These checks should also be made when the utility is notified of a violation that could endanger public health.

When a customer is not convinced of the seriousness of the potential hazard or refuses to cooperate, the water purveyor must have the legal authority to shut off the customer's water. Before the water is shut off, the purveyor needs to make a final effort to gain the customer's cooperation and warn of the consequences of noncompliance. Discontinuing water service should be a last-resort measure, and it is recommended that legal counsel be consulted first. All warnings and notifications must be made exactly as specified by ordinance.

One possible corrective action resulting from inspections is the actual elimination of cross-connections. These actions will require only follow-up inspections to ensure that the old connections are not reconnected. A second corrective action is the installation of backflow-prevention devices. If records of testing and maintenance are not maintained by the agency performing the field surveys, repeated inspections should include a check of the owner's records to ensure that the backflow equipment is being tested and maintained.

Backflow-Prevention Devices

For backflow-prevention devices to operate properly for long periods of time, two primary conditions must be met.

1. Good quality backflow-prevention units must be installed.
2. Inspection and maintenance must be performed periodically.

The first condition is met by setting minimum standards (e.g., ANSI/AWWA C510, C511; ANSI/NSF Standard 61) for design, construction, and performance, followed by tests to evaluate conformance. Testing and approval are done by special laboratories and in some cases by the state or utility. A customer should consult a list of state-approved devices before installing any backflow preventer.

The second condition—periodic inspection and maintenance—is met by systematic inspection and testing of all units that have been installed. Inspection consists of visually checking a unit weekly or biweekly for leaks and external damage. The owner should be required to make the inspections and keep a record of the findings. If an inspection shows defects, qualified repair personnel should be called to perform further examination and repair. Each unit should be tested for correct operation annually by an individual specifically qualified for such testing and repair. Any devices that fail must be repaired immediately, and a report of the test and repair should be made when the job is completed.

Education and training for cross-connection control should be performed at several levels. The first level is a continuous program of public education. This program should describe applicable codes and cross-connections and the hazards they pose. The goal is to obtain the consumer's cooperation and reduce the overall risk. At the next level, water utility personnel, plumbing inspectors, and health personnel need special training to develop and operate the control program. Finally, specialized training in maintenance and testing (certification of testers may be required by regulation) is needed for those individuals assigned to maintain backflow preventers. The names of schools with programs for training in backflow-preventer repair can be obtained from the manufacturers of the devices.

Records and Reports

Records and reports play an essential role in administering a cross-connection control program. Complete records are a utility's first line of defense against potential legal liability in case of public health problems resulting from a cross-connection.

Records should be kept on installing, inspecting, testing, and repairing backflow-prevention devices as a shared responsibility of the water customer, the water utility or agency operating the control program, and personnel performing tests and repairs.

Water Customer Reports

To fulfill their responsibility for maintaining safety within the premises, water customers need to be informed about their building plumbing systems, any hazardous conditions that are creating potential cross-connections, and the condition of any backflow-prevention devices that have been installed. Information about the plumbing system should be available from building plans. Information about cross-connections and their protection comes from field inspection surveys, periodic visual observations, and test reports.

Utility or Agency Operating Records

Effective control dictates a time-based formal record system—that is, records that are organized so that each inspection, survey, test, and corrective action is performed as scheduled. Records should consist of inspection reports, test and repair reports, reports of corrective actions, authorized testing and repair personnel lists, approved backflow-preventer lists, and backflow-preventer installation locations.

Testing and Repair Personnel Reports

To ensure continued satisfactory operation of each backflow preventer, only qualified, authorized (certified testers may be required in many locations) individuals should perform the periodic test or needed maintenance. A report of work performed should be recorded. The distribution of the report by the control agency is specified by local regulations and practices.

Another way to ensure that each backflow preventer is adequately maintained is to assign each unit to a single, qualified individual or firm for testing and repair. The assignment can take the form of a contract between the owner and an independent repair service or it may be a letter of instruction to a qualified individual who is an employee of the owner. A record of the assignment should be kept by the owner and made available to the field inspector.

The agency operating the cross-connection control program should be given the name of the individuals to whom each backflow preventer is assigned. To ensure continued satisfactory performance, repair personnel should have servicing instructions available for all backflow preventers they must maintain. Records of previous servicing and performance tests are also important and may be required by law.

Study Questions

1. The correct protective methods for backflow-prevention devices, in order of decreasing effectiveness, are
   a. air gap, vacuum breaker (VB), reduced pressure zone backflow preventer (RPZ), and double check valve assembly (DCVA).
   b. air gap, VB, DCVA, and RPZ.
   c. air gap, RPZ, VB, and DCVA.
   d. air gap, RPZ, DCVA, and VB.

2. What is the likelihood of a swimming pool creating a cross-connection that could contaminate a potable water supply?
   a. Impossible
   b. Not very likely
   c. Moderately likely
   d. Highly likely

3. What is the likelihood of a sewage pump creating a cross-connection that could contaminate a potable water supply?
   a. Impossible
   b. Not very likely
   c. Moderately likely
   d. Highly likely

4. What type of backflow preventer consists of two spring-loaded check valves with a pressure-regulated relief valve located between them?
   a. Vacuum breaker
   b. Atmospheric vacuum breaker
   c. Double check valve backflow preventer
   d. Reduced pressure zone backflow preventer

5. What is a utility's first line of defense against potential legal liability in case of public health problems resulting from a cross-connection?
   a. Notifying the public
   b. Maintaining complete records
   c. Testing the problem
   d. Correcting the issue

6. What is the best method to prevent backflow?

7. How frequently do most regulations require testing of backflow-prevention devices?

8. What is a cross-connection?

9. Give two examples of cross-connections.

10. Give some examples of causes of backflow.

11. What is the term for a cross-connection for which something must be done to complete the connection?

Chapter 20
Information Management and System Mapping

The different types of maps and records described in this chapter have been developed over the years and are the standards for well-run water systems. Most water systems use computerized records, but some systems still use hardcopy records. The same information is required, but it is just stored and available in a different format.

Distribution System Maps

Various types of maps are necessary to provide information on mains, hydrants, and services. Because most of the distribution system equipment is underground, taking time to keep careful records during construction and repair saves a lot of time in the long run by knowing where everything is and being able to quickly find it. Accurate distribution system records are also necessary to determine a valuation of the distribution system, and these maps are essential for engineers to design system improvements.

Comprehensive Maps

A map that provides an overall picture of the entire system is usually called a *comprehensive map*, or *wall map*, because a large copy of it is usually hung on a blank wall in the distribution center office for ready reference by everyone. Computerized maps often contain layers of detail that can be accessed by zooming in or out of the image. The following information should appear on the map:

- Street names
- Distribution water mains with the sizes noted
- Transmission mains shown in a code different than distribution mains
- Fire hydrants and valves with their designated numbers
- Reservoirs, tanks, and booster stations
- Water source connections and interconnection with other water systems
- Pressure zone limits
- Notation of the street-numbering grid

Part of a typical comprehensive map is shown in Figure 20-1. The map should be as large as possible and should not be cluttered with unnecessary information that would be distracting. A good map scale for a small water system is 500 ft to 1 in. (6 m to 1 mm). Larger systems often use 1,000 ft to 1 in. (12 m to 1 mm).

> **comprehensive map**
> A map that provides a clear picture of the entire distribution system. It usually indicates the locations of water mains, fire hydrants, valves, reservoirs and tanks, pump stations, pressure zone limits, and closed valves at pressure zone limits.

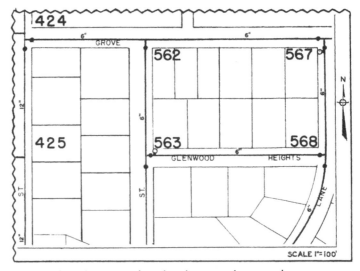

Figure 20-1 Comprehensive map showing intersection numbers
Courtesy of Water Works and Lighting Commission, Wisconsin Rapids, Wisconsin.

Ordinary commercial maps are sometimes available to use as basic layouts, but they should be carefully checked for accuracy before they are used. After any corrections are made, they can be enlarged by a professional reproduction firm, but care must be taken that the scale is not seriously distorted in the process. Copies of the comprehensive map are often provided to the municipal engineering department and the fire department.

Sectional Maps

A sectional map (also commonly called a plat) is a series of maps covering sections of the water system, usually stored in a plat book. Plat books are either computer files or hardcopy collections of system maps. The maps are usually at a scale of either 50 or 100 ft to 1 in. (0.6 or 1.2 m to 1 mm) for small and medium-sized communities, and 200 ft to 1 in. (2.4 m to 1 mm) for larger systems. In addition to the information on the comprehensive map, the following details are usually included:

- Type of main material (i.e., cast iron [CI], ductile iron [DI], polyvinyl chloride [PVC], or asbestos–cement [A–C])
- Main installation date
- Main distance from property line if other than standard
- Block numbers

If the scale of the map allows, details such as house numbers and curb boxes may be shown. However, if all this information would cause undue clutter on the map, it is best provided on other maps or records.

Sectional maps should not overlap each other. Each should have a definite cut-off line on each side so that one map abuts the next one (Figure 20-2). The maps should be indexed, and, for quick reference, the number of the adjoining map in each direction should be indicated. A common method of numbering maps is shown in Figure 20-3. Water systems that may experience growth in any direction sometimes use a variation of this numbering, starting at a central location with four quadrants so that the map number would also have a designation such as SE or NW on it.

---

sectional map
A map that provides a detailed picture of a portion (section) of the distribution system. Reveals the locations and valving of existing mains, locations of fire hydrants, and locations of active service lines.

plat
A map showing street names, mains, main sizes, numbered valves, and numbered hydrants for the plat-and-list method of setting up valve and hydrant maps.

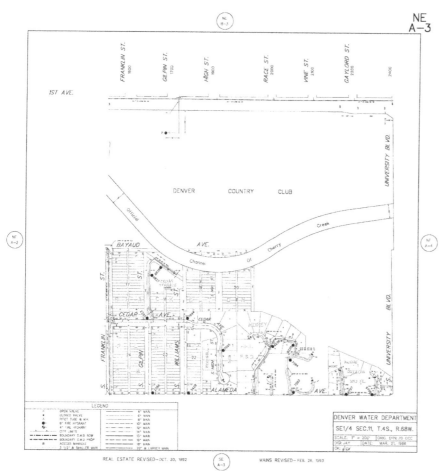

Figure 20-2 Sectional map
Courtesy of Denver Water.

1E	1D	1C	1B	1A
2E	2D	2C	2B	2A
3E	3D	3C	3B	3A

Figure 20-3 Small, comprehensive map divided into sections

Sectional maps can often be developed from tax assessment maps, insurance maps, subdivision maps, or city engineering maps, but they should be carefully checked for accuracy before being used. The historical method of preparing sectional maps was to have the original copies drawn in ink on tracing cloth, and many older systems are still using this method. The reasoning behind this method was that copies could be made for everyday use to reduce wear on the originals.

Computer file maps are preferred since they are not subject to wear. Backups of computer file maps, however, need to be kept in a safe and secure location, and these files need to contain the latest updates.

The originals should be kept in a safe place, such as a fireproof vault. Changes to the originals should preferably be made as soon as possible. New prints should be made and distributed whenever a significant number of changes have been made. One method of ensuring that the hardcopy information on sectional maps is never lost is to have microfilm copies made and stored at a separate, secure location.

Valve and Hydrant Maps

Many water systems also prepare valve and hydrant maps, either combined or as two separate maps. These maps are of sufficient detail to locate every valve and hydrant, and are primarily for use by field crews during maintenance or emergencies. Notes should also be included for special information, such as valves and hydrants that operate in reverse from most in the system.

Many systems use individual plat sheets of convenient size. Another alternative is the plat-and-list method, in which a relatively small map identifies the valves and hydrants by number. This is used in conjunction with a list that provides basic information, such as the type, make, size, street location, and location measurements. Both of these map systems can be accommodated by computerized mapping programs.

Another variation is to use the basic location maps in conjunction with computer maps that are detailed to the level of showing street intersections (Figure 20-4). (Where hardcopy systems are still used, cards, which are usually kept in a loose-leaf book, could be used in place of the computer maps). One advantage of this system is that the location map rarely needs to be corrected because it shows only an overview of part of the system. The intersection plats have plenty of room for all required field information and are usually updated immediately in the field as changes are made. With this system, each intersection must be given an identifying number that is shown on the basic location map.

valve-and-hydrant map
A mapped record that pinpoints the location of valves throughout the distribution system. Generally of plat-and-list or intersection type.

plat-and-list method
A method of preparing valve and hydrant maps. Plat is the map position, showing mains, valves, and hydrants. List is the text portion, which provides appropriate information for items on the plat.

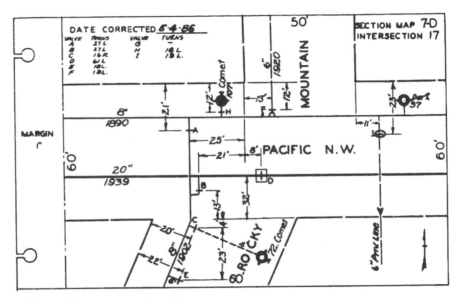

Figure 20-4 Valve intersection plat

Plan and Profile Drawings

Plan and profile drawings are usually provided by an engineer for all new construction. They show the *proposed* location of the new main and appurtenances, and indicate how they connect with the existing system, and how it is anticipated they will avoid all obstructions, such as sewers, gas mains, cables, trees, and telephone poles. The engineers often use abbreviations on these drawings that are not explained, including the following:

- POT—point on tangent
- POC—point on curve
- BC or PC—beginning of a curve
- EC or PT—end of a curve
- PI—point of intersection
- EL—elevation

Distances are usually provided in terms of stationing, starting with 0+00 at the beginning connection. For example, 2 + 42 on the map will indicate 242 ft from the starting point.

Installations are rarely constructed exactly as anticipated on the drawings. Some unanticipated obstructions are usually encountered that require deviation from the plan, and these changes must be quickly recorded on a copy of the plan so they are not forgotten. Good practice is then to prepare a set of "as-built" drawings at the completion of the job to keep as a permanent record.

Information from the construction drawings should be transferred to the other water system drawings, and the plans should be identified so they may be easily found if needed in the future. The plans should be filed in a safe, dry, clean location.

Map Symbols

Every water utility should adopt a set of map symbols that are used on all maps and records. They must be simple, clear, and preferably the same as symbols used by other water systems. Commonly used symbols are shown in Figure 20-5.

Equipment Records

Computerized records have replaced card files and are often kept for details that cannot be included on maps for each valve, hydrant, water service, and water meter.

Valve Records

A valve record usually has information on the make, size, type, and location of the valve on one side, and maintenance information on the other (Figure 20-6). Valves may be assigned numbers, which are referenced on the distribution maps, and are then filed numerically. It is wise to keep as much location information as possible so that the valve can be located quickly in an emergency. For example, although distances to trees are handy reference points in winter when there is snow on the ground, there should also be other measurements to more permanent markers, such as lot stakes, because trees are sometimes cut down.

---

plan and profile drawings
Engineering drawings showing depth of pipe, pipe location (both horizontal and vertical displacements), and the distance from a reference point.

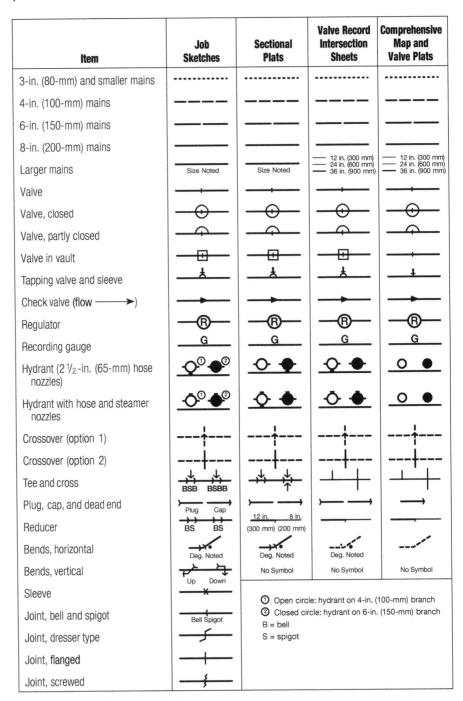

Figure 20-5 Typical map symbols

Hydrant Records

Like valve records, hydrant records should include information on both the hydrant and auxiliary valve on one side, and maintenance and repair information on the other (Figure 20-7). Although it is a job that almost everyone would like to put off, carefully kept hydrant maintenance records are particularly important. If a hydrant should ever fail to operate properly when there is a fire, and there is serious property loss from the fire, there is a good chance that the property owner will sue the water utility for damages. The best defense the

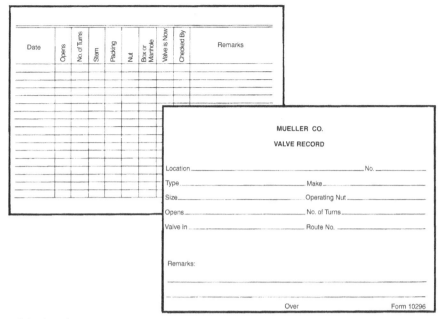

Figure 20-6 Valve record information input document
Courtesy of Mueller Company, Decatur, Illinois.

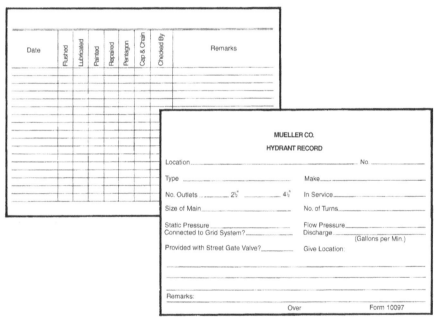

Figure 20-7 Hydrant record information input document
Courtesy of Mueller Company, Decatur, Illinois.

utility can have is records showing that there is a continuous, planned hydrant maintenance program.

Service Records

Many utilities also maintain records with details of each water service. Information includes installation date, name of the installing plumber or contractor, and details of all materials. The information often includes a sketch of the installation with dimensions, burial depth, point of entering the building, meter location, and anything else that may be of interest in future years. Details of the installation of plastic service lines should be particularly thorough because they are difficult to locate once they are covered. The best time to prepare the service information is during the plumbing inspection, before the contractor is allowed to backfill the trench.

Meter Records

Meter records are also often maintained by specialized system management programs. Information included identifies the meter make, size, model, and purchase date. Test data and installation location are also included.

Technical Information

Every water distribution system should develop a workable file system for technical information. An example would be to establish a file section or electronic information storage for fire hydrants. Within that section there should be file folders with literature on equipment from each manufacturer, and additional files for purchases, warranties, correspondence, and factory and local representative phone numbers.

New catalogs should be dated when they are received, and as they are filed, old ones should be discarded unless they have some special historical value. Most files gradually grow to unreasonable size because nobody takes the time to delete old material.

Geographic Information Management Systems

The array of data needed to efficiently operate and manage a water distribution system is immense. Almost all data used for this purpose are related to a geographic location. Computerized programs are used to assemble these data and organize them for practical use. These systems are known as automated mapping/facility management/geographic information systems (AM/FM/GIS).

An AM/FM/GIS shows locations and areas on a computerized map. This map has many layers with each containing related data such as surface elevations and features, streets, water distribution components, and even electric power facilities. There can be hundreds of factors associated with each location.

The objective of the AM/FM/GIS is to provide a fully integrated database to support operation, management, and maintenance functions. Some of the important applications are water resource management, such as water quality information from various entry points into the distribution system; customer services, like the location and frequency of complaints; operations to assist with directing water flow for firefighting; water demand forecasting by including weather information; and water system modeling where new information can be used to refine the hydraulic model.

Maintenance management information can be included in the AM/FM/GIS, or these records can be kept separately, either manually or in a maintenance management system. The maintenance management process is greatly aided by an automated system that is integrated with other information. This system provides for better planning and scheduling of work and allows quick access to historical records.

---

automated mapping/ facility management/ geographic information system (AM/FM/GIS)

A computerized system for collecting, storing, and analyzing water system components for which geographic location is an important characteristic.

Supervisory control and data acquisition (SCADA) system data can be very useful for the AM/FM/GIS. Operational data from the distribution system can be integrated for use in many ways. As indicated earlier, these data can help refine the system hydraulic model. They can also provide valuable information to analyze system performance and target quality improvement investigations.

Several other information systems are important to a distribution system operation. These include source of supply systems, treatment plant process control systems, laboratory information management systems, leakage control data, emergency response data, and customer information. Ideally, all of these systems are integrated so that the most important information is shared and available for analysis and system operation. Many utilities have chosen to integrate only selected data from various systems to make the integration easier and still get the benefit of the most important information.

WATCH THE VIDEO
GIS for Water Systems (www.awwa.org/wsovideoclips)

Study Questions

1. Which type of maps should not overlap each other?
   a. Index maps
   b. Comprehensive maps
   c. Construction maps
   d. Sectional maps

2. Comprehensive maps of medium to large systems generally have scales ranging from
   a. 250–500 feet to 1 inch.
   b. 500–1,000 feet to 1 inch.
   c. 1,000–1,500 feet to 1 inch.
   d. 1,500–2,000 feet to 1 inch.

3. Sectional maps generally have scales ranging from
   a. 50–100 feet to 1 inch.
   b. 100–200 feet to 1 inch.
   c. 200–250 feet to 1 inch.
   d. 250–400 feet to 1 inch.

4. A comprehensive map should be
   a. compact enough to fit in a folder.
   b. as large as possible.
   c. as detailed as possible.
   d. written in technical language so that only engineers can read it.

5. On a plan and profile drawing, what does the abbreviation EL mean?
   a. English language
   b. Estimated length
   c. Electric
   d. Elevation

6. What type of map is also referred to as a wall map?

7. What type of map, commonly called a plat, is a series of maps covering sections of the water system?

8. What is the term for a computerized program used to manage data relating to geographic locations within a water distribution system?

Chapter 21
Safety, Security, and Emergency Response

Personal Safety Considerations

Everyone is responsible for maintaining safe working conditions, including water utility upper management, supervisors, and operations personnel. Management is responsible both for maintaining safe working conditions and for supporting a policy that encourages safe performance of work duties. Supervisors at all levels are responsible for the direct control of work conditions. This means that each crew chief has a responsibility to see that all work is done in compliance with safety practices and regulations. Supervisors have a duty to correct safety violations.

The employees have a special position with regard to safety. Safety practices and safe equipment must be used or the safety program will not be successful. The employees must help guard against unsafe acts and conditions. Supervisors are required to discipline employees who do not comply with safety policies.

Regulatory Requirements

The obvious reason for exercising safe practices is to eliminate suffering, injury, and possible death of individuals. Beyond that is the cost to the utility from lost time, medical costs, and possibly even legal judgments. Other considerations are damage to equipment and property with their resulting repair costs.

Another reason for safe practices is the federal Occupational Safety and Health Act. Under this act, the federal government has established minimum health and safety standards that are applicable to every industry. The act mandates that every employer must furnish employees with a workplace that is free from recognized hazards that are likely to cause death or serious physical harm. The act provides for 6 months in prison and/or a $10,000 penalty on conviction for violations.

Causes of Accidents

An accident can almost always be traced to either an unsafe act or an unsafe condition. All supervisors and workers should learn to recognize unsafe conditions and unsafe acts.

Personal Protection Equipment

Most water utilities now provide basic personal protection equipment for workers. The equipment should be issued to all workers who will need it; thus they will have no excuse for not using it. Each person must be responsible for maintaining his or her equipment in good condition and having it available when it is needed.

373

A hardhat should be worn whenever a worker is in a trench or has someone working above him or her, or when he or she is near electrical equipment. Metal hardhats should never be used near an electrical hazard.

Gloves are necessary for protection from rough, sharp, hot materials and cold weather. Special long gloves are available to provide wrist and forearm protection. Workers should also wear gloves when handling oils, solvents, and other chemicals. Gloves should not, however, be worn around revolving machinery because of the danger of a glove being caught and pulled into the machine.

Respiratory equipment must also be available for use in some situations. For example, masks should always be worn when working with asbestos–cement pipe because of the danger of inhaling asbestos fibers. A self-contained breathing apparatus should be available for emergency use wherever chlorine gas is being used.

Other personal safety equipment that should be used when a related danger exists includes safety goggles, face shields, steel-toed shoes, aprons, and ear protection.

Equipment Safety

Material Handling

One of the most common and most debilitating types of injury related to material handling is back injury. Lifting heavy objects can be done safely and easily if common sense and a few basic guidelines are followed. These recommended procedures should be followed when lifting heavy objects:

1. Bend at the knees to grasp the weight, keeping the back straight.
2. Get a firm hold on the object.
3. Maintain good footing with feet about shoulder-width apart.
4. Keeping the back as straight and upright as possible, lift slowly by straightening the legs.
5. If the object is too heavy to lift alone, do not hesitate to ask for assistance.

Other general safety guidelines for workers who are handling material or doing manual labor are listed here.

- Do not lift or shove sharp, heavy, or bulky objects without the help of other workers or the use of tools.
- To change direction when carrying a heavy load, turn the whole body, including the feet, rather than just twisting the back.
- Never try to lift a load that is too heavy or too large to lift comfortably. Use a mechanical device to assist in lifting heavy objects.
- Even though pipes and fittings look tough, they should be handled carefully. They should be lifted or lowered from the truck to the ground—not dropped. In addition to potentially damaging the equipment, dropping it can be dangerous in several ways.
- When pipe is being unloaded from a truck, there is considerable danger of it rolling. All workers must be warned to stay clear of the load at all times. Figure 21-1 shows a sign warning of the dangers involved in unloading pipe.
- If several people are moving or placing a pipe, they must all work together. Only one person should give directions and signals.

Figure 21-1 Warning of the dangers of injury during pipe unloading
Courtesy of US Pipe and Foundry Company.

- When a crane is handling pipe, only one person should direct the machine operator. No one should ever stand or walk under the suspended pipe or crane boom.
- If pipe is to be lowered by skids, it is important to make sure the skids are strong enough to hold the weight and are firmly secured. Snubbing ropes should be both strong enough to support the weight and large enough for workers to maintain a good grip.
- Individuals working with ropes should wear gloves to prevent rope burns.
- For lifting and moving large valves and fire hydrants, slings placed around the body should be carefully secured so as not to slip. Special lifting clamps for valves and hydrants should be used if possible.
- Equipment for transporting objects by hand, such as wheelbarrows and hand trucks, should be properly maintained and not overloaded.
- Workers using jackhammers must wear goggles, ear protection, and protective foot gear.
- Horseplay has been the cause of many serious accidents and should be absolutely prohibited on the job.

Trench Safety

Trenches can be made safe if the excavation is properly made and appropriate equipment is used. Proper trench shoring cannot be reduced to a standard formula. Each job is an individual problem and must be considered in relation to local conditions.

If an excavation is 5 ft (1.5 m) or more deep, cave-in protection is required under any soil conditions. Where soil is unstable, protection may be necessary in shallower trenches.

Confined-Space Safety

Dangers of confined spaces include injury, acute illness, disability, and death. The National Institute for Occupational Safety and Health estimates that, until recent years, an average of at least 174 confined-space deaths were occurring each year. Many of the incidents are extremely tragic because they involve multiple deaths. A common scenario is that a worker enters a confined space without proper safety preparations and equipment and is overcome, then coworkers attempt rescue and are also overcome. A great number of those who have died were water and wastewater system workers. Among the dangers that may cause injury or death in a confined space are oxygen-deficient or oxygen-enriched atmosphere, toxic gases, flammable gases, temperature extremes, flooding potential, slick or wet surfaces, falling objects, and electrical hazards.

The Occupational Safety and Health Administration (OSHA) established standards for confined-space entry (29 CFR Part 1910.146) in 1993. The number of yearly deaths is now decreasing but is far from acceptable. A *confined space* is defined by OSHA as any space that has the following characteristics:

- Has limited or restricted means of entry or exit
- Is large enough for an employee to enter and perform work
- Is not designated for continuous work occupancy

Examples of confined spaces in the water and wastewater industries include access holes for valves, meters, and air vents; sewer access holes; tanks; wet wells; digesters; and reservoirs.

Confined spaces, as further defined by OSHA, fall into two categories: permit-required and nonpermit-required. A permit-required space (or *permit space*) is a space that has one or more of the following characteristics:

- Contains, or has a potential to contain, a hazardous atmosphere
- Contains a material that has the potential for engulfing an entrant
- Has an internal configuration such that an entrant could be trapped or asphyxiated by inwardly converging walls or by a floor that slopes downward and tapers to a smaller cross-section
- Contains any other recognized serious safety or health hazard

A nonpermit-required space is a space that does not have any of the above hazards.

OSHA requires a written program for any permit-required, confined-space entry. This program includes identifying locations and making preparations prior to entry. Before an employee enters a permit space, the internal atmosphere must be tested with a calibrated, direct-reading instrument for oxygen content, flammable gases and vapors, and potentially toxic air contaminants. In addition, the permit space must be periodically tested during work inside the space to ensure that acceptable conditions exist.

The employer is required to supply and ensure that employees use the following equipment:

- Testing and monitoring equipment
- Ventilation equipment needed to obtain acceptable entry conditions
- Communications equipment
- Personal protective equipment
- Lighting equipment needed to enable employees to see well enough to work safely
- Barriers and shields as required
- Ladders needed for safe entry and exit
- Rescue and emergency equipment (Figure 21-2)

The OSHA standards include many other requirements that must be observed. These standards may include a requirement for a trained attendant outside the confined space that can assist with a rescue or operate emergency equipment. Additional information should be obtained from state and federal offices.

WATCH THE VIDEO
Confined Space Safety (www.awwa.org/wsovideoclips)

---

permit-required space
A space defined by the Occupational Safety and Health Administration as having one or more of the following characteristics: contains, or has a potential to contain, a hazardous atmosphere; contains a material that has the potential for engulfing an entrant; has an internal configuration such that an entrant could be trapped or asphyxiated by inwardly converging walls or by a floor that slopes downward and tapers to a smaller cross-section; and/or contains any other recognized serious safety or health hazard.

nonpermit-required space
A space defined by the Occupational Safety and Health Administration as not having any of the risks associated with a permit-required space.

Figure 21-2 Confined space rescue and retrieval system
Courtesy of DBI/SALA.

Hand Tool Safety

Some of the basic rules for safe use of hand tools are as follows:

- Always use an appropriate tool for the job. A very large percentage of on-the-job injuries are caused by the use of an improper tool. A screwdriver is not the same as a crowbar, a wrench should not be used as a hammer, and so forth.
- Check the condition of tools frequently. Repair or replace them if they are damaged or defective.
- Avoid using tools on machinery that is moving. It is best to shut off the machine and lock it out before making adjustments.
- Check clearance at the workplace to make sure there is sufficient space to recover a tool if it should slip.
- Maintain good support underfoot to reduce the hazard of slipping, stumbling, or falling when working.
- If at all possible, do not wear rings while doing mechanical work.
- Carry sharp or pointed tools in covers and pointed away from the body. Do not carry sharp-edged tools in trouser pockets.
- Wear eye protection when using impact tools.
- After using tools, wipe them clean and put them away in a safe place. Keep the workplace orderly.
- Do not lay tools on tops of stepladders or other elevated places from which they could fall on someone below.
- Learn and follow the correct methods for using hand tools.
- Try not to hurry unduly under emergency conditions. When hurrying, one tends to forget good safety practices and often takes dangerous shortcuts.

Portable Power Tools

An electric power tool should be grounded unless it has an all-plastic case. A ground-fault interrupter circuit should be used whenever a power tool is used outdoors or in a damp situation. Power cords should be inspected frequently and replaced if any breaks are noted. Power tools should not be lifted by the cords. A worker must make sure to have a firm footing before starting to use a power tool to avoid being thrown off balance when it starts.

Air tools can be dangerous if the hoses and connections are not properly maintained. Workers should not point air tools at anyone and should not use compressed air to clean off their bodies or clothing.

Traffic Control

Barricades, traffic cones, warning signs, and flashing lights should be used to warn the public of construction work that is taking place. These devices should be placed far enough ahead of the work site so that the public has ample opportunity to stop or avoid the obstructions. If necessary, one or more flaggers should be used to slow and direct traffic. Everyone involved in work near roadways should wear bright, reflective vests.

Approved traffic safety control devices and procedures for various classes of roadways are detailed in *Manual on Uniform Traffic Control Devices for Streets and Highways*, prepared by the US Department of Transportation, Federal Highway Administration. Most states have also prepared simplified booklets describing work area protection.

Water utility operators should be aware that they could be liable for damages if an accident occurs as a result of work on a street or highway that was not guarded in conformance with state-directed procedures. Figure 21-3 illustrates the

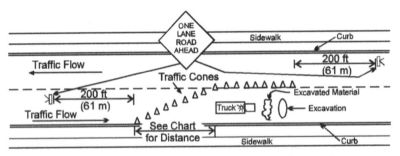

(If traffic is heavy or construction work causes interference in the open lane, one or more flaggers should be used.)

Figure 21-3 Recommended barricade placement for working in a roadway

Speed Limit mph (km/h)	Lane Width						Number of Cones Required
	10 ft (3 m) Taper Length,		11 ft (3.5 m) Taper Length,		12 ft (3.7 m) Taper Length,		
	ft	(m)	ft	(m)	ft	(m)	
20 (32)	70	(21)	75	(23)	80	(24)	5
25 (40)	105	(32)	115	(35)	125	(38)	6
30 (48)	150	(46)	165	(50)	180	(55)	7
35 (56)	205	(62)	225	(69)	245	(75)	8
40 (64)	270	(82)	295	(90)	320	(98)	9
45 (72)	450	(137)	495	(151)	540	(165)	13
50 (81)	500	(152)	550	(168)	600	(183)	13
55 (89)	550	(168)	605	(184)	660	(201)	13

guarding procedures required for construction work that will obstruct one lane of a low-traffic-volume, two-lane roadway.

WATCH THE VIDEO
Traffic Control (www.awwa.org/wsovideoclips)

Chemical Safety

Three chemicals are commonly used to disinfect mains and other facilities. *Liquid chlorine* is used with a gas-flow chlorinator and ejector to disinfect water mains. This type of disinfection should be used only by experienced personnel with specific training in chlorine safety and emergency response for chlorine release. *Sodium hypochlorite*, a liquid, may be used to disinfect mains and other facilities. *Calcium hypochlorite* is generally found in tablet or powder form and is the most common means of disinfecting mains. It is a strong oxidizer, and while it is not flammable by itself, its heat of reaction with other materials may cause fires. Hypochlorites are corrosive, and personal protective equipment (such as eye goggles and gloves, at a minimum) should be used for worker safety. Refer to the Safety Data Sheet (SDS) provided for specific chemicals to access additional information about chemical hazards and safe handling procedures.

Vehicle Safety

Records indicate that most accidents in the water works industry involve vehicles. Workers should be particularly made aware of the potential dangers to themselves, fellow workers, and the public while they are operating large trucks and heavy construction equipment. Many utilities require all workers to periodically attend a safe-driving school.

Water Supply System Security

The types of threats that would affect water utilities include both natural and accidental disasters and intentional disasters, which are discussed below. Natural disasters include floods, windstorms, ice storms, snowstorms, fires, droughts, and earthquakes. Accidental disasters include chemical spills, fires, transportation accidents, and explosions. Intentional disasters include the terrorist activities that are covered in this chapter. Three main types of consequences may occur: (1) complete interruption of supply, (2) sufficient quantity, but compromised quality, and (3) sufficient quality, but insufficient quantity.

Water distribution systems are extensive, with many components and subcomponents. The components and subcomponents are relatively unprotected and accessible and are often isolated. The physical destruction of a water distribution system's components and subcomponents or the disruption of water supply could be more likely than the introduction of contaminants to a system. The actual probability of a terrorist threat to drinking water is probably very low; however, the consequences could be extremely severe for exposed populations.

Four major types of intentional threats to drinking water systems (Mays, 2004) are discussed here. Although the potential threats to a water distribution are many, utilities can take steps to minimize risks by taking proper physical/cyber security steps, applying vigilant water quality monitoring, and developing an appropriate emergency response plan. Training and practice is essential to maintaining workforce preparedness for emergency events. Refer to AWWA Standard J100 and AWWA Manual M19 for additional information.

Cyber Threats

The electronic control functions intended to make water treatment more efficient and effective can be turned against the treatment system when hacked or otherwise violated. Examples of cyber threats include the following:

- Physical disruption of a supervisory control and data acquisition (SCADA) network
- Attacks on the central control system to create simultaneous failures
- Software attacks using worms/viruses
- Network flooding
- Jamming of control system functions
- Data corruption to give the appearance of appropriate chlorination when in fact no disinfectant has been added, allowing the proliferation of microbes

Physical Threats

Physical destruction of a system's assets or disruption of water supply is perhaps more likely than is contamination of the water supply. A single terrorist or a small group of terrorists could easily cripple an entire city by destroying a critical component of the water system. For example, opening and closing major control valves and turning pumps off and on too quickly could create a water hammer effect and, consequently, simultaneous main breaks. The resulting loss of water pressure would compromise firefighting capabilities and possibly lead bacterial buildup in the system.

Chemical Threats

There are numerous chemical warfare agents and industrial chemical poisons. Some of the chemical warfare agents include hydrogen cyanide, tabun, sarin, VX, lewisite (arsenic fraction), sulfur mustard, 3-quinuclidinyl benzilate, and lysergic acid diethylamide. Some of the industrial chemical poisons include cyanides, arsenic fluoride, cadmium, mercury, dieldrin, sodium fluoroacetate, and parathion.

Biological Threats

Several pathogens and biotoxins exist that have been weaponized, are potentially resistant to disinfection by chlorination, and are stable for relatively long periods in water. The pathogens include *Clostridium perfringens*, plague, and others. Biotoxins include botulinum, aflatoxin, ricin, and others. Water does provide dilution potential; however a neutrally buoyant particle of any size could be used to disperse pathogens into drinking water systems. Water storage and distribution systems can facilitate the delivery of an effective dose of toxicant to a potentially large population. A more extensive discussion of microbiological contaminants and threats of concern is presented by Abbaszadegan and Alum (2004).

Study Questions

1. If an excavation on a road requires that one of the lanes be closed and the speed limit is 25 mph (40 km/h), how many cones are required to divert the traffic?
   a. 6
   b. 9
   c. 13
   d. 15

2. If an excavation on a road requires that one of the lanes be closed and the speed limit is 45 mph (72 km/h), how many cones are required to divert the traffic?
   a. 9
   b. 13
   c. 15
   d. 18

3. Who is responsible for maintaining safe working conditions?
   a. Upper management
   b. Operations personnel
   c. Supervisors
   d. All water utility personnel

4. What item should be worn whenever a worker is in a trench or has someone working above him or her, or when he or she is near electrical equipment?
   a. Gloves
   b. A hardhat
   c. Respiratory equipment
   d. A reflective vest

5. When lifting a load that is too heavy or too large to lift comfortably,
   a. get a firm hold on the object before lifting.
   b. bend at the knees and lift with the legs.
   c. use a mechanical device to assist in lifting heavy objects.
   d. maintain good footing, with feet about shoulder-width apart, while lifting.

6. Which of the following is *not* an example of an accidental disaster that might disrupt water utilities?
   a. Earthquake
   b. Chemical spill
   c. Fire
   d. Transportation accident

7. What is the most common injury of distribution system workers?

8. Would a meter pit be a permit-required confined space?

9. Under what conditions is cave-in protection always required?

10. What three chemicals are commonly used to disinfect mains and other facilities?

Chapter 22
Public Relations

The Importance of Public Relations

It is important to maintain the public's confidence in the quality of drinking water and the services provided by a utility. Satisfied customers will pay their bills promptly and will provide political support for necessary rate increases or bond issues. They will also be less likely to turn to bottled water or home treatment devices of questionable quality.

The public relations activities of a utility are directed at maintaining public confidence and customer satisfaction. Water distribution personnel generally have the greatest exposure to the public and can do more to ensure public confidence than all the formal media campaigns and large-scale public relations projects put together. Conversely, management's formal public relations campaigns can be undermined by an employee's poor attitude or unwillingness to be of service beyond the immediate requirements of the job.

The Role of Public Relations

The role of public relations is to help create and maintain public confidence in the utility's product and organization. The active cooperation of all utility personnel is essential. Every customer contact should be viewed as an opportunity to improve communications and build goodwill.

Utilities depend on customer support for new budgets and for the implementation of special projects that require additional charges. It can be goodwill that tips the balance in favor of a badly needed bond issue. Customers who see their utility in a favorable light will generally vote in a like manner. Customers who are favorably impressed with water system operations are most likely to support increased service charges when they are required. In addition, where there are good public relations, customers will also be more tolerant of problems such as temporary tastes and odors, voluntary conservation measures, or repairs that require disruption of service.

Conversely, a utility that fosters an "us versus them" attitude can expect little customer cooperation. In fact, this approach may prompt outright antagonism from customers who already envision the utility as an unfriendly institution with the power to withhold a product vital to life. Everybody suffers in such cases, including water distribution personnel, who may bear the brunt in terms of lay-offs, lower salaries, or budget reductions affecting the system.

Customers must never be taken for granted. Although a utility rarely finds itself in the situation of a retail store, with competition offering lower prices or

public relations
The methods and activities employed to promote a favorable relationship with the public.

better service on the next corner, the same "best value for the dollar" approach to customer service should be applied. Higher prices must be justified by better products and services. Customer satisfaction must be a top priority.

The Role of Water Distribution Personnel

As noted earlier, water distribution personnel are often in much closer contact with customers than is anyone else in the utility. They may be the only contact between the utility and some customers. It may seem unfair, but a customer will remember the meter reader who took time at the doorway to wipe mud from his or her shoes much longer than today's favorable newspaper article about the utility. Likewise, the customer will recall the repair crew that heckled the family's cocker spaniel more vividly than the last rate hike.

Field personnel can enhance or detract from a utility's public image. On-the-job behavior can tell customers that the utility values their business or simply does not care. Once in the field, the utility employee *is* the utility. If utility employees do not respect the customer, in the customer's mind, the utility itself does not respect the customer.

Having effective public relations with customers requires three basic ingredients:

1. *Good communications* means really listening to what the customer has to say, then explaining policy, answering questions, or pointing out how the customer can save water or money.
2. *Caring* involves employees taking pride in themselves, their appearance, and the well-being of the customer.
3. *Courtesy* requires that field personnel follow the commonsense rules of polite behavior.

Meter Readers

Meter readers find themselves in the unenviable position of being associated with the water bill. Whether the customer blames a high bill on a misreading of the meter or simply feels that rates are too high, the meter reader is the most available target for complaint. The meter reader must also deal with unfriendly dogs, bad weather, and the general reluctance of homeowners to allow a stranger into their homes.

Irate customers are generally the exception, but an overdose of complaints can make any job difficult. However, the meter reader should remain informative and polite. Meter readers are not expected to be superhuman, but a cheerful and helpful outlook goes a long way toward effective public relations and makes the job much easier.

Here are some basic behavioral guidelines that a meter reader can follow in performing day-to-day tasks:

- For a good first impression, maintain a neat appearance (Figure 22-1). Most water utilities now feel it is well worth the cost to provide uniforms so that the employee is easily identifiable. Issuing uniforms also ensures that the employee has no excuse not to wear appropriate clothes. The clothes should be clean and well pressed.
- Meter readers should display name tags and carry credentials in the event they are asked to verify their employment with the water system. Do not

make customers take your word that you are who you say you are. Many customers are wary about admitting strangers into their homes. Anyone can claim to be a meter reader. Be prepared to prove it.

- Meter readers must be polite. They don't have to answer detailed questions, but they can inform customers about inquiry or complaint procedures. Many customers merely want sympathy. A person who listens well and is courteous can often calm an irate customer.
- Short, succinct answers should be given whenever possible. A long, drawn-out explanation is confusing—and it also takes up time when the meter reader is supposed to be reading meters. If the utility furnishes informative brochures, a few of them can be carried to hand out when appropriate.
- Any leaks found on the premises should be reported. Make sure customers know that the water utility does not want them to pay for unnecessary water use.
- If customers read their own meters, time should be taken to explain the procedure. This helps prevent errors or future bill adjustments.
- Meter readers should show enthusiasm and keep a smile in their voice.
- Customers should be addressed properly. Use *Miss*, *Sir*, or *Mrs.*, rather than *Lady* or *Hey you!*
- Good judgment must be used. For example, wipe off muddy shoes; don't kick the dog; don't smoke in homes, and preferably not in public; don't swear; don't chat too long with a customer; walk on sidewalks, not gardens or lawns; obey all driving and parking rules.

In short, the meter reader should always try to be helpful and polite and let customers know that he or she and the utility are on the customer's side.

Maintenance and Repair Crews

Like meter readers, maintenance and repair crews are highly visible to the public. Field personnel who are conducting routine inspections or performing minor repairs should stay as well groomed as possible. In most cases, maintenance or

Figure 22-1 Meter reader appearance is important

Courtesy of Colorado Springs Utilities.

repair crews will have little face-to-face contact with customers. Personnel should answer questions politely and to the best of their ability or refer the customer to a supervisor or the appropriate customer relations representative.

A common questions is, What are you doing? The inclination by workers who are busy and don't want to be bothered is to give a curt reply, or worse, an inaccurate reply. It must be remembered that most people who ask questions really are interested and want a correct answer. Although the worker can't take the time to provide a long dissertation, a reasonable response should be provided.

As a matter of routine, customers should be notified when service is to be temporarily discontinued. Shutoffs should be scheduled to coincide with low-water-use hours. Figure 22-2 shows a sample doorknob card that is often used for this purpose. Be sure the card isn't hung on a door that residents rarely or never use. In most cases, it is best to hang such cards on the back door. Other water systems print letters on letterhead paper and have workers hand them to homeowners or slip them under the door if no one is home.

If water must be shut off immediately, customers should be given at least a few minutes' notice to finish a shower or collect a pitcher of drinking water. At the same time, customers should be warned that water may be cloudy for a short time after it is turned back on. They should be informed that this presents no health threat but that clothes should not be washed until the water clears.

Property should be respected and the worksite kept as clean as possible. Utilities usually have policies governing lunch and breaks. Under no circumstances should field crews litter an area with wastepaper or soda cans. In general, customers don't like having workers lounging on their lawn while eating lunch. Smokers should not use the customer's lawn as an ashtray.

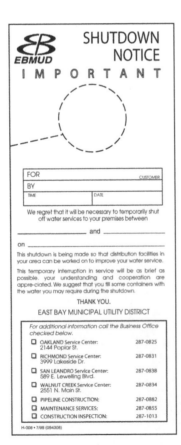

Figure 22-2 Example of a customer service card
Courtesy of EBMUD.

Damage to lawns, sidewalks, gardens, or streets should be kept to a minimum, and customers should be forewarned if property damage is expected to occur. Repairs should be completed and property restored to its original state before workers leave the site.

Vehicles should be clean and parked out of the way. They should not block driveways or alleys, nor should they hamper the flow of traffic in an area. Workers should never nap in a utility vehicle; doing so gives the impression that they are permitted to sleep on the job.

If streets are to be dug up, the neighborhood should be warned well in advance. The utility should suggest that those people affected move their vehicles. Good safety habits also promote public relations. Road barriers and warning signs communicate the utility's concern about customer and employee safety.

During the course of the job, workers should maintain a friendly attitude and a genuine desire to accomplish their work quickly and in the most inconspicuous manner possible.

Public Relations Behind the Wheel

At some time, most water distribution personnel will use a company vehicle. Public image is enhanced by good driving habits. If the driver of a utility vehicle drives over the speed limit or tailgates the driver in front, it will not be long before a customer telephones the utility to complain. Careless driving is dangerous, makes people angry, and gives the utility a bad image.

Driving rules should always be obeyed. Vehicles should be parked so as not to impede the flow of traffic. Parked vehicles should not block driveways, intersections, or alleys unless absolutely necessary. If blocking cannot be avoided, warning signals should be used on the vehicle. Where work must be performed in the street, appropriate cones, barricades, and other work-area protection methods should be employed to channel traffic smoothly around the jobsite.

Utility personnel must use good judgment regarding parking vehicles in front of coffee and doughnut shops. Although workers may be taking a legitimate coffee break, customers who see trucks regularly parked there get the impression that workers are spending most of their working day there.

Some water systems have developed a program with the local police department whereby every few years, all service personnel take a 4-hour safe driving course. Workers don't usually like having to go through the training, but it is still worth the time in terms of reduced accidents and improved public relations.

Public Relations and the Media

Although water distribution personnel may never be confronted with news reporters, a situation could arise—a main break, for instance, or several months of water conservation restrictions—that draws the attention of local newspapers and/or television stations. The general rule for talking to reporters is, *Don't!*

Large utilities maintain a public relations department whose job it is to prepare news releases and meet with the media. In other utilities, a manager or someone else is designated to take that responsibility. If approached by a reporter, distribution system operators should courteously but firmly state that they are not qualified to answer questions. Comments to the press will do little to help the situation. They may, on the contrary, do a great deal of damage.

Study Questions

1. What are the basic elements to good customer relations?

2. What should meter readers do when asked a question to which they don't know the answer?

3. What should customers be told when water service must be temporarily stopped?

4. _____ generally presents the greatest exposure to the public and can have the greatest impact in ensuring public confidence.
    a. Formal media campaigns
    b. Water distribution personnel
    c. Large-scale public relations projects
    d. Political campaigns

5. _____ should be viewed as an opportunity to improve communications and build goodwill.
    a. The initial customer contact
    b. A response to a customer complaint
    c. Every customer contact
    d. A meter reading

6. If damage to a customer's property occurs,
    a. customers should be told how to repair the damage.
    b. customers should be put in touch with a professional who can repair the damage.
    c. it is the customer's responsibility to fix the damage.
    d. property must be restored to its original state before workers leave the site.

7. Shutoffs should be scheduled
    a. at the utility's convenience.
    b. during holidays.
    c. on weekends.
    d. to coincide with low-water-use hours.

8. What is the general rule for talking to reporters?
    a. Don't do it.
    b. Keep answers brief.
    c. Make only positive statements.
    d. Validate all statements with relevant statistics.

Appendix A
Conversion Tables

Table 19-1 Conversion factors

Conversions		Procedure		Approximations (Actual answer will be within 25% of approximate answer.)
From	To	Multiply number of	by	To get number of
acres	hectares (ha)	acres	0.4047	ha	1 acre ≈ 0.4 ha
acres	square feet (ft²)	acres	43,560	ft²	1 acre ≈ 40,000 ft²
acres	square kilometers (km²)	acres	0.004047	km²	1 acre ≈ 0.004 km²
acres	square meters (m²)	acres	4,047	m²	1 acre ≈ 4,000 m²
acres	square miles (mi²)	acres	0.001563	mi²	1 acre ≈ 0.0015 mi²
acres	square yards (yd²)	acres	4,840	yd²	1 acre ≈ 5,000 yd²
acre-feet (acre-ft)	cubic feet (ft³)	acre-ft	43,560	ft³	1 acre-ft ≈ 40,000 ft³
acre-feet (acre-ft)	cubic meters (m³)	acre-ft	1,233	m³	1 acre-ft ≈ 1,000 m³
acre-feet (acre-ft)	gallons (gal)	acre-ft	325,851	gal	1 acre-ft ≈ 300,000 gal
centimeters (cm)	feet (ft)	cm	0.03281	ft	1 cm ≈ 0.03 ft
centimeters (cm)	inches (in.)	cm	0.3937	in.	1 cm ≈ 0.4 in.
centimeters (cm)	meters (m)	cm	0.01	m	—
centimeters (cm)	millimeters (mm)	cm	10	mm	—
centimeters per second (cm/s)	meters per minute (m/min)	cm/s	0.6	m/min	—
cubic centimeters (cm³)	cubic feet (ft³)	cm³	0.00003531	ft³	1 cm³ ≈ 0.00004 ft³
cubic centimeters (cm³)	cubic inches (in.³)	cm³	0.06102	in.³	1 cm³ ≈ 0.06 in.³
cubic centimeters (cm³)	cubic meters (m³)	cm³	0.000001	m³	—
cubic centimeters (cm³)	cubic yards (yd³)	cm³	0.000001308	yd³	1 cm³ ≈ 0.0000015 yd³
cubic centimeters (cm³)	gallons (gal)	cm³	0.0002642	gal	1 cm³ ≈ 0.0003 gal
cubic centimeters (cm³)	liters (L)	cm³	0.001	L	—

Table 19-1 Conversion factors (continued)

From	Conversions To	Procedure Multiply number of	by	To get number of	Approximations (Actual answer will be within 25% of approximate answer.)
cubic feet (ft^3)	acre-feet (acre-ft)	ft^3	0.00002296	acre-ft	1 ft^3 ≈ 0.00002 acre-ft
cubic feet (ft^3)	cubic centimeters (cm^3)	ft^3	28,320	cm^3	1 ft^3 ≈ 30
cubic feet (ft^3)	cubic inches (in.^{3})	ft^3	1,728	in.^{3}	1 ft^3 ≈ 1,500 in.^{3}
cubic feet (ft^3)	cubic meters (m^3)	ft^3	0.02832	m^3	1 ft^3 ≈ 0.03 m^3
cubic feet (ft^3)	cubic yards (yd^3)	ft^3	0.03704	yd^3	1 ft^3 ≈ 0.04 yd^3
cubic feet (ft^3)	gallons (gal)	ft^3	7.481	gal	1 ft^3 ≈ 7 gal
cubic feet (ft^3)	kiloliters (kL)	ft^3	0.02832	kL	1 ft^3 ≈ 0.03 kL
cubic feet (ft^3)	liters (L)	ft^3	28.32	L	1 ft^3 ≈ 30 L
cubic feet (ft^3)	pounds (lb) of water	ft^3	62.4	lb of water	1 ft^3 ≈ 60 lb of water
cubic feet per second (ft^3/s)	cubic meters per second (m^3/s)	ft^3/s	0.02832	m^3/s	1 ft^3/s ≈ 0.03 m^3/s
cubic feet per second (ft^3/s)	million gallons per day (mgd)	ft^3/s	0.6463	mgd	1 ft^3/s ≈ 0.6 mgd
cubic feet per second (ft^3/s)	gallons per minute (gpm)	ft^3/s	448.8	gpm	1 ft^3/s ≈ 400 gpm
cubic feet per minute (ft^3/min)	gallons per second (gps)	ft^3/min	0.1247	gps	1 ft^3/min ≈ 0.1 gps
cubic feet per minute (ft^3/min)	liters per second (L/s)	ft^3/min	0.4720	L/s	1 ft^{13}/min ≈ 0.5 L/s
cubic inches (in.^{3})	cubic centimeters (cm^3)	in.^{3}	16.39	cm^3	1 in.^{3} = 15 cm^3
cubic inches (in.^{3})	cubic feet (ft^3)	in.^{3}	0.0005787	ft^3	1 in.^{3} = 0.0006 ft^3
cubic inches (in.^{3})	cubic meters (m^3)	in.^{3}	0.00001639	m^3	1 in.^{3} = 0.00015 m^3
cubic inches (in.^{3})	cubic millimeters (mm^3)	in.^{3}	16,390	mm^3	1 in.^{3} = 15,000 mm^3
cubic inches (in.^{3})	cubic yards (yd^3)	in.^{3}	0.00002143	yd^3	1 in.^{3} = 0.00002 yd^3
cubic inches (in.^{3})	gallons (gal)	in.^{3}	0.004329	gal	1 in.^{3} = 0.004 gal

(continued)

Table 19-1 Conversion factors (continued)

	Conversions	Procedure			Approximations (Actual answer will be within 25% of approximate answer.)
From	To	Multiply number of	by	To get number of	
cubic inches (in.^{3})	liters (L)	in.^{3}	0.01639	L	1 in.^{3} = 0.015 L
cubic meters (m^3)	acre-feet (acre-ft)	m^3	0.0008107	acre-ft	1 m^3 = 0.0008 acre-ft
cubic meters (m^3)	cubic centimeters (cm^3)	m^3	1,000,000	cm^3	—
cubic meters (m^3)	cubic feet (ft^3)	m^3	35.31	ft^3	1 m^3 = 40 ft^3
cubic meters (m^3)	cubic inches (in.^{3})	m^3	61,020	in.^{3}	1 m^3 = 60,000 in.^{3}
cubic meters (m^3)	cubic yards (yd^3)	m^3	1.308	yd^3	1 m^3 = 1.5 yd^3
cubic meters (m^3)	gallons (gal)	m^3	264.2	gal	1 m^3 = 300 gal
cubic meters (m^3)	kiloliters (kL)	m^3	1.0	kL	—
cubic meters (m^3)	liters (L)	m^3	1,000	L	—
cubic meters per day (m^3/d)	gallons per day (gpd)	m^3/d	264.2	gpd	1 m^3/d = 300 gpd
cubic meters per second (m^3/s)	cubic feet per second (ft^3/s)	m^3/s	35.31	ft^3/s	1 m^3/s = 40 ft^3/s
cubic millimeters (mm^3)	cubic inches (in.^{3})	mm^3	0.00006102	in.^{3}	1 mm^3 = 0.00006 in.^{3}
cubic yards (yd^3)	cubic centimeters (cm^3)	yd^3	764,600	cm^3	1 yd^3 ≈ 800,000 cm^3
cubic yards (yd^3)	cubic feet (ft^3)	yd^3	27	ft^3	1 yd^3 ≈ 30 ft^3
cubic yards (yd^3)	cubic inches (in.^{3})	yd^3	46,660	in.^{3}	1 yd^3 ≈ 50,000 in.^{3}
cubic yards (yd^3)	cubic meters (m^3)	yd^3	0.7646	m^3	1 yd^3 ≈ 0.8 m^3
cubic yards (yd^3)	gallons (gal)	yd^3	202.0	gal	1 yd^3 ≈ 200 gal
cubic yards (yd^3)	liters (L)	yd^3	764.6	L	1 yd^3 ≈ 800 L
feet (ft)	centimeters (cm)	ft	30.48	cm	1 ft ≈ 30 cm
feet (ft)	inches (in.)	ft	12	in.	—

Table 19-1 Conversion factors (continued)

From	Conversions To	Procedure Multiply number of	by	To get number of	Approximations (Actual answer will be within 25% of approximate answer.)
feet (ft)	kilometers (km)	ft	0.0003048	km	1 ft ≈ 0.0003 km
feet (ft)	meters (m)	ft	0.3048	m	1 ft ≈ 0.3 m
feet (ft)	miles (mi)	ft	0.0001894	mi	1 ft ≈ 0.0002 mi
feet (ft)	millimeters (mm)	ft	304.8	mm	1 ft ≈ 300 mm
feet (ft)	yards (yd)	ft	0.3333	yd	1 ft ≈ 0.3 yd
feet (ft) of hydraulic head	kilopascals (kPa)	ft of head	2.989	kPa	1 ft of head ≈ 3 kPa
feet (ft) of hydraulic head	meters (m) of hydraulic head	ft of head	0.3048	m of head	1 ft of head ≈ 0.3 m of head
feet (ft) of hydraulic head	pascals (Pa)	ft of head	2,989	Pa	1 ft of head ≈ 3,000 Pa
feet (ft) of water	inches of mercury (in. Hg)	ft of water	0.8826	in. Hg	1 ft of water ≈ 0.9 in. Hg
feet (ft) of water	pounds per square foot (lb/ft^2)	ft of water	62.4	lb/ft^2	1 ft of water ≈ 60 lb/ft^2
feet (ft) of water	pounds per square inch gauge (psig)	ft of water	0.4332	psig	1 ft of water ≈ 0.4 psig
feet per hour (ft/h)	meters per second (m/s)	ft/h	0.00008467	m/s	1 ft/h ≈ 0.00008 m/s
feet per minute (ft/min)	feet per second (ft/s)	ft/min	0.01667	ft/s	1 ft/min ≈ 0.015 ft/s
feet per minute (ft/min)	kilometers per hour (km/h)	ft/min	0.01829	km/h	1 ft/min ≈ 0.02 km/h
feet per minute (ft/min)	meters per minute (m/min)	ft/min	0.3048	m/min	1 ft/min ≈ 0.3 m/min
feet per minute (ft/min)	meters per second (m/s)	ft/min	0.005080	m/s	1 ft/min ≈ 0.005 m/s
feet per minute (ft/min)	miles per hour (mph)	ft/min	0.01136	mph	1 ft/min ≈ 0.01 mph
feet per second (ft/s)	feet per minute (ft/min)	ft/s	60	ft/min	—
feet per second (ft/s)	kilometers per hour (km/h)	ft/s	1.097	km/h	1 ft/s ≈ 1 km/h
feet per second (ft/s)	meters per minute (m/min)	ft/s	18.29	m/min	1 ft/s ≈ 20 m/min

(continued)

Table 19-1 Conversion factors (continued)

From	Conversions To	Procedure Multiply number of	by	To get number of	Approximations (Actual answer will be within 25% of approximate answer.)
feet per second (ft/s)	meters per second (m/s)	ft/s	0.3048	m/s	1 ft/s ≈ 0.3 m/s
feet per second (ft/s)	miles per hour (mph)	ft/s	0.6818	mph	1 ft/s ≈ 0.7 mph
foot-pounds per minute (ft-lb/min)	horsepower (hp)	ft-lb/min	0.00003030	hp	1 ft-lb/min ≈ 0.00003 hp
foot-pounds per minute (ft-lb/min)	kilowatts (kW)	ft-lb/min	0.00002260	kW	1 ft-lb/min ≈ 0.00002 kW
foot-pounds per minute (ft-lb/min)	watts (W)	ft-lb/min	0.02260	W	1 ft-lb/min ≈ 0.02 W
gallons (gal)	acre-feet (acre-ft)	gal	0.000003069	acre-ft	1 gal ≈ 0.000003 acre-ft
gallons (gal)	cubic centimeters (cm^3)	gal	3,785	cm^3	1 gal ≈ 4000 cm^3
gallons (gal)	cubic feet (ft^3)	gal	0.1337	ft^3	1 gal ≈ 0.15 ft^3
gallons (gal)	cubic inches (in.^{3})	gal	231.0	in.^{3}	1 gal ≈ 200 in.^{3}
gallons (gal)	cubic meters (m^3)	gal	0.003785	m^3	1 gal ≈ 0.004 m^3
gallons (gal)	cubic yards (yd^3)	gal	0.004951	yd^3	1 gal ≈ 0.005 yd^3
gallons (gal)	kiloliters (kL)	gal	0.003785	kL	1 gal ≈ 0.004 kL
gallons (gal)	liters (L)	gal	3.785	L	1 gal ≈ 4 L
gallons (gal)	pounds (lb) of water	gal	8.34	lb of water	1 gal ≈ 8 lb of water
gallons (gal)	quarts (qt)	gal	4	qt	
gallons per capita per day (gpcd)	liters per capita per day (L/d per capita)	gpcd	3.785	L/d per capita	1 gpcd ≈ 4 L/d per capita
gallons per day (gpd)	cubic meters per day (m^3/d)	gpd	0.003785	m^3/d	1 gpd ≈ 0.004 m^3/d
gallons per day (gpd)	liters per day (L/d)	gpd	3.785	L/d	1 gpd ≈ 4 L/d

Table 19-1 Conversion factors (continued)

Conversions			Procedure		Approximations (Actual answer will be within 25% of approximate answer.)
From	To	Multiply number of	by	To get number of	
gallons per day per foot (gpd/ft)	square meters per day (m²/d)	gpd/ft	0.01242	m²/d	1 gpd/ft ≈ 0.01 m²/d
gallons per day per foot (gpd/ft)	square millimeters per second (mm²/s)	gpd/ft	0.1437	mm²/s	1 gpd/ft ≈ 0.15 mm²/s
gallons per day per square foot (gpd/ft²)	millimeters per second (mm/s)	gpd/ft²	0.0004716	mm/s	1 gpd/ft² ≈ 0.0005 mm/s
gallons per hour (gph)	liters per second (L/s)	gph	0.001052	L/s	1 gph ≈ 0.001 L/s
gallons per minute (gpm)	cubic feet per second (ft³/s)	gpm	0.002228	ft³/s	1 gpm ≈ 0.0002 ft³/s
gallons per minute (gpm)	liters per second (L/s)	gpm	0.06309	L/s	1 gpm ≈ 0.06 L/s
gallons per minute per square foot (gpm/ft²)	millimeters per second (mm/s)	gpm/ft²	0.6790	mm/s	1 gpm/ft² ≈ 0.7 mm/s
gallons per second (gps)	cubic feet per minute (ft³/min)	gps	8.021	ft³/min	1 gpm ≈ 8 ft³/min
gallons per second (gps)	liters per minute (L/min)	gps	227.1	L/min	1 gps ≈ 200 L/min
grains (gr)	grams (g)	gr	0.06480	g	1 gr ≈ 0.06 g
grains (gr)	pounds (lb)	gr	0.0001428	lb	1 gr ≈ 0.00015 lb
grams (g)	grains (gr)	g	15.43	gr	1 g ≈ 15 gr
grams (g)	kilograms (kg)	g	0.001	kg	—
grams (g)	milligrams (mg)	g	1,000	mg	—
grams (g)	ounces (oz), avoirdupois	g	0.03527	oz	1 g ≈ 0.04 oz
grams (g)	pounds (lb)	g	0.002205	lb	1 g ≈ 0.002 lb
hectares (ha)	acres	ha	2.471	acres	1 ha ≈ 2 acres

(continued)

Table 19-1 Conversion factors (continued)

Conversions		Procedure		Approximations
				(Actual answer will be within 25% of approximate answer.)
From	To	Multiply number of	by	To get number of
hectares (ha)	square meters (m²)	ha	10,000	m²	—
hectares (ha)	square miles (mi²)	ha	0.003861	mi²	1 ha ≈ 0.004 mi²
horsepower (hp)	foot-pounds per minute (ft-lb/min)	hp	33,000	ft-lb/min	1 hp ≈ 30,000 ft-lb/min
horsepower (hp)	kilowatts (kW)	hp	0.7457	kW	1 hp ≈ 0.7 kW
horsepower (hp)	watts (W)	hp	745.7	W	1 hp ≈ 700 W
inches (in.)	centimeters (cm)	in.	2.540	cm	1 in. ≈ 3 cm
inches (in.)	feet (ft)	in.	0.08333	ft	1 in. ≈ 0.08 ft
inches (in.)	meters (m)	in.	0.02540	m	1 in. ≈ 0.03 m
inches (in.)	millimeters (mm)	in.	25.40	mm	1 in. ≈ 30 mm
inches (in.)	yards (yd)	in.	0.02778	yd	1 in. ≈ 0.03 yd
inches of mercury (in. Hg)	feet (ft) of water	in. Hg	1.133	ft of water	1 in. Hg ≈ 1 ft of water
inches of mercury (in. Hg)	inches (in.) of water	in. Hg	13.60	in. of water	1 in. Hg ≈ 15 in. of water
inches of mercury (in. Hg)	pounds per square foot (lb/ft²)	in. Hg	70.73	lb/ft²	1 in. Hg ≈ 70 lb/ft³
inches of mercury (in. Hg)	pounds per square inch (psi)	in. Hg	0.4912	psi	1 in. Hg ≈ 0.5 psi
inches per minute (in./min)	millimeters per second (mm/s)	in./min	0.4233	mm/s	1 in./min ≈ 0.4 mm/s
inches (in.) of water	inches of mercury (in. Hg)	in. of water	0.07355	in. Hg	1 in. of water ≈ 0.07 in. Hg
inches (in.) of water	pounds per square foot (lb/ft²)	in. of water	5.198	lb/ft²	1 in. of water ≈ 5 lb/ft²

Table 19-1 Conversion factors (continued)

From	Conversions To	Procedure Multiply number of	by	To get number of	Approximations (Actual answer will be within 25% of approximate answer.)
inches (in.) of water	pounds per square inch gauge (psig)	in. of water	0.03610	psig	1 in. of water ≈ 0.04 psig
kilograms (kg)	grams (g)	kg	1,000	g	—
kilograms (kg)	pounds (lb)	kg	2.205	lb	1 kg ≈ 2 lb
kiloliters (kL)	cubic feet (ft³)	kL	35.31	ft³	1 kL ≈ 40 ft³
kiloliters (kL)	cubic meters (m³)	kL	1.0	m³	—
kiloliters (kL)	gallons (gal)	kL	264.2	gal	1 kL ≈ 300 gal
kiloliters (kL)	liters (L)	kL	1,000	L	—
kilometers (km)	feet (ft)	km	3,281	ft	1 km ≈ 3,000 ft
kilometers (km)	meters (m)	km	1,000	m	—
kilometers (km)	miles (mi)	km	0.6214	mi	1 km ≈ 0.6 mi
kilometers (km)	yards (yd)	km	1,094	yd	1 km ≈ 1,000 yd
kilometers per hour (km/h)	feet per minute (ft/min)	km/h	54.68	ft/min	1 km/h ≈ 50 ft/min
kilometers per hour (km/h)	feet per second (ft/s)	km/h	0.9113	ft/s	1 km/h ≈ 1 ft/s
kilometers per hour (km/h)	meters per minute (m/min)	km/h	16.67	m/min	1 km/h ≈ 15 m/min
kilometers per hour (km/h)	meters per second (m/s)	km/h	0.2778	m/s	1 km/h ≈ 0.3 m/s
kilometers per hour (km/h)	miles per hour (mph)	km/h	0.6214	mph	1 km/h ≈ 0.6 mph
kilopascals (kPa)	feet (ft) of hydraulic head	kPa	0.3346	ft of head	1 kPa ≈ 0.3 ft of head
kilowatts (kW)	foot-pounds per minute (ft-lb/min)	kW	44,250	ft-lb/min	1 kW ≈ 40,000 ft-lb/min
kilowatts (kW)	horsepower (hp)	kW	1.341	hp	1 kW ≈ 1.5 hp
kilowatts (kW)	watts (W)	kW	1,000	W	—

(continued)

Table 19-1 Conversion factors (continued)

From	Conversions To	Procedure Multiply number of	by	To get number of	Approximations (Actual answer will be within 25% of approximate answer.)
liters (L)	cubic centimeters (cm³)	L	1,000	cm³	—
liters (L)	cubic feet (ft³)	L	0.03531	ft³	1 L ≈ 0.04 ft³
liters (L)	cubic inches (in.³)	L	61.03	in.³	1 L ≈ 60 in.³
liters (L)	cubic meters (m³)	L	0.001	m³	—
liters (L)	cubic yards (yd³)	L	0.001308	yd³	1 L ≈ 0.0015 yd³
liters (L)	gallons (gal)	L	0.2642	gal	1 L ≈ 0.3 gal
liters (L)	kiloliters (kL)	L	0.001	kL	—
liters (L)	milliliters (mL)	L	1,000	mL	—
liters (L)	ounces (oz), fluid	L	33.81	oz (fluid)	1 L ≈ 30 oz (fluid)
liters (L)	quarts (qt), fluid	L	1.057	qt (fluid)	1 L ≈ 1 qt (fluid)
liters per capita per day (L/d per capita)	gallons per capita per day (gpcd)	L/d per capita	0.2642	gpcd	1 L/d per capita ≈ 0.3 gpcd
liters per day (L/d)	gallons per day (gpd)	L/d	0.2642	gpd	1 L/d ≈ 0.3 gpd
liters per minute (L/min)	gallons per second (gps)	L/min	0.004403	gps	1 L/min ≈ 0.004 gps
liters per second (L/s)	cubic feet per minute (ft³/min)	L/s	2.119	ft³/min	1 L/s ≈ 2 ft³/min
liters per second (L/s)	gallons per hour (gph)	L/s	951.0	gph	1 L/s ≈ 1000 gph
liters per second (L/s)	gallons per minute (gpm)	L/s	15.85	gpm	1 L/s ≈ 15 gpm
megaliters per day (ML/d)	million gallons per day (mgd)	ML/d	0.2642	mgd	1 ML/d ≈ 0.3 mgd
meters (m)	centimeters (cm)	m	100	cm	—
meters (m)	feet (ft)	m	3.281	ft	1 m ≈ 3 ft

Table 19-1 Conversion factors (continued)

From	Conversions To	Procedure Multiply number of	by	To get number of	Approximations (Actual answer will be within 25% of approximate answer.)
meters (m)	inches (in.)	m	39.37	in.	1 m ≈ 40 in.
meters (m)	kilometers (km)	m	0.001	km	—
meters (m)	miles (mi)	m	0.0006214	mi	1 m ≈ 0.0006 mi
meters (m)	millimeters (mm)	m	1,000	mm	—
meters (m)	yards (yd)	m	1.094	yd	1 m ≈ 1 yd
meters (m) of hydraulic head	feet (ft) of hydraulic head	m of head	3.281	ft of head	1 m of head ≈ 3 ft of head
meters (m) of hydraulic head	pounds per square inch gauge (psig)	m of head	1.422	psig	1 m of head ≈ 1.5 psig
meters per minute (m/min)	centimeters per second (cm/s)	m/min	1.667	cm/s	1 m/min ≈ 1.5 cm/s
meters per minute (m/min)	feet per minute (ft/min)	m/min	3.281	ft/min	1 m/min ≈ 3 ft/min
meters per minute (m/min)	feet per second (ft/s)	m/min	0.05468	ft/s	1 m/min ≈ 0.05 ft/s
meters per minute (m/min)	kilometers per hour (km/h)	m/min	0.06	km/h	—
meters per minute (m/min)	miles per hour (mph)	m/min	0.03728	mph	1 m/min ≈ 0.04 mph
meters per second (m/s)	feet per hour (ft/h)	m/s	11,810	ft/h	1 m/s ≈ 10,000 ft/h
meters per second (m/s)	feet per minute (ft/min)	m/s	196.8	ft/min	1 m/s ≈ 200 ft/min
meters per second (m/s)	feet per second (ft/s)	m/s	3.281	ft/s	1 m/s ≈ 3 ft/s
meters per second (m/s)	kilometers per hour (km/h)	m/s	3.6	km/h	1 m/s ≈ 4 km/h
meters per second (m/s)	miles per hour (mph)	m/s	2.237	mph	1 m/s ≈ 2 mph
miles (mi)	feet (ft)	mi	5,280	ft	1 mi ≈ 5,000 ft
miles (mi)	kilometers (km)	mi	1.609	km	1 mi ≈ 1.5 km
miles (mi)	meters (m)	mi	1,609	m	1 mi ≈ 1,500 m

(continued)

Table 19-1 Conversion factors (continued)

From	Conversions To	Multiply number of	Procedure by	To get number of	Approximations (Actual answer will be within 25% of approximate answer.)
miles (mi)	yards (yd)	mi	1,760	yd	1 mi ≈ 2,000 yd
miles per hour (mph)	feet per minute (ft/min)	mph	88	ft/min	1 mph ≈ 90 ft/min
miles per hour (mph)	feet per second (ft/s)	mph	1.467	ft/s	1 mph ≈ 1.5 ft/s
miles per hour (mph)	kilometers per hour (km/h)	mph	1.609	km/h	1 mph ≈ 1.5 km/h
miles per hour (mph)	meters per minute (m/min)	mph	26.82	m/min	1 mph ≈ 30 m/min
miles per hour (mph)	meters per second (m/s)	mph	0.4470	m/s	1 mph ≈ 0.4 m/s
milligrams (mg)	grams (g)	mg	0.001	g	—
milliliters (mL)	liters (L)	mL	0.001	L	—
millimeters (mm)	centimeters (cm)	mm	0.1	cm	—
millimeters (mm)	feet (ft)	mm	0.003281	ft	1 mm ≈ 0.003 ft
millimeters (mm)	inches (in.)	mm	0.03937	in.	1 mm ≈ 0.04 in.
millimeters (mm)	meters (m)	mm	0.001	m	—
millimeters (mm)	yards (yd)	mm	0.001094	yd	1 mm ≈ 0.001 yd
millimeters per second (mm/s)	gallons per day per square foot (gpd/ft^2)	mm/s	2,121	gpd/ft^2	1 mm/s ≈ 2,000 gpd/ft^2
millimeters per second (mm/s)	gallons per minute per square foot (gpm/ft^2)	mm/s	1.473	gpm/ft^2	1 mm/s ≈ 1.5 gpm/ft^2
millimeters per second (mm/s)	inches per minute (in./min)	mm/s	2.362	in./min	1 mm/s ≈ 2 in./min
million gallons per day (mgd)	cubic feet per second (ft^3/s)	mgd	1.547	ft^3/s	1 mgd ≈ 1.5 ft^3/s

Table 19-1 Conversion factors (continued)

From	Conversions		Procedure		Approximations (Actual answer will be within 25% of approximate answer.)
	To	Multiply number of	by	To get number of	
million gallons per day (mgd)	megaliters per day (ML/d)	mgd	3.785	ML/d	1 mgd ≈ 4 ML/d
ounces (oz), avoirdupois	grams (g)	oz	28.35	g	1 oz ≈ 30 g
ounces (oz), avoirdupois	pounds (lb)	oz	0.0625	lb	1 oz ≈ 0.06 lb
ounces (oz), fluid	liters (L)	oz	0.02957	L	1 oz ≈ 0.03 L
pascals (Pa)	feet (ft) of hydraulic head	Pa	0.0003346	ft of head	1 Pa ≈ 0.0003 ft of head
pascals (Pa)	pounds per square inch (psi)	Pa	0.0001450	psi	1 Pa ≈ 0.00015 psi
pounds (lb)	grains (gr)	lb	7,000	gr	
pounds (lb)	grams (g)	lb	453.6	g	1 lb ≈ 500 g
pounds (lb)	kilograms (kg)	lb	0.4536	kg	1 lb ≈ 0.5 kg
pounds (lb)	ounces (oz), avoirdupois	lb	16	oz	—
pounds (lb) of water	cubic feet (ft³)	lb of water	0.01603	ft³	1 lb of water ≈ 0.015 ft³
pounds (lb) of water	gallons (gal)	lb of water	0.1199	gal	1 lb of water ≈ 0.1 gal
pounds per square foot (lb/ft²)	feet (ft) of water	lb/ft²	0.01603	ft of water	1 lb/ft² ≈ 0.015 ft of water
pounds per square foot (lb/ft²)	inches of mercury (in. Hg)	lb/ft²	0.01414	in. Hg	1 lb/ft² ≈ 0.015 in. Hg
pounds per square foot (lb/ft²)	inches (in.) of water	lb/ft²	0.1924	in. of water	1 lb/ft² ≈ 0.2 in. of water
pounds per square inch gauge (psig)	feet (ft) of water	psig	2.31	ft of water	1 psig ≈ 2 ft of water

(continued)

Table 19-1 Conversion factors (continued)

Conversions		Procedure			Approximations (Actual answer will be within 25% of approximate answer.)
From	To	Multiply number of	by	To get number of	
pounds per square inch (psi)	inches of mercury (in. Hg)	psi	2.036	in. Hg	1 psi ≈ 2 in. Hg
pounds per square inch gauge (psig)	inches (in.) of water	psig	27.70	in. of water	1 psig ≈ 30 in. of water
pounds per square inch gauge (psig)	meters (m) of hydraulic head	psig	0.7034	m of head	1 psig ≈ 0.7 m of head
pounds per square inch (psi)	pascals (Pa)	psi	6,895	Pa	1 psi ≈ 7,000 Pa
quarts (qt)	gallons (gal)	qt	0.25	gal	—
quarts (qt)	liters (L)	qt	0.9464	L	1 qt ≈ 0.9 L
square centimeters (cm^2)	square inches ($in.^2$)	cm^2	0.1550	$in.^2$	1 cm^2 ≈ 0.15 $in.^2$
square centimeters (cm^2)	square millimeters (mm^2)	cm^2	100	mm^2	—
square feet (ft^2)	acres	ft^2	0.00002296	acres	1 ft^2 ≈ 0.00002 acre
square feet (ft^2)	square inches ($in.^2$)	ft^2	144	$in.^2$	1 ft^2 ≈ 150 $in.^2$
square feet (ft^2)	square meters (m^2)	ft^2	0.09290	m^2	1 ft^2 ≈ 0.09 m^2
square feet (ft^2)	square millimeters (mm^2)	ft^2	92,900	mm^2	1 ft^2 ≈ 90,000 mm^2
square feet (ft^2)	square yards (yd^2)	ft^2	0.1111	yd^2	1 ft^2 ≈ 0.1 yd^2
square inches ($in.^2$)	square centimeters (cm^2)	$in.^2$	6.452	cm^2	1 $in.^2$ ≈ 6 cm^2
square inches ($in.^2$)	square feet (ft^2)	$in.^2$	0.006944	ft^2	1 $in.^2$ ≈ 0.007 ft^2
square inches ($in.^2$)	square meters (m^2)	$in.^2$	0.0006452	m^2	1 $in.^2$ ≈ 0.0006 m^2
square inches ($in.^2$)	square millimeters (mm^2)	$in.^2$	645.2	mm^2	1 $in.^2$ ≈ 600 mm^2
square inches ($in.^2$)	square yards (yd^2)	$in.^2$	0.0007716	yd^2	1 $in.^2$ ≈ 0.0008 yd^2

Table 19-1 Conversion factors (continued)

Conversions		Procedure			Approximations (Actual answer will be within 25% of approximate answer.)
From	To	Multiply number of	by	To get number of	
square kilometers (km^2)	acres	km^2	247.1	acres	1 km^2 ≈ 200 acres
square kilometers (km^2)	square miles (mi^2)	km^2	0.3861	mi^2	1 km^2 ≈ 0.4 mi^2
square meters (m^2)	acres	m^2	0.0002471	acres	1 m^2 ≈ 0.0002 acre
square meters (m^2)	hectares (ha)	m^2	0.0001	ha	—
square meters (m^2)	square feet (ft^2)	m^2	10.76	ft^2	1 m^2 ≈ 10 ft^2
square meters (m^2)	square inches ($in.^2$)	m^2	1,550	$in.^2$	1 m^2 ≈ 1,500 $in.^2$
square meters (m^2)	square miles (mi^2)	m^2	0.0000003861	mi^2	1 m^2 ≈ 0.0000004 mi^2
square meters (m^2)	square yards (yd^2)	m^2	1.196	yd^2	1 m^2 ≈ 1 yd^2
square meters per day (m^2/d)	gallons per day per foot (gpd/ft)	m^2/d	80.53	gpd/ft	1 m^2/d ≈ 80 gpd/ft
square miles (mi^2)	acres	mi^2	640	acres	1 mi^2 ≈ 600 acres
square miles (mi^2)	square kilometers (km^2)	mi^2	2.590	km^2	1 mi^2 ≈ 3 km^2
square miles (mi^2)	square meters (m^2)	mi^2	2,590,000	m^2	1 mi^2 ≈ 3,000,000 m^2
square millimeters (mm^2)	square centimeters (cm^2)	mm^2	0.01	cm^2	—
square millimeters (mm^2)	square feet (ft^2)	mm^2	0.00001076	ft^2	1 mm^2 ≈ 0.00001 ft^2
square millimeters (mm^2)	square inches ($in.^2$)	mm^2	0.001550	$in.^2$	1 mm^2 ≈ 0.0015 $in.^2$
square millimeters per second (mm^2/s)	gallons per day per foot (gpd/ft)	mm^2/s	6.958	gpd/ft	1 mm^2/s ≈ 7 gpd/ft
square yards (yd^2)	acres	yd^2	0.0002066	acres	1 yd^2 ≈ 0.0002 acre
square yards (yd^2)	square feet (ft^2)	yd^2	9	ft^2	—
square yards (yd^2)	square inches ($in.^2$)	yd^2	1,296	$in.^2$	1 yd^2 ≈ 1,500 $in.^2$

(continued)

Table 19-1 Conversion factors (continued)

Conversions		Procedure		Approximations
From	To	Multiply number of	by	To get number of	(Actual answer will be within 25% of approximate answer.)
square yards (yd^2)	square meters (m^2)	yd^2	0.8361	m^2	1 yd^2 ≈ 0.8 m^2
watts (W)	foot-pounds per minute (ft-lb/min)	W	44.25	ft-lb/min	1 W ≈ 40 ft-lb/min
watts (W)	horsepower (hp)	W	0.001341	hp	1 W ≈ 0.0015 hp
watts (W)	kilowatts (kW)	W	0.001	kW	—
yards (yd)	feet (ft)	yd	3	ft	—
yards (yd)	inches (in.)	yd	36	in.	1 yd ≈ 40 in.
yards (yd)	kilometers (km)	yd	0.0009144	km	1 yd ≈ 0.0009 km
yards (yd)	meters (m)	yd	0.9144	m	1 yd ≈ 0.9 m
yards (yd)	miles (mi)	yd	0.0005681	mi	1 yd ≈ 0.0006 mi
yards (yd)	millimeters (mm)	yd	914.4	mm	1 yd ≈ 900 mm

Table 19-2 Temperature conversions, Celsius to Fahrenheit: °F = (9/5)°C + 32

°C to °F		°C to °F		°C to °F		°C to °F		°C to °F	
−29	−20.2	−9	15.8	11	51.8	31	87.8	51	123.8
−28	−18.4	−8	17.6	12	53.6	32	89.6	52	125.6
−27	−16.6	−7	19.4	13	55.4	33	91.4	53	127.4
−26	−14.8	−6	21.2	14	57.2	34	93.2	54	129.2
−25	−13.0	−5	23.0	15	59.0	35	95.0	55	131.0
−24	−11.2	−4	24.8	16	60.8	36	96.8	56	132.8
−23	−9.4	−3	26.6	17	62.6	37	98.6	57	134.6
−22	−7.6	−2	28.4	18	64.4	38	100.4	58	136.4
−21	−5.8	−1	30.2	19	66.2	39	102.2	59	138.2
−20	−4.0	0	32.0	20	68.0	40	104.0	60	140.0
−19	−2.2	1	33.8	21	69.8	41	105.8	61	141.8
−18	−0.4	2	35.6	22	71.6	42	107.6	62	143.6
−17	+1.4	3	37.4	23	73.4	43	109.4	63	145.4
−16	3.2	4	39.2	24	75.2	44	111.2	64	147.2
−15	5.0	5	41.0	25	77.0	45	113.0	65	149.0
−14	6.8	6	42.8	26	78.8	46	114.8	66	150.8
−13	8.6	7	44.6	27	80.6	47	116.6	67	152.6
−12	10.4	8	46.4	28	82.4	48	118.4	68	154.4
−11	12.2	9	48.2	29	84.2	49	120.2	69	156.2
−10	14.0	10	50.0	30	86.0	50	122.0	70	158.0

°C to °F	
71	159.8
72	161.6
73	163.4
74	165.2
75	167.0
76	168.8
77	170.6
78	172.4
79	174.2
80	176.0
81	177.8
82	179.6
83	181.4
84	183.2
85	185.0
86	186.8
87	188.6
88	190.4
89	192.2
90	194.0

(continued)

Table 19-2 Temperature conversions, Celsius to Fahrenheit: °F = (9/5)°C + 32 (continued)

°C	to °F	°C	to °F	°C	to °F	°C	to °F	°C	to °F	°C	to °F
91	195.8	101	213.8	111	231.8	121	249.8	131	267.8	141	285.8
92	197.6	102	215.6	112	233.6	122	251.6	132	296.6	142	287.6
93	199.4	103	217.4	113	235.4	123	253.4	133	271.4	143	289.4
94	201.2	104	219.2	114	237.2	124	255.2	134	273.2	144	291.2
95	203.0	105	221.0	115	239.0	125	257.0	135	275.0	145	293.0
96	204.8	106	222.8	116	240.8	126	258.8	136	276.8	146	294.8
97	206.6	107	224.6	117	242.6	127	260.6	137	278.6	147	296.6
98	208.4	108	226.4	118	244.4	128	262.4	138	280.4	148	298.4
99	210.2	109	228.2	119	246.2	129	264.2	139	282.2	149	300.2
100	212.0	110	230.0	120	248.0	130	266.0	140	284.0	150	302.0

Table 19-3 Temperature conversions, Fahrenheit to Celsius: °C = (5/9)°F − 32

°F	°F to °C	°F	°F to °C	°F	°F to °C	°F	°F to °C	°F	°F to °C
−19	−28.3	1	−17.2	21	−6.1	41	5.0	61	16.1	81	27.2
−18	−27.8	2	−16.7	22	−5.6	42	5.6	62	16.7	82	27.8
−17	−27.2	3	−16.1	23	−5.0	43	6.1	63	17.2	83	28.3
−16	−26.7	4	−15.6	24	−4.4	44	6.7	64	17.8	84	28.9
−15	−26.1	5	−15.0	25	−3.9	45	7.2	65	18.3	85	29.4
−14	−25.6	6	−14.4	26	−3.3	46	7.8	66	18.9	86	30.0
−13	−25.0	7	−13.9	27	−2.8	47	8.3	67	19.4	87	30.6
−12	−24.4	8	−13.3	28	−2.2	48	8.9	68	20.0	88	31.1
−11	−23.9	9	−12.8	29	−1.7	49	9.4	69	20.6	89	31.7
−10	−23.3	10	−12.2	30	−1.1	50	10.0	70	21.1	90	32.2
−9	−22.8	11	−11.7	31	−0.6	51	10.6	71	21.7	91	32.8
−8	−22.2	12	−11.1	32	0.0	52	11.1	72	22.2	92	33.3
−7	−21.7	13	−10.6	33	+0.6	53	11.7	73	22.8	93	33.9
−6	−21.1	14	−10.0	34	1.1	54	12.2	74	23.3	94	34.4
−5	−20.6	15	−9.4	35	1.7	55	12.8	75	23.9	95	35.0
−4	−20.0	16	−8.9	36	2.2	56	13.3	76	24.4	96	35.6
−3	−19.4	17	−8.3	37	2.8	57	13.9	77	25.0	97	36.1
−2	−18.9	18	−7.8	38	3.3	58	14.4	78	25.6	98	36.7
−1	−18.3	19	−7.2	39	3.9	59	15.0	79	26.1	99	37.2
0	−17.8	20	−6.7	40	4.4	60	15.6	80	26.7	100	37.8

(continued)

Table 19-3 Temperature conversions, Fahrenheit to Celsius: °C = (5/9)°F − 32 (continued)

°F	to °C	°F	to °C	°F	to °C	°F	to °C	°F	to °C
101	38.3	121	49.4	141	60.6	161	71.7	181	82.8	201	93.9
102	38.9	122	50.0	142	61.1	162	72.2	182	83.3	202	94.4
103	39.4	123	50.6	143	61.7	163	72.8	183	83.9	203	95.0
104	40.0	124	51.1	144	62.2	164	73.3	184	84.4	204	95.6
105	40.6	125	51.7	145	62.8	165	73.9	185	85.0	205	96.1
106	41.1	126	52.2	146	63.3	166	74.4	186	85.6	206	96.7
107	41.7	127	52.8	147	63.9	167	75.0	187	86.1	207	97.2
108	42.2	128	53.3	148	64.4	168	75.6	188	86.7	208	97.8
109	42.8	129	53.9	149	65.0	169	76.1	189	87.2	209	98.3
110	43.3	130	54.4	150	65.6	170	76.7	190	87.8	210	98.9
111	43.9	131	55.0	151	66.1	171	77.2	191	88.3	211	99.4
112	44.4	132	55.6	152	66.7	172	77.8	192	88.9	212	100.0
113	45.0	133	56.1	153	67.2	173	78.3	193	89.4	213	100.6
114	45.6	134	56.7	154	67.8	174	78.9	194	90.0	214	101.1
115	46.1	135	57.2	155	68.3	175	79.4	195	90.6	215	101.7
116	46.7	136	57.8	156	68.9	176	80.0	196	91.1	216	102.2
117	47.2	137	58.3	157	69.4	177	80.6	197	91.7	217	102.8
118	47.8	138	58.9	158	70.0	178	81.1	198	92.2	218	103.3
119	48.3	139	59.4	159	70.6	179	81.7	199	92.8	219	103.9
120	48.9	140	60.0	160	71.1	180	82.2	200	93.3	220	104.4
221	105.0	241	116.1	261	127.2	281	138.3	301	149.4	321	160.6
222	105.6	242	116.7	262	127.8	282	138.9	302	150.0	322	161.1

Table 19-3 Temperature conversions, Fahrenheit to Celsius: °C = (5/9)°F − 32 (continued)

°F	to °C	°F	to °C	°F	to °C	°F	to °C
223	106.1	243	117.2	263	128.3	283	139.4	303	150.5	323	161.7
224	106.7	244	117.8	264	138.9	284	140.0	304	151.1	324	162.2
225	107.2	245	118.3	265	129.4	285	140.6	305	151.7	425	162.8
226	107.8	246	118.9	266	130.0	286	141.1	306	152.2	326	163.3
227	108.3	247	119.4	267	130.6	287	141.7	307	152.8	327	163.9
228	108.9	248	120.0	268	131.1	288	142.2	308	153.3	328	164.4
229	109.4	249	120.6	269	131.7	289	142.8	309	153.9	329	165.0
230	110.0	250	121.1	270	132.2	290	143.3	310	154.4	330	165.6
231	110.6	251	121.7	271	132.8	291	143.9	311	155.0	331	166.1
232	111.1	252	122.2	272	133.3	292	144.4	312	155.6	332	166.7
233	111.7	253	122.8	273	133.9	293	145.0	313	156.1	333	167.2
234	112.2	254	123.3	274	134.4	294	145.6	314	156.7	334	167.8
235	112.8	255	123.9	275	135.0	295	146.1	315	157.2	335	168.3
236	113.3	256	124.4	276	135.6	296	146.7	316	157.8	336	168.9
237	113.9	257	125.0	277	136.1	297	147.2	317	158.3	337	169.4
238	114.4	258	125.6	278	136.7	298	147.8	318	158.9	338	170.0
239	115.0	259	126.1	279	137.2	299	148.3	319	159.4	339	170.6
240	115.6	260	126.7	280	137.8	300	148.9	320	160.0	340	171.1

Appendix B
Specifications and Approval of Treatment Chemicals and System Components

Drinking Water Additives

Many chemicals are added to water during the various water treatment processes. Chemical additives are commonly used in the following processes:

- Coagulation and flocculation
- Control of corrosion and scale
- Chemical softening
- Sequestering of iron and manganese
- Chemical precipitation
- pH adjustment
- Disinfection and oxidation
- Algae and aquatic plant control

It is important that the chemicals added to water during treatment do not add any objectionable tastes, odors, or color to the water; otherwise, customers will likely complain. It is of even greater importance that the chemicals, or side effects that they cause, do not create any danger to public health.

Coatings and Equipment in Contact with Water

In addition to chemicals that are added to water for treatment, many products come in direct contact with potable water and could, under some circumstances, cause objectionable contamination. The general categories of products that could cause an adverse effect because of their contact with water are

- pipes, faucets, and other plumbing materials;
- protective materials such as paints, coatings, and linings;
- joining and sealing materials;
- lubricants;
- mechanical devices that contact drinking water; and
- process media, such as filter and ion exchange media.

Objectionable effects that can be caused by construction and maintenance materials include unpleasant tastes and odors, support of microbiological growth, and the liberation of toxic organic and inorganic chemicals.

NSF International Standards and Approval

For many years, the water industry relied on American Water Works Association (AWWA) standards and on approval by the US Environmental Protection Agency (USEPA) of individual products to ensure that harmful chemicals were not unknowingly added to potable water. However, there was no actual testing of the products, and the water treatment industry had to rely on the manufacturers' word that their products did not contain toxic materials.

In recent years, increasing numbers of products have been offered for public water supply use, many of them made with new manufactured chemicals that have not been thoroughly tested. In addition, recent toxicological research has revealed potential adverse health effects due to rather low levels of continuous exposure to many chemicals and substances previously considered safe.

In the early 1980s, it became evident that more exacting standards were needed, as well as more definite assurance that products positively meet safety standards. In view of the growing complexity of testing and approving water treatment chemicals and components, the USEPA awarded a grant in 1985 for the development of private-sector standards and a certification program. The grant was awarded to a consortium of partners. National Science Foundation (NSF) International was designated the responsible lead. Other cooperating organizations were the Association of State Drinking Water Administrators (ASDWA), AWWA, the Water Research Foundation (formerly AwwaRF), and the Conference of State Health and Environmental Managers (COSHEM).

Two standards were developed by these organizations, along with the help of many volunteers from the water supply and manufacturing industries who served on development committees. These standards have now been adopted by the American National Standards Institute (ANSI), so they bear an ANSI designation in addition to the NSF reference.

ANSI/NSF Standard 60 essentially covers treatment chemicals for drinking water. The standard sets up testing procedures for each type of chemical, provides limits on the percentage of the chemical that can safely be added to potable water, and places limitations on any harmful substances that might be present as impurities in the chemical.

ANSI/NSF Standard 61 covers materials that are in contact with water, such as coatings, construction materials, and components used in processing and distributing potable water. The standard sets up testing procedures for each type of product to ensure that they do not unduly contribute to microbiological growth, leach harmful chemicals into the water, or otherwise cause problems or adverse effects on public health.

Manufacturers of chemicals and other products that are sold for the purpose of being added to water or that will be in contact with potable water must now submit samples of their product to NSF International or another qualifying laboratory for testing based on Standards 60 and 61. If a product qualifies, it is then "listed." There are also provisions for periodic retesting and inspection of the manufacturer's processes by the testing laboratory.

The listings of certified products provided by NSF International are used in particular by three groups in the water supply industry.

1. In the design of water treatment facilities, engineers must specify that pipes, paints, caulks, liners, and other products that will be used in construction, as well as the chemicals to be used in the treatment process, are

listed under one of the standards. This ensures in a very simple manner that only appropriate materials will be used. It also provides contractors with specific information on what materials qualify for use, without limiting competition.

2. Individual states and local agencies have the right to impose more stringent requirements or to allow the use of products based on other criteria, but most states have basically agreed to accept NSF International standards. When state authorities approve plans and specifications for the construction of new water systems or improvements to older systems, they will generally specify that all additives, coatings, and components be listed as having been tested for compliance with the standards.

3. Water system operators can best protect themselves and their water systems from customer complaints or possibly even lawsuits by insisting that only listed products be used for everything that is added to or in contact with potable water. Whether it concerns purchasing paint for plant maintenance or taking bids for supplies of chemicals, the manufacturer or representative should be asked to provide proof that the exact product has been tested and is listed.

Copies of the current listing of products approved based on Standards 60 and 61 should be available at state drinking water program offices. A copy of the current listing can also be obtained by contacting the nearest NSF International regional office or the following address:

NSF International
P.O. Box 130140
789 N. Dixboro Road
Ann Arbor, MI 48113-0140
(734) 769-8010
www.nsf.org

AWWA Standards

Since 1908, AWWA has developed and maintained a series of voluntary consensus standards for products and procedures used in the water supply community. These standards cover products such as pipe, valves, and water treatment chemicals. They also cover procedures such as disinfection of storage tanks and design of pipe. As of 2016, AWWA had over 170 standards in existence and many new standards under development. A list of standards currently available may be obtained from AWWA at any time.

AWWA offers and encourages use of its standards by anyone on a voluntary basis. AWWA has no authority to require the use of its standards by any water utility, manufacturer, or other person. Many individuals in the water supply community, however, choose to use AWWA standards. Manufacturers often produce products complying with the provisions of AWWA standards. Water utilities and consulting engineers frequently include the provisions of AWWA standards in their specifications for projects or purchase of products. Regulatory agencies require compliance with AWWA standards as part of their public water supply regulations. All of these uses of AWWA standards can establish a mandatory relationship, for example, between a buyer and a seller, but AWWA is not part of that relationship.

AWWA recognizes that others use its standards extensively in mandatory relationships and takes great care to avoid provisions in the standards that could give one party a disadvantage relative to another. Proprietary products are avoided whenever possible in favor of generic descriptions of functionality or construction. AWWA standards are not intended to describe the highest level of quality available, but rather describe minimum levels of quality and performance expected to provide long and useful service in the water supply community.

While maintaining product and procedural standards, AWWA does not endorse, test, approve, or certify any product. No product is or ever has been AWWA approved. Compliance with AWWA standards is encouraged, and demonstration of such compliance is entirely between the buyer and seller, with no involvement by AWWA.

AWWA standards are developed by balanced committees of persons from the water supply community who serve on a voluntary basis. Product users and producers, as well as those with general interest, are all involved in AWWA committees. Persons from water utilities, manufacturing companies, consulting engineering firms, regulatory agencies, universities, and others gather to provide their expertise in developing the content of the standards. Agreement of such a group is intended to provide standards, and thereby products, that serve the water supply community well.

Appendix C
Other Sources of Information

Many trade associations, publishers, and other groups have information available that can be useful to water system operators. Depending on the organization, they may have product literature, association standards, installation manuals, slide presentations, and video programs available. Some groups will also provide technical advice on their area of expertise.

The following are names and addresses of some of these organizations. Note that this list is not exhaustive. The local state water agency may be contacted for more sources.

American Concrete Pressure Pipe Association
3900 University Dr., Ste. 110
Fairfax, VA 22030-2513
(703) 273-PCCP (7227)
www.acppa.org

American Public Works Association
2345 Grand Blvd., Ste. 700
Kansas City, MO 64108-2625
(816) 472-6100
www.apwa.net

American Water Works Association
6666 W. Quincy Ave.
Denver, CO 80235-3098
(303) 794-7711
www.awwa.org

Association of State Drinking Water Administrators
1401 Wilson Blvd., Ste. 1225
Arlington, VA 22209
(703) 812-9505
www.asdwa.org

The Chlorine Institute Inc.
1300 Wilson Blvd.
Arlington, VA 22209
(703) 741-5760
www.chlorineinstitute.org

Ductile Iron Pipe Research Association
245 Riverchase Parkway East, Ste. O
Birmingham, AL 35244
(205) 402-8700
www.dipra.org

Hydraulic Institute
6 Campus Dr., First Floor North
Parsippany, NJ 07054-4406
(973) 267-9700
www.pumps.org

International Desalination Association
P.O. Box 387
94 Central St., Ste. 200
Topsfield, MA 01983
(978) 887-0410
www.idadesal.org

International Ozone Association
Pan American Group
P.O. Box 28873
Scottsdale, AZ 85255
(480) 529-3787
www.io3a.org

National Association of Water Companies
2001 L Street NW, Ste. 850
Washington, DC 20036
(202) 833-8383
www.nawc.org

National Fire Protection Association
1 Batterymarch Park
Quincy, MA 02169-7471
(617) 770-3000
www.nfpa.org

National Ground Water Association
601 Dempsey Rd.
Westerville, OH 43081
(800) 551-7379
www.ngwa.org

National Lime Association
200 N. Glebe Rd., Ste. 800
Arlington, VA 22203
(703) 243-5463
www.lime.org

National Rural Water Association
2915 S. 13th St.
Duncan, OK 73533
(580) 252-0629
www.nrwa.org

National Small Flows Clearinghouse
West Virginia University/NRCCE
P.O. Box 6064
Morgantown, WV 26506-6064
(304) 293-4191
www.nesc.wvu.edu

North American Society for Trenchless Technology
1655 N. Ft. Myer Dr., Ste. 700
Arlington, VA 22209
(703) 351-5252
www.nastt.org

NSF International
P.O. Box 130140
789 N. Dixboro Rd.
Ann Arbor, MI 48113-0140
(734) 769-8010
www.nsf.org

Plastics Pipe Institute
1825 Connecticut Ave., NW, Ste. 680
Washington, DC 20009
(202) 462-9607
www.plasticpipe.org

Salt Institute
700 N. Fairfax St., Ste. 600
Alexandria, VA 22314-2040
(703) 549-4648
www.saltinstitute.org

Steel Tube Institute of North America
2516 Waukegan Rd., Ste. 172
Glenview, IL 60025
(847) 461-1701
www.steeltubeinstitute.org

Trench Shoring & Shielding Association
A Product Specific Group of the Association of Equipment Manufacturers
6737 W. Washington St., Ste. 2400
Milwaukee, WI 53214-5647
(414) 272-0943
www.aem.org/CBC/ProdSpec/TSSA/

Uni-Bell PVC Pipe Association
2711 LBJ Freeway, Ste. 1000
Dallas, TX 75234
(972) 243-3902
www.uni-bell.org

Valve Manufacturers Association of America
1050 17th St., NW, Ste. 280
Washington, DC 20036-5521
(202) 331-8105
www.vma.org

Water Quality Association
4151 Naperville Rd.
Lisle, IL 60532-3696
(630) 505-0160
www.wqa.org

The following are some published sources of current information on water system equipment:

- *American Water Works Association Sourcebook*, sourcebook.awwa.org (covering the field of water supply and treatment). Published annually by the American Water Works Association, 6666 W. Quincy Ave., Denver, CO 80235.
- *Pollution Equipment News Buyer's Guide* (covering water supply and pollution fields). Published annually by Rimbach Publishing Inc., 8650 Babcock Blvd., Pittsburgh, PA 15237-5821
- *Public Works Manual* (covering the fields of general operations, streets and highways, water supply and treatment, and water pollution control). Hanley Wood Business Media, 8725 W. Higgins Rd., Suite 600, Chicago, IL 60631.

Appendix D
Sample Material Safety Data for Chlorine

PRODUCT NAME: CHLORINE

1. Chemical Product and Company Identification

BOC Gases,
Division of
The BOC Group, Inc.
575 Mountain Avenue
Murray Hill, NJ 07974

BOC Gases
Division of
BOC Canada Limited
5975 Falbourne Street, Unit 2
Mississauga, Ontario L5R 3W6
TELEPHONE NUMBER: (905) 501-1700

TELEPHONE NUMBER: (908) 464-8100
24-HOUR EMERGENCY TELEPHONE NUMBER:
CHEMTREC (800) 424-9300

24-HOUR EMERGENCY TELEPHONE NUMBER:
(905) 501-0802
EMERGENCY RESPONSE PLAN NO: 2-0101

PRODUCT NAME: CHLORINE
CHEMICAL NAME: Chlorine
COMMON NAMES/SYNONYMS: Bertholite, Molecular Chlorine
TDG (Canada) CLASSIFICATION: 2.3 (5.1)
WHMIS CLASSIFICATION: A, D1A, D2B, E, C

PREPARED BY: Loss Control (908)464-8100/(905)501-1700
PREPARATION DATE: 6/1/95
REVIEW DATES: 4/16/02

2. Composition, Information on Ingredients

EXPOSURE LIMITS[1]:

INGREDIENT	% VOLUME	PEL-OSHA[2]	TLV-ACGIH[3]	LD_{50} or LC_{50} Route/Species
Chlorine FORMULA: Cl_2 CAS: 7782-50-5 RTECS #: FO2100000	100.0	1 ppm Ceiling	0.5 ppm TWA 1 ppm STEL	LC_{50}: 293 ppm inhalation/rat (1H)

[1] Refer to individual state or provincial regulations, as applicable, for limits which may be more stringent than those listed here.
[2] As stated in 29 CFR 1910, Subpart Z (revised July 1, 1993)
[3] As stated in the ACGIH 2002 Threshold Limit Values for Chemical Substances and Physical Agents.

OSHA Regulatory Status: This material is classified as hazardous under OSHA regulations.
IDLH: 10 ppm

3. Hazards Identification

EMERGENCY OVERVIEW

Greenish yellow gas with bleach-like choking odor. Corrosive and poison gas. Contact may cause severe irritation or corrosive burns to the eyes, skin and mucous membranes. Inhalation may result in chemical pneumonitis, retention of body fluid in the lungs (pulmonary edema), and respiratory collapse. Nonflammable. Oxidizer. May react violently with reducing agents. Can accelerate combustion and increase risk of fire and explosion in flammable and combustible materials. Contents under pressure. Use and store below 125°F.

Source: BOC Gases, a Division of The BOC Group Inc.

ROUTE OF ENTRY:

Skin Contact	Skin Absorption	Eye Contact	Inhalation	Ingestion
Yes	No	Yes	Yes	No

HEALTH EFFECTS:

Exposure Limits		Irritant		Sensitization	
Yes		Yes		No	
Teratogen		Reproductive Hazard		Mutagen	
No		No		No	
Synergistic Effects					
Other agents that irritate the respiratory system					

Carcinogenicity: -- NTP: No   IARC: No   OSHA: No

EYE EFFECTS:
Corrosive and irritating to the eyes. Contact with the liquid or vapor causes painful burns and ulcerations. Burns to the eyes result in lesions and possible loss of vision.

SKIN EFFECTS:
Corrosive and irritating to the skin and all living tissue. It hydrolyzes very rapidly yielding hydrochloric acid. Skin burns and mucosal irritation are like that from exposure to volatile inorganic acids. Chlorine burns result in severe pain, redness, possible swelling and early necrosis.

INGESTION EFFECTS:
Ingestion is unlikely.

INHALATION EFFECTS:
Corrosive and irritating to the upper and lower respiratory tract and all mucosal tissue. Symptoms include lacrimation, cough, labored breathing, and excessive salivary and sputum formation. Excessive irritation of the lungs causes acute pneumonitis, pulmonary edema, and respiratory collapse which could be fatal. Residual pulmonary malfunction may also occur. Chemical pneumonitis and pulmonary edema may result from exposure to the lower respiratory tract and deep lung.

MEDICAL CONDITIONS AGGRAVATED BY EXPOSURE: May aggravate pre-existing eye, skin, and respiratory conditions.

POTENTIAL ENVIRONMENTAL EFFECTS: Toxic to fish and wildlife. Chlorine is designated as a marine pollutant by DOT. The LC_{50} in the fathead minnow has been cited as 0.1 mg/l/96 H and an LC_{50} of 0.097 mg/L/30 min has been cited for the *Daphnia magna*.

4. First Aid Measures

EYES:
PERSONS WITH POTENTIAL EXPOSURE SHOULD NOT WEAR CONTACT LENSES. Flush contaminated eye(s) with copious quantities of water. Part eyelids to assure complete flushing. Continue for a minimum of 30 minutes. Seek immediate medical attention.

SKIN:
Flush affected area with copious quantities of water while removing contaminated clothing. Seek immediate medical attention.

INGESTION:
None required.

INHALATION:
PROMPT MEDICAL ATTENTION IS MANDATORY IN ALL CASES OF OVEREXPOSURE. RESCUE PERSONNEL SHOULD BE EQUIPPED WITH SELF-CONTAINED BREATHING APPARATUS. Conscious persons should be assisted to an uncontaminated area and inhale fresh air. If breathing is difficult, administer oxygen. Unconscious persons should be moved to an uncontaminated area and given artificial resuscitation and supplemental oxygen. Assure that mucus or vomited material does not obstruct the airway by use of positional drainage. Delayed pulmonary edema may occur. Keep the patient under medical observation for at least 24 hours.

5. Fire Fighting Measures

Conditions of Flammability: Not flammable		
Flash point: None	Method: Not Applicable	Autoignition Temperature: None
LEL(%): None	UEL(%): None	
Hazardous combustion products: None		
Sensitivity to mechanical shock: None		
Sensitivity to static discharge: None		

FIRE AND EXPLOSION HAZARDS:
Strong oxidizer. Most combustible materials burn in chlorine as they do in oxygen producing irritating and poisonous gases. Flame impingment upon steel chlorine container will result in iron/chlorine fire causing rupture of the container. Cylinder may vent rapidly or rupture violently from pressure when involved in a fire situation.

EXTINGUISHING MEDIA:
Use media suitable for surrounding materials. If it can be done without risk, stop the flow of chlorine which is accelerating the fire.

FIRE FIGHTING INSTRUCTIONS:
Firefighters should wear respiratory protection (SCBA) and full turnout or Bunker gear with additional chemical protective clothing to prevent exposure to chlorine. Use water spray to keep fire exposed containers cool. Continue to cool fire exposed cylinders until well after flames are extinguished. Control runoff and isolate discharged material for proper disposal.

6. Accidental Release Measures

Evacuate all personnel from affected area. Deny entry to unauthorized and unprotected individuals. Extinguish all ignition sources. No smoking, sparks, flames, or flares in hazard area. Appropriate protective equipment is essential to prevent exposure (See Section 8). Stop the flow of gas or remove cylinder to outdoor location if this can be done without risk. Ventilate enclosed areas. A leak near incompatible, flammable or combustible materials may create a fire or explosion hazard. Consult a HAZMAT specialist and the appropriate emergency telephone number in Section 1 or your closest BOC location. If leak is in user's equipment, be certain to purge piping with inert gas prior to attempting repairs.

7. Handling and Storage

Electrical classification: Nonhazardous.

Most metals corrode rapidly with wet chlorine. Systems must be kept dry. Lead, gold, tantalum and Hastelloy are most resistant to wet chlorine.

Do not inhale. Prevent contact with skin and eyes. Use only in well-ventilated areas. Valve protection caps must remain in place unless container is secured with valve outlet piped to use point. Do not drag, slide or roll cylinders. Use a suitable hand truck for cylinder movement. Use a pressure reducing regulator when connecting cylinder to lower pressure piping or systems. Do not heat cylinder by any means to increase rate of product from the cylinder. Use a check valve or trap in the discharge line to prevent hazardous back flow into cylinder. Do not insert any object (i.e.: screwdriver) into valve cap openings as this can damage the valve causing leakage.

Protect cylinders from physical damage. Store in cool, dry, well-ventilated areas of non-combustible construction away from heavily trafficked areas and emergency exits. Do not allow the temperature where cylinders are stored to exceed 125°F (52°C). Cylinders should be stored upright and firmly secured to prevent falling or being knocked over. Full & empty cylinders should be segregated. Use a "first in-first out" inventory system to prevent full cylinders from being stored for excessive periods of time. Separate from combustibles, organic, and easily oxidizable materials. Isolate from acetylene, ammonia, hydrogen, hydrocarbons, ether, turpentine, finely divided metals, and other incompatible materials. Outside or detached storage is preferred.

Never carry a compressed gas cylinder or a container of a gas in cryogenic liquid form in an enclosed space such as a car trunk, van or station wagon. A leak can result in a fire, explosion, asphyxiation or a toxic exposure.

For additional storage recommendations, consult Compressed Gas Association's Pamphlet P-1.

8. Exposure Controls, Personal Protection

ENGINEERING CONTROLS:
Hood with forced ventilation may be used for small quantities. Use local exhaust ventilation in combination with enclosed processes as needed to prevent accumulation above the exposure limit.

EYE/FACE PROTECTION:
Gas-tight safety goggles and full faceshield or full-face respirator.

SKIN PROTECTION:
Protective gloves or fully encapsulated vapor protective clothing. (Butyl rubber, neoprene, and Teflon ® provide adequate protection for exposures to chlorine greater than 8 hours.)

RESPIRATORY PROTECTION:
For emergency release use a positive pressure NIOSH approved air-supplying respirator systems (SCBA or airline/escape bottle) using a full-face mask and at a minimum Grade D air.

For normal conditions below fifty times the exposure limit but where engineering can not control exposures below the applicable limits, than appropriately selected air-purifying respirators with full-face mask can be used.

OTHER/GENERAL PROTECTION:
Safety shoes, safety shower, eyewash "fountain"

9. Physical and Chemical Properties

PARAMETER	VALUE	UNITS
Physical state (gas, liquid, solid)	: Gas	
Vapor pressure at 70 °F	: 100.2	psia
Vapor density at STP (Air = 1)	: 2.47	
Evaporation point	: Not Available	
Boiling point	: -29.3	°F
	: -34.1	°C
Freezing point	: -149.8	°F
	: -101	°C
pH	: Not Available	
Specific gravity	: Not Available	
Oil/water partition coefficient	: Not Available	
Solubility (H_2O)	: Very Soluble	
Odor threshold	: Not Available	
Odor and appearance	: Greenish-yellow gas with sharp suffocating odor. Liquid is amber colored.	

10. Stability and Reactivity

STABILITY: Stable

INCOMPATIBLE MATERIALS: Strong oxidizer. Will react with organic and other oxidizable materials. Reacts explosively or forms explosive compounds with many common substances including acetylene, ether, turpentine, ammonia, fuel gas, hydrogen and finely divided metals. Reacts with water to form corrosive acidic solution.

HAZARDOUS DECOMPOSITION PRODUCTS: Hydrochloric acid on contact with water.

HAZARDOUS POLYMERIZATION: Will not occur.

11. Toxicological Information

INHALATION:
Inhalation of chlorine concentrations as low as 1 ppm may cause nose, throat and conjunctiva irritation. Irritation becomes more pronounced at concentrations of 1.3 ppm and above with coughing and labored breathing. Death may occur after a few breaths at 1000 ppm. Delayed effects following high exposure may include bronchitis, edema, and pneumonia.

SKIN AND EYE:
Extremely irritating to the skin, eyes, and mucous membranes. Can cause corrosive burns. May cause corrosion of the teeth. Prolonged exposure to low concentrations may cause chloracne.

CHRONIC:
Repeated contact with low concentrations may cause dermatitis.

OTHER:
Equivocal evidence of carcinogenicity for chlorine was noted in an IARC review and a 2-year drinking water study in F344/N rats and B6C3F1 mice by the NTP. Literature references suggest the possibility of mutagenic and teratogenic effects from hypochlorites (a hydrolysis product of chlorine).

12. Ecological Information

Product does not contain Class I or Class II ozone depleting substances. Chlorine is highly toxic to all forms of aquatic life (See Section 3). There is no potential for bioaccumulation or bioconcentration. Chlorine is designated as a hazardous substance under section 311(b)(2)(A) of the Federal Water Pollution Control Act and further regulated by the Clean Water Act Amendments of 1977 and 1978. Listed as a hazardous air pollutant (HAP) and a marine pollutant. Chlorine is listed as an extremely hazardous substance (EHS) subject to state and local reporting under Section 304 of SARA Title III (EPCRA) with a Threshold Planning Quantity (TPQ) of 100 pounds. The CERCLA reportable quantity (RQ) for chlorine is 10 pounds.

13. Disposal Considerations

Do not attempt to dispose of residual waste or unused quantities. Return in the shipping container PROPERLY LABELED, WITH ANY VALVE OUTLET PLUGS OR CAPS SECURED AND VALVE PROTECTION CAP IN PLACE to BOC Gases or authorized distributor for proper disposal.

14. Transport Information

PARAMETER	United States DOT	Canada TDG
PROPER SHIPPING NAME:	Chlorine	Chlorine
HAZARD CLASS:	2.3 (8)	2.3 (8)
IDENTIFICATION NUMBER:	UN 1017	UN 1017
SHIPPING LABEL:	POISON GAS, CORROSIVE	TOXIC GAS, CORROSIVE

* Effective August 15, 2002

Additional Marking Requirement: "Inhalation Hazard"
    If net weight of product ≥ 10 pounds, the container must be also marked with the letters "RQ".
    "Marine Pollutant" – For vessel transportation the Marine Pollutant Mark shall be placed in association with the hazard warning labels, or in the absence of any labels, in association with the marked proper shipping name.

Additional Shipping Paper Description Requirement: "Poison-Inhalation Hazard, Zone B"
    If net weight of product ≥ 10 pounds, the shipping papers must be also marked with the letters "RQ".
    The words "Marine Pollutant" shall be entered in association with the basic description for a material which is a marine pollutant.

15. Regulatory Information

SARA TITLE III NOTIFICATIONS AND INFORMATION
SARA TITLE III - HAZARD CLASSES:
Acute Health Hazard
Chronic Health Hazard
Fire Hazard
Sudden Release of Pressure Hazard
Reactivity Hazard

SARA TITLE III - SECTION 313 SUPPLIER NOTIFICATION:
This product contains the following toxic chemicals subject to the reporting requirements of section 313 of the Emergency Planning and Community Right-To-Know Act (EPCRA) of 1986 and of 40 CFR 372:

CAS NUMBER	INGREDIENT NAME	PERCENT BY VOLUME
7782-50-5	CHLORINE	100.0

This information must be included on all MSDSs that are copied and distributed for this material.

U.S. TSCA/Canadian DSL: All ingredients are listed on the U.S. Toxic Substances Control Act (TSCA) inventory or exempt from listing and on the Canadian Domestic Substance List (DSL).

California Proposition 65: This product does not contain ingredient(s) known to the State of California to cause cancer or reproductive toxicity.

16. Other Information

NFPA HAZARD CODES	HMIS HAZARD CODES	RATINGS SYSTEM
Health: 4	Health: 3	0 = No Hazard
Flammability: 0	Flammability: 0	1 = Slight Hazard
Instability: 0	Physical Hazard: 2	2 = Moderate Hazard
OXIDIZER		3 = Serious Hazard
		4 = Severe Hazard

ACGIH	American Conference of Governmental Industrial Hygienists
DOT	Department of Transportation
IARC	International Agency for Research on Cancer
NTP	National Toxicology Program
OSHA	Occupational Safety and Health Administration
PEL	Permissible Exposure Limit
SARA	Superfund Amendments and Reauthorization Act
STEL	Short Term Exposure Limit
TDG	Transportation of Dangerous Goods
TLV	Threshold Limit Value
WHMIS	Workplace Hazardous Materials Information System

Compressed gas cylinders shall not be refilled without the express written permission of the owner. Shipment of a compressed gas cylinder which has not been filled by the owner or with his/her (written) consent is a violation of transportation regulations.

DISCLAIMER OF EXPRESSED AND IMPLIED WARRANTIES:
Although reasonable care has been taken in the preparation of this document, we extend no warranties and make no representations as to the accuracy or completeness of the information contained herein, and assume no responsibility regarding the suitability of this information for the user's intended purposes or for the consequences of its use. Each individual should make a determination as to the suitability of the information for their particular purpose(s).

Study Question Answers

Chapter 1 Answers

1. **b.** United States Environmental Protection Agency
2. **c.** based on population served.
3. **c.** 0.080 mg/L
4. **a.** coliform samples.
5. **d.** 0.30 mg/L.
6. **a.** state
7. **d.** To promote routine surveillance of distribution system water quality to search for fecal matter and/or disease-causing bacteria
8. US Environmental Protection Agency
9. Tier 1 is an acute MCL violation requiring public notification within 24 hours. Tier 2 notification must occur within 30 days. Tier 3 requires notification within 1 year.
10. 24 hours
11. Population served
12. 0.015 mg/L
13. 15 or more
14. The state primacy agency
15. 0.015 mg/L

Chapter 2 Answers

1. **d.** 8.3 hr

   First, convert pipe diameter from inches to feet:

   Pipe diameter, ft = (14 in.)(1 ft/12 in.) = 1.167 ft

   Next, find the number of gallons in the pipeline:

   Number of gal = (0.785)(diameter, ft)^2(length, ft)(7.48 gal/ft^3)

   Number of gal = (0.785)(1.167 ft)(1.167 ft)(549 ft)(7.48 gal/ft^3) = 4,390 gal

   Add the pipe and tank volume to get the total number of gallons:

   Pipe and tank volume, gal = 4,390 gal + 2,310,000 gal = 2,314,390 gal

   Then convert mgd to gallons per day:

   Number of gal = (6.72 mgd)(1,000,000) = 6,720,000 gal/day

Using the following equation, solve for the detention time:

$$\text{Equation: Detention time, hr} = \frac{(\text{Total volume})(24 \text{ hr/day})}{\text{Flow, gal/day}}$$

Substitute known values and solve:

$$\frac{(2{,}314{,}390 \text{ gal})(24 \text{ hr/day})}{6{,}720{,}000 \text{ gal/day}} = 8.266 \text{ hr, round to } \mathbf{8.3 \text{ hr}}$$

2. **b.** 180 lb/day

   Set up a ratio and solve for the unknown, x

   $$\frac{x \text{ lb/day}}{16 \text{ cfs}} = \frac{260 \text{ lb/day}}{23 \text{ cfs}}$$

   $$x = \frac{(260 \text{ lb/day})(16 \text{ cfs})}{23 \text{ cfs}} = \mathbf{180 \text{ lb/day}}$$

3. **c.** 1.3 gal sodium hypochlorite

   First, find the volume of the pipe:

   $$\text{Volume, gal} = (0.785)(\text{diameter, ft})^2(\text{length, ft})(7.48 \text{ gal/ft}^3)$$
   $$\text{Volume} = (0.785)(1.5)(1.5)(283 \text{ ft})(7.48 \text{ gal/ft}^3) = 3{,}739 \text{ gal}$$

   Next, find the number of million gallons (mil gal):

   $$\text{mil gal} = (3{,}739 \text{ gal})(1 \text{ M}/1{,}000{,}000) = 0.003739 \text{ mil gal}$$

   Then use the "pound" equation:

   $$\text{Sodium hypochlorite solution, lb} = \frac{(\text{mil gal})(\text{Dosage, mg/L})(8.34 \text{ lb/gal})}{\% \text{ available chlorine} \div 100\%}$$

   $$= \frac{(0.003739 \text{ mil gal})(50.00 \text{ mg/L})(8.34 \text{ lb/gal})}{12.1\% \text{ available chlorine} \div 100\%}$$

   $$= 12.89 \text{ lb}$$

   Lastly, calculate the number of gallons of sodium hypochlorite:

   $$\text{Sodium hypochlorite, gal} = 12.89 \text{ lb} \div 9.92 \text{ lb/gal} =$$
   $$1.299 \text{ gal, rounded to } \mathbf{1.3 \text{ gal}}$$

4. **b.** 17.8 hr

   First, find the diameter:

   $$\text{Diameter, ft} = 2(\text{radius}) = 2(60.0 \text{ ft}) = 120 \text{ ft}$$

   Then, determine the volume of water in the storage tank:

   $$\text{Volume, gal} = (0.785)(\text{diameter, ft})^2(\text{depth, ft})(7.48 \text{ gal/ft}^3)$$

   $$\text{Average Tank Volume, gal} = (0.785)(120 \text{ ft})(120 \text{ ft})(25.5 \text{ ft})(7.48 \text{ gal/ft}^3)$$
   $$= 2{,}156{,}125 \text{ gal}$$

Next, convert mgd to gallons per day:

Flow through tank, gal/day = (2.91 mgd)(1,000,000 gal) = 2,910,000 gal/day

Next, solve for the detention:

Detention time, hr = [(tank volume)(24 hr/day)] ÷ (flow, gal/day)

Substitute known values and solve:

Detention time, hr = [(2,156,125 gal)(24 hr/day)] ÷ (2,910,000 gal/day) = 17.8 hr

5. d. 54 mg/L chlorine

First, convert the diameter of the pipeline from inches to feet:

Number of feet = 24.0 in. ÷ 12 in./ft = 2.0 ft

Next, find the number of gallons by determining the volume of the pipeline:

Volume of pipe, gal = (0.785)(diameter, ft)^2(length, ft)

Volume of pipe, gal = (0.785)(2.0 ft)(2.0 ft)(427 ft)(7.48 gal/ft^3) = 10,029 gal

Then convert number of gallons to mil gal:

Number of mil gal = 10,029 gal ÷ 1,000,000 = 0.010029 mil gal

Finally, calculate the dosage by rearranging the "pounds" equation:

$$\text{Calcium hypochlorite, lb/day} = \frac{(\text{mgd})(\text{Dosage, mg/L})(8.34 \text{ lb/gal})(100\%)}{\% \text{ available chlorine}}$$

Rearrange the equation and drop the day on each side of the equation, as it is not needed:

$$\text{Dosage, mg/L} = \frac{(\text{calcium hypochlorite, lb})(65.0\% \text{ available chlorine})}{(\text{mil gal})(8.34 \text{ lb/gal})(100\% \text{ calcium hypochlorite})}$$

$$= \frac{(7.0)(65.0\%)}{(0.010029)(8.34)(100\%)}$$

= 54.4 mg/L, round to **54 mg/L**

6. b. 142 gpm

$$\text{Equation: Well yield, gpm} = \frac{\text{gallons produced}}{\text{test duration, min}}$$

$$= \frac{2,840 \text{ gal}}{20 \text{ min}} = \textbf{142 gpm}$$

7. c. 8,170 ft^2

Equation: Area = πr^2, where π = 3.14

First find the radius: Radius = diameter/2 = 102/2 = 51 ft

Area of tank = (3.14)(51 ft)(51 ft) = 8,167.14 ft^2, rounded to **8,170 ft^2**

8. **a.** 11.6 psi

   Equation: First subtract water level from the level in question, 1.85 feet.

   Number of feet in question = 28.7 ft − 1.85 ft = 26.85 ft

   Pressure, psi = (26.85 ft)(0.433 psi/ft) = 11.626 psi, rounded to **11.6 psi**

9. **c.** 15,400 gal

   First convert the diameter from inches to feet:

   $$\text{Number of feet} = \frac{18.0 \text{ in.}}{12 \text{ in./ft}} = 1.50 \text{ ft}$$

   Next, calculate the volume:

   $$\begin{aligned}\text{Pipe volume, gal} &= (0.785)(\text{diameter, ft})^2(\text{length, ft})(7.48 \text{ gal/ft}^3) \\ &= (0.785)(1.50 \text{ ft})(1.50 \text{ ft})(1{,}165 \text{ ft})(7.48 \text{ gal/ft}^3) \\ &= 15{,}391 \text{ gal, rounded to } \mathbf{15{,}400 \text{ gal}}\end{aligned}$$

10. **a.** 3.0°C

    Equation: °C = (°F − 32) × 5/9

    First: 37.4 − 32 = 5.4

    Then: 5.4 × 5 = 270 ÷ 9 = 3.0 °C

11. **a.** 410

    $$\text{Equation: 100\% number} = \frac{(\text{number given})(100\%)}{\text{percent of given number}}$$

    $$= \frac{(288)(100\%)}{70.3\%} = 409.67, \text{ round to } \mathbf{410}$$

12. Conversion table
13. 7.48
14. 5
15. $$\text{ADF} = \frac{\text{sum of all daily flows}}{\text{total number of daily flows used}}$$

16. Multiply by 100, or move the decimal two places to the right.

Chapter 3 Answers

1. **c.** feet.
2. **a.** a map.
3. **a.** Early morning
4. **d.** adequate fire flow at an appropriate pressure.
5. **c.** 50 to 75 psi (345 to 517 kPa).
6. **b.** commercial districts
7. **b.** reasonable sharing."
8. Summer

9. Tree system
10. Shutoff valves

Chapter 4 Answers

1. **c.** feet.
2. **d.** 58 psi

   Equation: Pressure head, ft = (Pressure, psi)(2.31 ft/psi)

   Rearrange to solve for pressure in psi:

   Pressure, psi = (Pressure head, ft) ÷ (2.31 ft/psi)

   Pressure, psi = 134 ft ÷ 2.31 ft/psi = **58 psi**

3. **a.** 820 gal

   First, convert the flow in ft³/min to gallons per minute (gpm):

   gpm = (5.5 ft³/min)(7.48 gal/ft³) = 41.14 gpm

   Then determine the number of gallons that flowed through the fire hydrant:

   Gallons = (41.14 gpm)(20 min) = 822.8 gal, rounded to **820 gal**

4. **c.** 30,355 gal/day

   $$\text{Equation: Pumped, gal/day} = \frac{(\text{last read of gallons pumped} - \text{first read of gallons pumped})}{\text{number of days}}$$

   $$= \frac{(72{,}487{,}008 \text{ gal} - 71{,}576{,}344 \text{ gal})}{30 \text{ days}}$$

   $$= \frac{910{,}664 \text{ gal}}{30 \text{ days}}$$

   = 30,355.467, round to **30,355 gal/day**

5. **a.** divided by the cross-sectional area of the pipe.
6. Gauge pressure
7. Head is expressed in units of foot-pounds per pound, written

   $$\text{head} = \frac{\text{ft-lb}}{\text{lb}}$$

8. Head
9. Elevation head, velocity head, and pressure head

Chapter 5 Answers

1. **a.** cathodic protection.
2. **d.** roughness that retards flow due to friction.

3. **b.** Push-on joint
4. **b.** Water hammer
5. **c.** Riveted joint
6. External load
7. Permeation
8. Reinforced concrete cylinder pipe

Chapter 6 Answers

1. **b.** 3.0 feet
2. **c.** Not having the joint completely clean
3. **a.** moisture.
4. **b.** 1.0 : 1.0
5. **b.** Always store gaskets in a clean, secure location until they are needed.
6. **b.** 18 in. (0.45 m)
7. Truck delivery
8. Excavation
9. Sheeting

Chapter 7 Answers

1. **a.** 6
2. **d.** Sifting
3. **b.** 5-psi (34-kPa)
4. **a.** The stream method
5. **c.** 5 ft (1.5 m)
6. 25 mg/L
7. 6–12 in. (150–300 mm), 12–24 in. (300–610 mm).
8. The amount is given in AWWA standards and manuals.
9. The pipeline must be flushed of chlorinated water and refilled with water from the system.
10. Jetting

Chapter 8 Answers

1. **a.** service line.
2. **c.** a compressed beveled gasket.
3. **c.** cold climates.
4. **c.** In areas where there is deep frost, meter pits are much more expensive to construct.
5. **a.** the meter is equipped with a remote reading device.
6. Direct connection to brass and other metal fittings
7. Turning off service for repair or as the result of nonpayment
8. ¾-in. (19-mm)
9. Corporation stop
10. At an angle of about 45 degrees down from the top of the pipe

Chapter 9 Answers

1. c. Blowoff valve
2. b. Globe valve
3. a. Plug valve
4. c. check valve.
5. d. Needle valve
6. c. Gate valve
7. c. Globe valve
8. d. Air-and-vacuum relief valve
9. Any five of the following types of valve: gate, globe, pressure-relief, air-and-vacuum relief, diaphragm, pinch, rotary, butterfly, check
10. Isolation valve
11. Pressure-reducing valves and altitude valves
12. Electric, hydraulic, and pneumatic
13. Three

Chapter 10 Answers

1. a. 6 in. (150 mm)
2. d. 2.5-in. (64-mm); 4.5-in. (114-mm)
3. c. is generally discouraged but may be authorized in a controlled context.
4. a. Flow hydrant
5. d. 2 in. (50 mm)
6. Dry-barrel hydrant
7. Turning off the hydrant for repair or maintenance
8. To prevent operation using a standard socket wrench
9. Flush hydrant
10. Green

Chapter 11 Answers

1. d. It will be the same in all three tanks.
2. b. 6 hours
3. d. 24 hours
4. d. Operating storage
5. c. Federal Aviation Administration
6. More efficient use of energy and lower operating costs
7. Standpipe
8. Riser
9. Reservoir

Chapter 12 Answers

1. b. Ohm
2. a. air

3. **d.** RTUs, communications, master station, and HMI
4. **a.** 4 to 20 mA DC
5. **c.** an analog (uses a needle) meter.
6. **c.** pH monitor
7. Information provided by sensors
8. Thermocouples and thermistors
9. Direct manual control
10. Feedforward proportional control

Chapter 13 Answers

1. **b.** Squirrel-cage induction motor
2. **a.** larger than fractional horsepower.
3. **a.** Synchronous motor
4. **c.** soon after starting the engine.
5. 104°F (40°C)
6. Three-phase motors
7. Power emergency generators for pump stations and water treatment plants
8. Exercise under load on a predetermined schedule
9. Reduced-voltage controller

Chapter 14 Answers

1. **b.** impeller.
2. **c.** Mechanical seal
3. **c.** Seals
4. **a.** Packing gland
5. **d.** Stuffing box
6. **c.** Gate valve
7. **d.** wear and deteriorate with normal use.
8. **d.** To prevent cavitation from occurring
9. Any three of the following: high initial cost, high repair costs, the need to lubricate support bearings located within the casing, inability to pump water containing any suspended matter, an efficiency that is at best limited to a very narrow range of discharge flow and head conditions
10. Vertical turbine pump
11. Rotary pump
12. Wear rings
13. Either packing rings or mechanical seals

Chapter 15 Answers

1. Positive-displacement meter
2. Low friction loss, accurate, low maintenance
3. When flow is dirty or corrosive

4. Multiple meters can be tested simultaneously.
5. **d.** Inverse
6. **b.** underregister
7. **c.** strainer
8. **b.** Detector-check meter
9. **d.** compound meter

Chapter 16 Answers

1. **c.** has a persistent residual.
2. **b.** 2.5
3. **c.** Coliform test
4. **b.** 5–20%
5. **d.** somewhere in the distribution system.
6. **b.** 150 lb (68 kg)
7. To maintain a more consistent chlorine residual throughout the system
8. A known concentration of disinfectant
9. Chlorine residual
10. Injector
11. Hypochlorination

Chapter 17 Answers

1. **a.** slightly scale forming.
2. **d.** maintain adequate pressure and deliver adequate flow.
3. **a.** not changed
4. **a.** Fire flow requirements
5. **c.** Staying under budget to allow for design modifications
6. Finished water storage facilities
7. A computer simulation of the behavior of physical facilities and water use within the distribution system
8. May form dead-end boundaries
9. Inspection, maintenance, monitoring, and mixing practices
10. Lead and Copper Rule

Chapter 18 Answers

1. **c.** sodium thiosulfate.
2. **a.** 100 mL.
3. **b.** 15 minutes.
4. **a.** faucet without threads.
5. **a.** DPD colorimetric method
6. **b.** grab sample
7. Continuous

8. Raw-water supply, treatment plant, and distribution system
9. USEPA
10. Reject the samples

Chapter 19 Answers

1. **d.** air gap, RPZ, DCVA, and VB.
2. **c.** Moderately likely
3. **d.** Highly likely
4. **d.** Reduced pressure zone backflow preventer
5. **b.** Maintaining complete records
6. Air gap
7. Annually
8. Any connection between the potable water supply and a source of contamination
9. Examples of correct answers include cooling towers, boilers, service sinks.
10. Examples of correct answers include high-pressure boiler, pressurized chemical storage tank, and hot-water recirculation system.
11. Potential cross-connection

Chapter 20 Answers

1. **d.** Sectional maps
2. **b.** 500–1,000 feet to 1 inch.
3. **a.** 50–100 feet to 1 inch.
4. **b.** as large as possible
5. **d.** Elevation
6. Comprehensive map
7. Sectional map
8. Automated mapping/facility management/geographic information system (AM/FM/GIS)

Chapter 21 Answers

1. **a.** 6
2. **b.** 13
3. **d.** All water utility personnel
4. **b.** A hardhat
5. **c.** use a mechanical device to assist in lifting heavy objects.
6. **a.** Earthquake
7. Back injury
8. Yes. It may contain a hazardous atmosphere.
9. All excavations at least 5 ft (1.5 m) deep
10. Calcium hypochlorite, liquid chlorine, and sodium hypochlorite

Chapter 22 Answers

1. Good communications, caring, and courtesy
2. Tell the customer that they don't know but will find out and get back to them.
3. Notify them with enough time so they can prepare.
4. **b.** Water distribution personnel
5. **c.** Every customer contact
6. **d.** property must be restored to its original state before workers leave the site.
7. **d.** to coincide with low-water-use hours.
8. **a.** Don't do it.

References

Abbaszadegan, M. and A. Alum. 2004. Microbiological Contaminants and Threats of Concern. In *Water Supply Systems Security*, ed. L. W. Mays. New York: McGraw-Hill.

Fire Hydrants: Installation, Field Testing, and Maintenance. AWWA Manual M17. Denver, CO: American Water Works Association (AWWA).

Mays, L. W., ed. 2004. *Water Supply Systems Security*. New York: McGraw-Hill.

Rice, E. W., Baird, R. B., and L. S. Eaton, eds. 2012. *Standard Methods for the Examination of Water and Wastewater*. American Public Health Association, AWWA, and Water Environment Federation.

Risk and Resilience Management of Water and Wastewater Systems. AWWA Standard J100. Denver, CO: AWWA.

US Environmental Protection Agency. April 1999. *Guidance Manual for Conducting Sanitary Surveys of Public Water Systems: Surface Water and Ground Water Under the Direct Influence (GWUDI)*. Washington DC: USEPA.

US Environmental Protection Agency. 2003. *Long Term 2 Enhanced Surface Water Treatment Rule: Toolbox Guidance Manual*. EPA 815-D-03-009. Washington DC: USEPA.

Water Treatment Plant Operation and Management. ANSI/AWWA Standard G100. Denver, CO: AWWA.

Ysusi, M. A. 2000. System Design: An Overview. In *Water Distribution Systems Handbook*, ed. by L. W. Mays. New York: McGraw-Hill.

Glossary

absolute ownership A water rights term referring to water that is completely owned by one person.

absolute pressure The total pressure in a system, including both the pressure of water and the pressure of the atmosphere (about 14.7 psi, at sea level). Compare with *gauge pressure*.

actual cross-connection Any arrangement of pipes, fittings, or devices that connects a potable water supply directly to a nonpotable source at all times. Also known as a *direct cross-connection*.

actuator A device, usually electrically or pneumatically powered, that is used to operate valves.

air-and-vacuum relief valve A dual-function air valve that (1) permits entrance of air into a pipe being emptied, thus preventing a vacuum, and (2) allows air to escape in a pipe while being filled or under pressure.

air gap In plumbing, the unobstructed vertical distance through the free atmosphere between (1) the lowest opening from any pipe or outlet supplying water to a tank, plumbing fixture, or other container and (2) the overflow rim of that container.

air-relief valve An air valve placed at a high point in a pipeline to release air automatically, thereby preventing air binding and pressure buildup.

alternating current (AC) Electrical current that reverses its direction in a periodic manner.

altitude valve A valve that automatically shuts off water flow when the water level in an elevated tank reaches a preset elevation, then opens again when the pressure on the system side is less than that on the tank side.

appropriation–permit system A water use system in which permits to use water are regulated so that overdraft cannot occur.

arterial-loop system A distribution system layout involving a complete loop of arterial mains (sometimes called trunk mains or feeders) around the area being served, with branch mains projecting inward. Such a system minimizes dead ends.

atmospheric vacuum breaker A mechanical device consisting of a float check valve and an air-inlet port designed to prevent backsiphonage.

automated mapping/facility management/geographic information system (AM/FM/GIS) A computerized system for collecting, storing, and analyzing water system components for which geographic location is an important characteristic.

automatic control A system in which equipment is controlled entirely by machines or computers, without human intervention, under normal conditions.

average daily flow (ADF) A measurement of the amount of water treated by a plant each day. It is the average of the actual daily flows that occur within a period of time, such as a week, a month, or a year. Mathematically, it is the sum of all daily flows divided by the total number of daily flows used.

backfill (1) The operation of refilling an excavation, such as a trench, after the pipeline or other structure has been placed into the excavation. (2) The material used to fill the excavation in the process of backfilling.

backflow A hydraulic condition, caused by a difference in pressures, in which nonpotable water or other fluids flow into a potable water system.

backpressure A condition in which a pump, boiler, or other equipment produces a pressure greater than the water supply pressure.

backsiphonage A condition in which the pressure in the distribution system is less than atmospheric pressure, which allows contamination to enter a water system through a cross-connection.

ball-and-socket joint A special-purpose pipe joint that provides for a large deflection (up to 15 degrees) and is positively connected so it won't come apart.

ball valve A valve consisting of a ball resting in a cylindrical seat. A hole is bored through the ball to allow water to flow when the valve is open. When the ball is rotated 90°, the valve is closed.

base The inlet structure of a fire hydrant. An elbow-shaped piece that is usually constructed as a gray cast-iron casting. Also known as the *shoe*, *inlet*, *elbow*, or *foot piece*.

bearing Antifriction device used to support and guide pump and motor shafts.

beneficial use A water rights term indicating that the water is being used for good purposes.

bonnet The top cover or closure on the hydrant upper section. It is removable for the purpose of repairing or replacing the internal parts of the hydrant.

breakaway hydrant A two-part, dry-barrel post hydrant with a coupling or other device joining the upper and lower sections. The coupling and barrel are designed to break cleanly when the hydrant is struck by a vehicle, preventing water loss and allowing easy repair.

breakpoint The point at which the chlorine dosage has satisfied the chlorine demand.

bubbler tube A level-sensing device that forces a constant volume of air into the liquid for which the level is being measured.

butterfly valve A valve in which a disc rotates on a shaft as the valve opens or closes. In the fully open position, the disc is parallel to the axis of the pipe.

bypass valve A small valve installed in parallel with a larger valve. Used to equalize the pressure on both sides of the disc of the larger valve before the larger valve is opened.

C × T value The product of the residual disinfectant concentration C, in milligrams per liter, and the corresponding disinfectant contact time T, in minutes. Minimum $C \times T$ values are specified by the Surface Water Treatment Rule as a means of ensuring adequate kill or inactivation of pathogenic microorganisms in water.

cathodic protection An electrical system for preventing corrosion to metals, particularly metallic pipe and tanks.

cavitation A condition that can occur when pumps are run too fast or water is forced to change direction quickly. During cavitation, a partial vacuum forms near the pipe wall or impeller blade, causing potentially rapid pitting of the metal.

centrifugal pump A pump consisting of an impeller on a rotating shaft enclosed by a casing that has suction and discharge connections. The spinning impeller throws water outward at high velocity, and the casing shape converts this high velocity to a high pressure.

chain of custody A written record of the sample's history from the time of collection to the time of analysis and subsequent disposal.

check valve A valve designed to open in the direction of normal flow and close with reversal of flow. An approved check valve has substantial construction and suitable materials, is positive in closing, and permits no leakage in a direction opposite to normal flow.

chlorination The process of disinfecting water through the controlled use of chlorine; usually accomplished by adding gaseous chlorine, liquid sodium hypochlorite, or solid calcium hypochlorite.

chlorinator Any device that is used to add chlorine to water.

chlorine cylinder A container that holds 150 lb (68 kg) of chlorine and has a total filled weight of 250–285 lb (110–130 kg).

chlorine demand The amount of chlorine that will combine with organic and inorganic materials to form chlorine compounds when added to water; once the demand is satisfied, additional chlorine will not combine with the organic and inorganic materials.

chlorine evaporator A heating device used to convert liquid chlorine to chlorine gas.

chlorine residual The total of all the compounds with disinfecting properties, plus any remaining uncombined chlorine.

closed-loop control A form of computerized control that automatically adjusts for changing conditions to produce the correct output, so that operator intervention may be minimized.

commutator A device that is part of the rotor of certain designs of motors and generators. The motor unit's brushes rub against the surface of the spinning commutator, allowing current to be transferred between the rotor and the external circuits.

composite sample A series of individual or grab samples taken at different times from the same sampling point and mixed together.

compound meter A water meter consisting of two single meters of different capacities and a regulating valve that automatically diverts all or part of the flow from one meter to the other. The valve senses flow rate and shifts the flow to the meter that can most accurately measure it.

comprehensive map A map that provides a clear picture of the entire distribution system. It usually indicates the locations of water mains, fire hydrants, valves, reservoirs and tanks, pump stations, pressure zone limits, and closed valves at pressure zone limits.

continuous feed method A method of disinfecting new or repaired mains in which chlorine is continuously added to the water being used to fill the pipe, so that a constant concentration can be maintained.

control system A means of controlling of equipment in a variety of ways, from completely manual to completely automatic.

control terminal unit (CTU) The receiving device in a digital telemetry system.

corporation stop A valve for joining a service line to a street water main. Cannot be operated from the surface. Also known as *corporation cock*.

correlative rights A rule that contends that the *overlying use* rule is not absolute but is related to the rights of other overlying users. This rule is used when there is not enough water to satisfy all overlying uses.

coupling A device that connects the pump shaft to the motor shaft.

coupon In tapping, the section of the main cut out by the drilling machine.

cross-connection Any arrangement of pipes, fittings, fixtures, or devices that connects a nonpotable system to a potable water system.

curb box A cylinder placed around the curb stop and extending to the ground surface to allow access to the valve.

curb stop A shutoff valve attached to a water service line from a water main to a customer's premises, usually placed near the customer's property line. It may be operated by a valve key to start or stop flow to the water supply line. Also known as *curb valve*.

current (1) The flow rate of electricity, measured in amperes. (2) In telemetry, a signal whose amperage varies as the parameter being measured varies.

current meter A device for determining flow rate by measuring the velocity of moving water. Turbine meters, propeller meters, and multijet meters are common types. Compare with *positive-displacement meter*.

cutting-in valve A specially designed gate valve used with a sleeve that allows the valve to be placed in an existing main.

D'Arsonval meter An electrical measuring device, consisting of an indicator needle attached to a coil of wire, placed within the field of a permanent magnet. The needle moves when an electric current is passed through the coil.

detector-check meter A meter that measures daily flow but allows emergency flow to bypass the meter. Consists of a weight-loaded check valve in the main line that remains closed under normal usage and a bypass around the valve containing a positive-displacement meter.

diffuser (1) A section of a perforated pipe or porous plates used to inject a gas, such as carbon dioxide or air, under pressure into water. (2) A type of pump.

direct current (DC) Current that flows continuously in one direction.

disinfection The water treatment process that kills disease-causing organisms in water, usually by the addition of chlorine.

dissolved oxygen (DO) The oxygen dissolved in water, wastewater, or other liquid, usually expressed in milligrams per liter, parts per million, or percent of saturation.

distribution main Any pipe in the distribution system other than a service line.

double check valve backflow preventer A mechanical designed basically the same way as the reduced pressure zone backflow preventer but without the relief valve.

double-suction pump A centrifugal pump in which the water enters from both sides of the impeller. Also known as a *split-case pump*.

dry-top hydrant A dry-barrel hydrant in which the threaded end of the main rod and the revolving or operating nut is sealed from water in the barrel when the main valve of the hydrant is in use.

dry tap A connection made to a main that is empty. Compare with *wet tap*.

dry-barrel hydrant A hydrant for which the main valve is located in the base. The barrel is pressurized with water only when the main valve is opened. When the main valve is closed, the barrel drains. This type of hydrant is especially appropriate for use in areas where freezing weather occurs.

duplexing A means by which an operator sends control signals back to the site of a transmitting sensor using a single transmission line. See *half-duplex*, *full duplex*, or *simplex*.

dynamic pressure Pressure that exists in water as moving energy.

electromagnetics The study of the combined effects of electricity and magnetism.

elevated storage In any distribution system, storage of water in a tank supported on a tower above the surface of the ground.

elevated tank A water distribution storage tank that is raised above the ground and supported by posts or columns.

elevation head The energy possessed per unit weight of a fluid because of its elevation above some reference point (called the "reference datum"). Elevation head is also called position head or potential head.

emergency storage Storage volume reserved for catastrophic situations, such as a supply line break or pump-station failure.

energy grade line (EGL) (Sometimes called "energy-gradient line" or "energy line.") A line joining the elevations of the energy heads; a line drawn above the hydraulic grade line by a distance equivalent to the velocity head of the flowing water at each section along a stream, channel, or conduit.

external load Any load placed on the outside of the pipe from backfill, traffic, or other sources. Also known as *superimposed load*.

feedforward proportional control A control system that measures a variable and adjusts the equipment proportionally.

fire demand The required fire flow and the duration for which it is needed, usually expressed in gallons (or liters) per minute for a certain number of hours. Also used to denote the total quantity of water needed to deliver the required fire flow for a specified number of hours.

fire flow The rate of flow, usually measured in gallons per minute (gpm) or liters per minute (L/min), that can be delivered from a water distribution system at a specified residual pressure for fire fighting. When delivery is to fire department pumpers, the specified residual pressure is generally 20 psi (140 kPa).

fire hydrant A device connected to a water main and provided with the necessary valves and outlet nozzles to which a fire hose may be attached. The primary purpose of a fire hydrant is to fight fires, but it is also used for washing down streets, filling water-tank trucks, and flushing out water mains. Sometimes called a *fire plug*.

flanged joint A pipe joint that consists of two machined surfaces that are tightly bolted together with a gasket between them. They are primarily used in exposed locations where rigidity, self-restraint, and tightness are required, such as inside treatment plants and pump stations.

flexural strength The ability of a material to bend (flex) without breaking.

flow-proportional composite sample A composite sample in which individual sample volumes are proportional to the flow rate at the time of sampling.

flush hydrant A fire hydrant with the entire barrel and head below ground elevation. The head, with operating nut and outlet nozzles, is encased in a box with a cover that is flush with the ground line. Usually a dry-barrel hydrant.

foot valve A check valve placed in the bottom of the suction pipe of a pump, which opens to allow water to enter the suction pipe but closes to prevent water from passing out of it at the bottom end.

full duplex Capable of sending and receiving data at the same time. Compare with *half-duplex* and *simplex*.

galvanic corrosion A form of localized corrosion caused by the connection of two different metals in an electrolyte, such as water.

gate valve A valve in which the closing element consists of a disc that slides across an opening to stop the flow of water.

gauge pressure The water pressure as measured by a gauge. Gauge pressure is not the total pressure. Total water pressure (absolute pressure) also includes the atmospheric pressure (about 14.7 psi at sea level) exerted on the water. However, because atmospheric pressure is exerted everywhere (against the outside of the main as well as the inside, for example), it is generally not written into water system calculations. Gauge pressure in pounds per square inch is expressed as "psig."

globe valve A valve having a round, ball-like shell and horizontal disc.

grab sample A single water sample collected at one time from a single point.

grid system A distribution system layout in which all ends of the mains are connected to eliminate dead ends.

ground-level tank In a distribution system, storage of water in a tank whose bottom is at or below the surface of the ground.

groundwater system A water system using wells, springs, or infiltration galleries as its source of supply.

half-duplex Capable of sending or receiving data but not both at the same time. Compare with *full duplex* and *simplex*.

Hazen–Williams formula A formula for the velocity of flow in a pipe, expressed as

$$V = 1.318 C R^{0.63} S_f^{0.54}$$

where V is the flow velocity in feet per second, C is the Hazen–Williams coefficient, R is the hydraulic radius in ft, and S_f is the friction slope in feet per feet.

head (1) A measure of the energy possessed by water at a given location in the water system, expressed in feet. (2) A measure of the pressure or force exerted by water, expressed in feet.

horizontal valve A gate that valve that is designed such that the valve-operating mechanism does not have to lift the weight of the gate to open the valve.

hydraulic grade line (HGL) A line (hydraulic profile) indicating the piezometric level of water at all points along a conduit, open channel, or stream. In an open channel, the HGL is the free water surface.

hydraulics The study of fluids in motion or under pressure.

hydropneumatic system A system using an airtight tank in which air is compressed over water (separated from the air by a flexible diaphragm). The air imparts pressure to water in the tank and the attached distribution pipelines.

hydrostatic pressure The pressure exerted by water at rest (for example, in a nonflowing pipeline).

hypochlorination Chlorination using solutions of calcium hypochlorite or sodium hypochlorite.

impeller The rotating set of vanes that forces water through a pump.

indicator The part of an instrument that displays information about a system being monitored. Generally either an analog or digital display.

injector The portion of a chlorination system that feeds the chlorine solution into a pipe under pressure.

in-plant piping system The network of pipes in a particular facility, such as a water treatment plant, that carry the water or wastes for that facility.

inserting valve A shutoff valve that can be inserted by special apparatus into a pipeline while the line is in service under pressure.

internal pressure The hydrostatic pressure within a pipe.

isolation valve A valve installed in a pipeline to shut off flow in a portion of the pipe, for the purpose of inspection or repair. Such valves are usually installed in the mainlines.

lantern ring A perforated ring placed around the pump shaft in the stuffing box. Water from the pump discharge is piped to the lantern ring so that it will form a liquid seal around the shaft and lubricate the packing.

lower barrel The section of a hydrant that carries the water flow between the base and the upper section. Usually buried in the ground with the connection to the upper section approximately 2 in. (50 mm) above ground line.

lower section The part of a dry-barrel hydrant that includes the lower barrel, the main valve assembly, and the base.

magnetic meter A flow-measuring device in which the movement of water induces an electrical current proportional to the rate of flow.

main valve In a dry-barrel hydrant, the valve in the hydrant's base that is used to pressurize the hydrant barrel, allowing water to flow from any open outlet nozzle.

manual control A type of system control in which personnel manually operate the switches and levers to control equipment from the physical location of the equipment.

maximum contaminant level (MCL) The maximum permissible level of a contaminant in water as specified in the regulations of the Safe Drinking Water Act.

maximum contaminant level goal (MCLG) Nonenforceable health-based goals published along with the promulgation of an MCL. Originally called *recommended maximum contaminant levels (RMCLs)*.

mechanical joint A pipe joint for ductile-iron pipe that uses bolts, flanges, and a special gasket.

mechanical seal A seal placed on the pump shaft to prevent water from leaking from the pump along the shaft. Also prevents air from entering the pump. Mechanical seals are an alternative to packing rings.

meter box A pit-like enclosure that protects water meters installed outside of buildings and allows access for reading the meter. Also known as *meter pit*.

multijet meter A type of current meter in which a vertically mounted turbine wheel is spun by jets of water from several ports around the wheel.

multiplexing The use of a single wire or channel to carry the information for several instruments or controls.

needle valve A valve that is similar to a globe valve except that a tapered metal shaft fits into a metal seat when the valve is closed; available only in small sizes and are primarily used for precise throttling of flow.

nonpermit-required space A space defined by the Occupational Safety and Health Administration as not having any of the risks associated with a *permit-required space*.

nonrising-stem valve A gate valve in which the valve stem does not move up or down as it is rotated.

nutating-disk meter A type of positive-displacement meter that uses a hard rubber disc that wobbles (rotates) in proportion to the volume of water flowing through the meter. Also known as a *wobble meter*.

on–off differential control A mode of controlling equipment in which the equipment is turned fully on when a measured parameter reaches a preset value, then turned fully off when it returns to another preset value.

operating nut A nut, usually pentagonal or square, rotated with a wrench to open or close a valve or hydrant valve. May be a single component or it may be combined with a weather shield.

operating storage A tank supplying a given area and capable of storing water during hours of low demand, for use when demands exceed the pumps' capacity to deliver water to the district.

orifice meter A type of flowmeter consisting of a section of pipe blocked by a disc pierced with a small hole or orifice. The entire flow passes through the orifice, creating a pressure drop proportional to the flow rate.

outlet nozzle A threaded bronze outlet on the upper section of a fire hydrant, providing a point of hookup for hose lines or suction hose from hydrant to pumper truck.

overlying use The land use that occurs on top of an aquifer.

packing Rings of graphite-impregnated cotton, flax, or synthetic materials, used to control leakage along a valve stem or a pump shaft.

packing ring See *packing*.

permeation The process by which organic compounds pass through plastic pipe.

permit-required space A space defined by the Occupational Safety and Health Administration as having one or more of the following characteristics: contains, or has a potential to contain, a hazardous atmosphere; contains a material that has the potential for engulfing an entrant; has an internal configuration such that an entrant could be trapped or asphyxiated by inwardly converging walls or by a floor that slopes downward and tapers to a smaller cross-section; and/or contains any other recognized serious safety or health hazard.

pinch valve A valve that is closed by pinching shut a flexible interior liner. This type of valve is available only in relatively small sizes and is useful for throttling the flow of liquids that are corrosive or might clog other types of valves.

piston meter A water meter of the positive-displacement type, generally used for pipeline sizes of 2 in. (50 mm) or less, in which the flow is registered by the action of an oscillating piston.

plan and profile drawings Engineering drawings showing depth of pipe, pipe location (both horizontal and vertical displacements), and the distance from a reference point.

plat A map showing street names, mains, main sizes, numbered valves, and numbered hydrants for the plat-and-list method of setting up valve and hydrant maps.

plat-and-list method A method of preparing valve and hydrant maps. *Plat* is the map position, showing mains, valves, and hydrants. *List* is the text portion, which provides appropriate information for items on the plat.

plug valve A valve in which the movable element is a cylindrical or conical plug.

polling A technique of monitoring several instruments over a single communications channel with a receiver that periodically asks each instrument to send current status.

positive-displacement meter A type of meter consisting of a measuring chamber of known size that measures the volume of water flowing through it by means of a moving piston or disk.

positive-displacement pump A pump that delivers a precise volume of liquid for each stroke of the piston or rotation of the shaft.

potential cross-connection Any arrangement of pipes, fittings, or devices that indirectly connects a potable water supply to a nonpotable source. This connection may not be present at all times but it is always there potentially. Also known as an *indirect cross-connection*.

pounds per square inch (psi) A measure of pressure.

pounds per square inch gauge (psig) Pressure measured by a gauge and expressed in terms of pounds per square inch.

power A measure of the amount of work done per unit time by an electrical circuit, expressed in watts.

pressure The force on a unit area of water.

pressure head A measurement of the amount of energy in water due to water pressure.

pressure vacuum breaker A device designed to prevent backsiphonage, consisting of one or two independently operating, spring-loaded check valves and an independently operating, spring-loaded air-inlet valve.

pressure-reducing valve A valve with a horizontal disk for automatically reducing water pressures in a main to a preset value.

pressure-relief valve A valve that opens automatically when the water pressure reaches a preset limit to relieve the stress on a pipeline.

prestressed concrete Reinforced concrete placed in compression by highly stressed, closely spaced, helically wound wire. The prestressing permits the concrete to withstand tension forces.

primary instrumentation Instruments required to operate the monitoring system by obtaining information relating to water flow, pressure, level, and temperature.

prior appropriation doctrine A water rights doctrine in which the first user has the right to water before subsequent users.

priority in time The assigning of water rights based on who has been using the water the longest.

propeller meter A meter that measures flow rate by measuring the speed at which a propeller spins as an indication of the velocity at which the water is moving through a conduit of known cross-sectional area.

proportional control A mode of automatic control in which a valve or motor is activated slightly to respond to small variations in the system, but activated at a greater rate to respond to larger variations.

proportional meter Any flowmeter that diverts a small portion of the main flow and measures the flow rate of that portion as an indication of the rate of the main flow. The rate of the diverted flow is proportional to the rate of the main flow.

public relations The methods and activities employed to promote a favorable relationship with the public.

purchased water system A water system that purchases water from another water system and so generally provides only distribution and minimal treatment.

push-on joint A pipe joint consisting of a bell with a specially designed recess to accept a rubber ring gasket. The spigot end must have a beveled edge so it will slip into the gasket without catching or tearing it.

reasonable use A water rights term indicating that the water use is acceptable in general terms.

receiver (1) The part of a meter that converts the signal from the sensor into a form that can be read by the operator; also called the *receiver–indicator*. (2) In a telemetry system, the device that converts the signal from the transmission channel into a form that the indicator can respond to.

reciprocating pump A type of positive-displacement pump consisting of a closed cylinder containing a piston or plunger to draw liquid into the cylinder through an inlet valve and force it out through an outlet valve. When the piston acts on the liquid in one end of the cylinder, the pump is termed single-action; when the piston acts in both ends, the pump is termed double-action.

reduced pressure zone backflow preventer (RPZ) A mechanical device consisting of two independently operating, spring-loaded check valves with a reduced-pressure zone between the check valves. Designed to protect against both backpressure and backsiphonage. Also known as *reduced-pressure backflow preventer (RPBP)*.

reduced-voltage controller An electric controller that uses less than the line voltage to start the motor. Used when full line voltage may overload or damage the electrical system.

remote manual control A type of system control in which personnel in a central location manually operate the switches and levers to control equipment at a distant site.

remote terminal unit (RTU) A computer terminal used to monitor the status of control elements, monitor and transmit inputs from instruments, and respond to data requests and commands from the master station.

representative sample A collected sample that accurately reflects the composition of the water to be tested.

reservoir (1) Any tank or basin used for the storage of water. (2) A ground-level storage tank for which the diameter is greater than the height.

resilient-seated valve A gate valve with a disc that has a resilient material attached to it to allow leak-tight shutoff at high pressure.

resistance A characteristic of an electrical circuit that tends to restrict the flow of current, similar to friction in a pipeline. Measured in ohms.

restraining fitting A device for restraining joints that are particularly useful in locations where other existing utilities or structures are so numerous that thrust blocks are precluded.

riparian doctrine A water right that allows the owners of land abutting a stream or other natural body of water to use that water.

riser The vertical supply pipe to an elevated tank.

rotary pump A type of positive-displacement pump consisting of elements resembling gears that rotate in a close-fitting pump case. The rotation of these elements alternately draws in and discharges the water being pumped. Such pumps act with neither suction nor discharge valves, operate at almost any speed, and do not depend on centrifugal forces to lift the water.

rotary valve See *plug valve* and *ball valve*.

saddle A device attached around a main to hold the corporation stop. Used with mains that have thinner walls to prevent leakage. Also known as a *service clamp*.

scanning A technique of checking the value of each of several instruments, one after another. Used to monitor more than one instrument over a single channel.

secondary instrumentation Instruments that display information provided by sensors. The display may be mounted adjacent to the sensor, in a nearby control room, or in a distant control center.

sectional map A map that provides a detailed picture of a portion (section) of the distribution system. Reveals the locations and valving of existing mains, locations of fire hydrants, and locations of active service lines.

semiautomatic control A form of system control equipment in which many actions are taken automatically but some situations require human intervention.

service clamp See *saddle*.

service line The pipe (and all appurtenances) that runs between the utility's water main and the customer's place of use, including fire lines.

shaft (1) The bearing-supported rod in a pump, turned by the motor, on which the impeller is mounted. (2) The portion of a butterfly valve attached to the disc and a valve actuator. The shaft opens and closes the disc as the actuator is operated.

sheeting A method to protect workers against cave-ins by installing tightly spaced upright planks against each other to form a solid barrier against the faces of the excavation.

shielding A method to protect workers against cave-ins through the use of a steel box open at the top, bottom, and ends. Allows the workers to work inside the box while installing water mains.

shoring A framework of wood and/or metal constructed against the walls of a trench to prevent cave-in of the earth walls.

simplex Related to a telemetry or data transmission system that can move data through a single channel in only one direction. Compare with *half-duplex* and *full-duplex*.

single-phase power Alternating current (AC) power in which the current flow reaches a peak in each direction only once per cycle.

single-suction pump A centrifugal pump in which the water enters from only one side of the impeller. Also known as *end-suction pump*.

slide valve A gate valve that uses a relatively thin gate or blade that slides up and down in a recess to stop low-pressure flows where tight shutoff is not important.

slip (1) In a pump, the percentage of water taken into the suction end that is not discharged because of clearances in the moving unit. (2) In a motor, the difference between the speed of the rotating magnetic field produced by the stator and the speed of the rotor.

sloping A method of preventing cave-ins that involves excavating the sides of the trench at an angle (the angle of repose) so that the sides will be stable.

slug method A method of disinfecting new or repaired water mains in which a high dosage of chlorine is added to a portion of the water used to fill the pipe. This slug of water is allowed to pass through the entire length of pipe being disinfected.

split-tee fitting A special sleeve that is bolted around a main to allow a wet tap to be made. Also known as a *tapping sleeve*.

squirrel-cage induction motor The most common type of induction electric motor. The rotor consists of a series of aluminum or copper bars parallel to the shaft, resembling a squirrel cage. Also known as a *split-phase motor*.

standpipe A ground-level water storage tank for which the height is greater than the diameter.

starter A motor-control device that uses a small push-button switch to activate a control relay, which sends electrical current to the motor.

static electricity A state in which electrons have accumulated but are not flowing from one position to another. Static electricity is often referred to as electricity at rest.

static pressure Pressure that exists in water although the water does not flow.

stator The stationary member of an electric generator or motor.

stringing (hydrants) The practice of dropping a weighted string down the barrel of a hydrant to check if the barrel has fully drained.

supervisory control and data acquisition (SCADA) A methodology involving equipment that both acquires data on an operation and provides limited to total control of equipment in response to the data.

surface water system A water system using water from a lake or stream for its supply.

surge See *water hammer*.

synchronous motor An electric motor in which the rotor turns at the same speed as the rotating magnetic field produced by the stator. This type of motor has no slip.

tablet method A method of disinfecting new or repaired water mains in which calcium hypochlorite tablets are placed in a section of pipe. As the water fills the pipe, the tablets dissolve, producing a chlorine concentration in the water.

tachometer generator A sensor for measuring the rotational speed of a shaft.

tapping The process of connecting laterals and service lines to mains and/or other laterals.

tapping sleeve See *split-tee fitting*.

tapping valve A special shutoff valve used with a tapping sleeve.

telemetry A system of sending data over long distances, consisting of a transmitter, a transmission channel (wire, radio, or microwave), and a receiver. Used for remote instrumentation and control.

tensile strength A measure of the ability of pipe or other material to resist breakage when it is pulled lengthwise.

thermistor A semiconductor type of sensor that measures temperature.

thermocouple A sensor, made of two wires of dissimilar metals, that measures temperature.

three-phase power Alternating current (AC) power in which the current flow reaches three peaks in each direction during each cycle.

thrust (1) A force resulting from water under pressure and in motion. Thrust pushes against fittings, valves, and hydrants; it can cause couplings to leak or to pull apart entirely. (2) In general, any pushing force.

thrust anchor A block of concrete, often a roughly shaped cube, cast in place below a fitting to be anchored against vertical thrust, and tied to the fitting with anchor rods.

thrust block A mass of concrete cast in place between a fitting to be anchored against thrust and the undisturbed soil at the side or bottom of the pipe trench.

tie rod A device frequently used to restrain mechanical-joint fittings that are located close together when thrust blocks cannot be used.

time composite sample A composite sample consisting of several equal-volume samples taken at specified times.

ton container A reusable, welded tank that holds 2,000 lb (910 kg) of chlorine. Containers weigh about 3,700 lb (1,700 kg) when full and are generally 30 in. (0.76 m) in diameter and 80 in. (2.03 m) long.

tone-frequency multiplexing A method of sending several signals simultaneously over a single channel by converting the signals into sounds (tones) and assigning a specific tone to each signal.

tracer study A study using a substance that can readily be identified in water (such as a dye) to determine the distribution and rate of flow in a basin, pipe, or channel.

transmission line The pipeline or aqueduct used for water transmission, i.e., movement of water from the source to the treatment plant and from the plant to the distribution system.

travel-stop nut A nut, used in dry-barrel hydrants, that is screwed on the threaded section of the main rod. It bottoms at the base of the packing plate, or revolving nut, and terminates downward travel (opening) of the hydrant valve.

tree system A distribution system layout that centers around a single arterial main, which decreases in size with length. Branches are taken off at right angles, with subbranches from each branch.

tuberculation The growth of nodules (tubercles) on the pipe interior, which reduces the inside diameter and increases the pipe roughness.

turbine meter A meter that measures flow rates by measuring the speed at which a turbine spins in water, indicating the velocity at which the water is moving through a conduit of known cross-sectional area.

turbine pump (1) A centrifugal pump in which fixed guide vanes (diffusers) partially convert the velocity energy of the water into pressure head as the water leaves the impeller. (2) A regenerative turbine pump.

ultrasonic meter A meter that utilizes sound-generating and -receiving sensors (transducers) attached to the sides of the pipe.

upper section The upper part of the main hydrant assembly, including the outlet nozzles and outlet-nozzle caps. The upper section is usually constructed of gray cast iron. Also known as *nozzle section* or *head*.

vacuum breaker A mechanical device that allows air into the piping system, thereby preventing backflow that could otherwise be caused by the siphoning action created by a partial vacuum.

valve A mechanical device installed in a pipeline to control the amount and direction of water flow.

valve-and-hydrant map A mapped record that pinpoints the location of valves throughout the distribution system. Generally of plat-and-list or intersection type.

valve box A metal, concrete, or composite box or vault set over a valve stem at ground surface to allow access to the stem so that the valve can be opened and closed. A cover for the box is usually provided at the surface to keep out dirt and debris.

velocity The speed at which water moves; measured in ft/sec or m/sec.

velocity head A measurement of the amount of energy in water due to its velocity, or motion.

velocity pump The general class of pumps that use a rapidly turning impeller to impart kinetic energy or velocity to fluids. The pump casing then converts this velocity head, in part, to pressure head. Also known as *kinetic pump*.

venturi meter A pressure-differential meter used for measuring flow of water or other fluids through closed conduits or pipes, consisting of a Venturi tube and a flow-registering device. The difference in velocity head between the entrance and the contracted throat of the tube is an indication of the rate of flow.

vertical turbine pump A centrifugal pump, commonly of the multistage diffuser type, in which the pump shaft is mounted vertically.

voltage (1) A measure of electrical potential (electrical pressure), measure in volts. One volt will send a current of 1 ampere through a resistance of 1 ohm. (2) In telemetry, a type of signal in which the electromotive force (measured in volts) varies as the parameter being measured varies.

warm-climate hydrant A fire hydrant with a two-piece barrel that is the main valve located at ground level.

water hammer The potentially damaging slam, bang, or shudder that occurs in a pipe when a sudden change in water velocity (usually as a result of someone too-rapidly starting a pump or operating a valve) creates a great increase in water pressure.

water meter A device installed in a pipe under pressure for measuring and registering the quantity of water passing through.

wattmeter An instrument for measuring real power in watts, stated as kilowatt-hours (kW·h).

wear rings Rings made of brass or bronze placed on the impeller and/or casing of a centrifugal pump to control the amount of water that is allowed to leak from the discharge to the suction side of the pump.

wet-barrel hydrant A fire hydrant with no main valve. Under normal, nonemergency conditions, the barrel is full and pressurized (as long as the lateral piping to the hydrant is under pressure and the gate valve ahead of the hydrant is open). Each outlet has an independent valve that controls discharge from that outlet. The wetbarrel hydrant is used mainly in areas where temperatures do not drop below freezing. The hydrant has no drain mechanism.

wet tap A connection made to a main that is full or under pressure. Compare with *dry tap*.

wet-top hydrant A dry-barrel hydrant in which the threaded end of the main rod and the revolving or operating nut are not sealed from water in the barrel when the main valve of the hydrant is open and the hydrant is in use.

wound-rotor induction motor A type of electric motor, similar to a squirrel-cage induction motor but easier to start and capable of variable-speed operation.

Index

NOTE: *f* indicates a figure; *t* indicates a table.

A

absolute ownership 75–76
absolute pressure 93
   pounds per square inch absolute 93
A-C pipe
   See asbestos-cement pipe
AC
   See alternating current
acres to square feet 38
actuators 195–196
   electric 196
   hydraulic 196
   operating speed 196
   pneumatic 196
ADF
   See average daily flow
air gaps 348
air vents 149
air-and-vacuum relief valves 84, 192
air-relief valves 186
alkalinity 290, 339
alternating current 239–240
   transformers 242
   See also electricity
altitude valves 184–185
ammonia 289, 293
   testing for 338–339
amperometric test methods 337
angle of repose 133
ANSI/AWWA Standards
   distribution systems 100
ANSI/NSF Standard 61 100–101, 172
arterial-loop system 82
asbestos-cement pipe 105–106
   taps 175
atoms 239
   electrons 239
   neutrons 239
   protons 239
autoclave 338*f*
automatic mapping/facility management/
   geographic information systems
   (AM/FM/GIS) 370–371
*Automatically Controlled, Impressed-Current
   Cathodic Protection for the Interior of Steel
   Water Tanks* 232
average daily flow 53–57
   annual 53, 56
   formulas 53, 55–56
   monthly 53, 56
   weekly 53, 55–56
AWWA Manuals
   M3 164–165
   M9 119, 120
   M11 112
   M22 171
   M23 115
   M42 236
AWWA Standards 105
   C105 109
   C115 109
   C205 113
   C210 113
   C223 175
   C300 119
   C301 118
   C302 120
   C303 119
   C304 118
   C502 206
   C503 206
   C606 111
   C651 159
   C652 234
   C800 172, 173
   C900 115, 173
   C905 115
   C906 116
   C950 117
   D100 223
   D103 224
   D110 224
axial-flow pumps 262–263

B

backfilling 153–155
   compacting 153–155
   equipment 155
   granular 154–155
   jetting 154
   placement 153
   shoring removal 155
   tamping 153–154, 155
   trenches 162
   water settling 154
   See also site restoration
backflow
   air gaps 348

control devices and methods 347–356
defined 341
double check valve assemblies 351
due to backpressure 345, 345f
due to backsiphonage 345–347
prevention devices 185, 359–360
recirculated system 345, 345f
reduced pressure zone backflow preventers 348–350
vacuum breakers 351–353
backpressure
defined 341
due to backflow 345, 345f
backsiphonage
booster pumps 346
broken hydrant 346–347
broken main 346–347
defined 341
due to backflow 345–347
bacteria
harmful 338
nitrifying 339
trapping 338
bacteriological analysis 338
equipment for 338f
batteries
storage 242
beam breakage 102
bell holes 139
blowoff valves 84
booster disinfection 310
booster pumps 265–266
backsiphonage 346, 347f
box method 30–31
breakpoint chlorination 293–294, 293f
bromide 310
butterfly valves 193–194
bypass valves 189

C

$C \times T$ 309
C values 103
calcium hypochlorite 290, 292, 379
capacitors 242
carbon dioxide 339
cast-iron pipe
  See gray cast-iron pipe
cathodic protection
  elevated tanks 231–232
Cement–Mortar Protective Lining and Coating for Steel Water Pipe—4 In. (100 mm) and Larger—Shop Applied 113
centrifugal pumps 261, 262, 268–278
advantages and disadvantages 264
bearing temperature 270–271
bearings 277
casing 272
couplings 277
cycling 270
double-suction 272–273
flow control 270
general observations 271
impellers 273–274
inspection and maintenance 278
lantern rings 275–276
mechanical seals 276–277
monitoring operational variables 270–271
motor temperature 270–271
operation 268–271
packing rings 275
priming 268
shaft 274
shaft sleeves 275
single-suction 272
speed 271
starting 268
stopping 268, 269–270
suction and discharge heads 270
throttling 270
vibration 271
wear rings 274
chain of custody 336–337
field log sheet 336
record keeping 336
sampling
sampler's liability 337
sampler's responsibility 337
check valves 194
chemical reactions 291
chloramination 339
chloramine 293f, 294, 309
chlorination 290, 339
automatic controls 304–306
automatic switchover systems 305–306
auxiliary equipment 303–306
auxiliary tank valve 299–300, 300f
booster pump 303, 304f
breakpoint 293–294, 293f
calcium hypochlorite 307
chlorinator 301–302, 301f, 302f
chlorine
alarms 306
chlorine evaporator 305f
solution mixing 303
chlorine gas 294–306
feeding 298–303
handling and storing 295–298
chlorine liquid 295
cylinders 295f, 296–297
handling 296–297
moving 296f
scale 299f
storing 296–297
valves 300f
valves and piping 299–300
weighing 299f
weighing scales 298
diffuser 302–303, 303f
evaporators 304–305, 305f
flow proportional control 304
gas facilities 294–306
hypochlorination facilities 307–308
hypochlorite compounds 307–308
with hypochlorite 292–293, 307t
injector 302
pH and 290
practical aspects of 290

Index

residual flow control 304
safety equipment 306
sodium hypochlorite 308
switchover system 306f
tank cars 295f
ton container truck 295f
ton containers 295f
  scale 299f
  valves 300f
  valves and piping 299–300
  weighing scales 298
chlorine 309
  adding 293
  demand for 289
  dilute solutions of 292
  disinfection with 290
  dosage for 291
  feeding 294
  free 291
  gaseous 290
  inorganics and 289, 289t
  liquid 379
  monitoring/continuous 337
  pretreatment 294
  reactions 291–292
  testing for 337
  and trihalomethanes 12
chlorine cylinders 290f
  leaking/burning 291
chlorine demand 291, 337
chlorine gas 290, 292, 294
chlorine residual 291
  testing 337
circuit breakers
  shorting 242
  See also electrical equipment
clearwell 309
coagulants 339
coagulation 339
  enhanced 294
coliform
  testing for 337–338
colorimetric tests 339
compound meters 283
compression fittings 173
concentration 339
concrete pipe 117–120
  advantages and disadvantages 120
  bar-wrapped cylinder 119
  joints and bends 120
  lined-cylinder 118
  prestressed cylinder 117–118
  reinforced cylinder 119
  reinforced noncylinder 120
Concrete Pressure Pipe 119
Concrete Pressure Pipe, Bar-Wrapped, Steel-Cylinder Type 119
concrete tanks 224
connecting to existing mains 142–146
consumer confidence reports (CCRs) 8–9
contaminants
  chemical monitoring requirements 332
  chemicals 332
  inorganic 289

contamination 289
conversions 28–51
  acres to square meters 47
  area measurement 37–38
  box method 30–31
  concentration measurement 40
  cross-system 47–49
  cubic feet to gallons 31, 32
  cubic feet to gallons to pounds 31
  dimensional analysis 28
  factors 29
  feet to inches 28
  flow rate 33–36
  gallons to cubic feet 29, 31
  gallons to liters 47
  gallons to pounds 29
  inches to feet 28–29, 36
  kilograms to ounces 48–49
  linear measurement 36
  meters to decimeters 43
  metric system 42–47
  metric-to-metric 43
  milligrams per liter to grains per gallon 40–41
  milligrams per liter to parts per million 41–42
  tables 28–29
  temperature 49–51
  volume measurement 38–40
copper service lines 172
  leaks 179
corporation stops 173
correlative rights 76
corrosion
  resistance in pipes 102
corrosion control 339
  ground-level storage facilities 233
  and Lead and Copper Rule 16–17
cross-connection control and backflow prevention
  regulations 9
cross-connections 341–361
  actual 343
  atmospheric vacuum breaker 344, 344f
  backflow due to backpressure 345, 345f
  backflow due to backsiphonage 345–347
  backflow prevention 359–360
  booster pumps 346
  broken hydrant 346–347
  broken main 346–347
  chemical tanks 344
  control programs 357–360
  customer reports 360
  dock to ship 344, 344f
  examples 343–347
  garden hose 343, 344f
  hose filling sink 346
  locations 342
  operating records 360
  potential 343
  recirculated system 345, 345f
  records and reports 360–361
  repair personnel reports 361
  slop sink 343, 343f

testing 361
types 342–347
water softeners 344, 344f
water tank 343, 343f
Cryptosporidium
and Interim Enhanced Surface Water Treatment Rule 14
and Long-Term 2 Enhanced Surface Water Treatment Rule 15
curb stops 174–175
current 240f
alternating 239–240
direct 239
induced 240
See also electricity
current meters 283
for customers 283
mainline 285–286
cutting-in valves 190

D

DBPR
See Disinfectants and Disinfection By-products Rule
DBPs
See disinfection by-products
DC
See direct current
dead ends 82
water quality 317
Design of Prestressed Concrete Cylinder Pipe 118
detector-check meters 284
diaphragm valves 192
dimensional analysis 28
direct current 239
Disinfectants and Disinfection By-products Rule (DBPR)
Stage 1 12
Stage 2 13–14
Disinfecting Water Mains 159
disinfection 159–161, 309–310, 337
application point 159
booster 310
chemical 289–294
chlorine 289, 290
chlorine dosage 159–160
chlorine gas 161
chlorine residual analyzer 305f
contact period 161
continuous feed method 160
day tank 308
equipment 308
hypochlorination facilities 307–308
hypochlorite compounds 307–308
mix tank 308
procedures 160–161
slug method 161
tablet method 161
See also chlorination
disinfection by-products (DBPs) 10, 309–310
formation of 294, 309

disinfection of new pipelines
chlorination by continuous feed (and calculation) 58–59
chlorination by slug (and calculation) 57–59
chlorination by tablets or granules (and calculation) 58, 58t
flushing rate calculation 57
pipeline volume calculation 57
disinfection of storage facilities
chlorinate-and-fill method (chlorine amount calculation) 61
chlorination spray solution calculation 60
full storage facility chlorination method (chlorine amount calculation) 60
storage facility volume calculation 59
storage facility walls surface area calculation 59
Disinfection of Water-Storage Facilities 234
distribution mains
See mains
distribution system design
arterial-loop system 82
dead ends 82
fire flow requirements 85
grid systems 82
mapping 83
network analysis 86
pipe material selection 99–104
pressure requirements 85–86
quantity requirements 85–86
sizing mains 84–86
tree system 82
valving 83–84
velocity requirements 86
water quality 316–318
distribution system hydraulics
See system hydraulics
distribution system mapping
See system mapping
distribution systems 77–81, 289, 338
ANSI/AWWA Standards 100
biological threats 380
chemical threats 380
cyber threats 380
dead ends 82
elevated tanks 225–226
groundwater 78
isolation valves 183
layout 82–84
physical threats 380
problems caused by hydrant operation 203
purchased water 78–79
surface water 77–78
system threats 379–380
tracer studies of 309
types 77
water supply 79–80
diurnal effect 321
doctrines
See water rights

Index

double check valve assemblies 351
   installation 351
   residential use 351
double-suction pumps 272–273
Dry-Barrel Fire Hydrants 206
dry-barrel hydrants 204–206
   breakaway 206
   dry-top 204
   inspection 212, 213
   operation 212
   testing 211
   valve types 205
   wet-top hydrants 204
ductile-iron pipe 108–111
   advantages and disadvantages 109
   ball-and-socket joints 110
   cement-mortar lining 108
   fittings 111
   flanged joints 109
   grooved joints 111
   joints 109
   mechanical joints 110
   pressure classes 108
   push-on joints 110
   restrained joints 110
   shouldered joints 111
   taps 175
dynamic electricity 239
   induced 240
   transformers 240
   See also alternating current, direct current
dynamic pressure 89–90

E

EGL
   See energy grade line
electrical equipment 241–242
   AC transformers 242
   storage battery 242
   See also generators
electrical measurements 241–242
   current 241
   electron 241
   potential 241
   resistance 241
   voltage drop 241
   volts 241
electricity 239–241
   dynamic 239
   electromagnetics 239, 240–241
   resistance 241
   static 239
   See also alternating current, direct current, transformers
electromagnetics 239
electromagnetics 239
electrons 239, 240, 241
elevated tanks 217, 219, 221, 225–226, 227–232
   access hatches 229
   aesthetic concerns 226
   air vents 229

cathodic protection 231–232
coatings 230–231
cold-weather operation 233–234
drain connection 228
equipment 227–232
inlet and outlet pipes 227
ladders 229
maintenance 234
monitoring devices 228
obstruction lighting 232
overflow pipes 228
pumping costs 226
pumping head 226
system hydraulics 225–226
transmission costs 226
valving 228–229
energy grade line 97
Enhanced Surface Water Treatment Rule (ESWTR)
   Interim 14
   Long-Term 1 15
   Long-Term 2 15–16
equipment
   chlorine 294
equipment records
   filing equipment information 370
   for hydrants 368–369, 369f
   for meters 370
   for service lines 369–370
   for valves 367, 369f
Escherichia coli 11
ESWTR
   See Enhanced Surface Water Treatment Rule
excavation and trenches 127–138
   angle of repose 133
   avoiding other utilities 136–137
   bedding 137–138
   bell holes 138
   cave-in prevention 133–136
   cold climates 128–129
   danger signs 132–133
   equipment selection 127
   groundwater 131
   haunching 137–138
   notifying other utilities 127
   operations 129–131
   preparation 127–128
   problems 131
   public information 127
   safety 165
   sheeting 135–136
   shielding 133–134
   shoring 134–135
   sloping 133
   soil types 131–132
   special bedding 137
   trench depth 128–129
   trench failure causes 132
   trench width 129
   trench-wall failure 131–136
   warm climates 129
external load 101

F

Fabricated Steel & Stainless Steel Tapping Sleeves 175
Factory-Coated Bolted Carbon Steel Tanks for Water Storage 224
fecal coliform 11
fiberglass pressure pipe 116–117
    joints and fittings 117
Fiberglass Pressure Pipe 117
fire flow
    requirements 85
    and water quality 316
fire hydrants
    See hydrants
flare fittings 173
flexural strength 102
flow
    fire flow requirements 85
    requirements 317
flow rate 63
    formulas 63–64
    instantaneous 63
fluoride 309
flush hydrants 206

G

galvanized iron service lines 172
    leaks 179
gas chlorinator 290f
gate valves 187–190
    bypass 189
    cost v. butterfly valves 194
    cutting-in 190
    horizontal 188–189
    horizontal thrust 189
    inserting 190
    nonrising-stem 188
    resilient-seated 190
    rising-stem 188
    slide 190
    tapping 189–190
gauge pressure
    conversion equations 93
    definition 92
    pounds per square inch gauge 92
generators 240, 240f
    motor-driven 242
geographic information systems (GIS) 83, 370–371
Giardia lamblia 290
    and Surface Water Treatment Rule 14
globe valves 191
gray cast-iron pipe 106
    taps 175
grid systems 82
Grooved and Shouldered Joints 111
Ground Water Rule (GWR) 17, 323
ground-level storage facilities 226–227, 232–233
    cold-weather operation 233
    corrosion protection 233
    drains 233
    equipment 232–233
    inlet and outlet pipes 232
    maintenance 234
    valves 184–185

H

haloacetic acids (HAAs) 309
head 92, 95–97
    and column height 89, 90f
    calculating 61–63
    definition 92, 95
    elevation 96, 96f
    energy grade line 97
    formula 95
    pressure 95–96, 95f
    velocity 96–97, 97f
hydrants 201–215
    allowing miscellaneous uses 202–203
    as air vents 149
    auxiliary valves 183, 208–209, 211
    blocking 210
    bonnet 207
    breakaway 206
    color coding 211
    damage 202
    drainage 210
    dry-barrel 204–206, 211, 212
    dry-top 204
    fire fighting 201–202
    flow diffusers 215
    flow testing 214, 215
    flush 206
    flushing 159
    footing 210
    inspection 209, 212–213
    installation 210–211
    location 201–202, 210
    lower section 208
    maintenance 212
    miscellaneous uses 202–203
    nozzles 207–208
    operating nut 207
    operation 212
    parts 207–209
    records 214
    repair 213
    safety 214–215
    system problems 203
    testing 211
    types 204–207
    upper section 207–208
    uses 201–203
    valve types 205
    warm-climate 206
    wet-barrel 206
    wet-top 204
    wrenches 207
hydraulics
    elevated tanks and distribution systems 225–226
    and water quality modeling 316–317
    See also system hydraulics
hydrochloric acid 291
hydrogen ions 291

hydrogen sulfide 289
hydropneumatic systems 222
hydrostatic pressure 89
hypochlorite 290, 292–293
hypochlorous acid 291, 292

I

induced current 240
inorganics
   alkalinity consumption with 289t
   chlorine and 289, 289t
in-plant pipes 104
inserting valves 190
instantaneous flow rate
   See flow rate
instrumentation and control 243
   automatic control systems 250–251
   bellows sensors 243, 244f
   Bourdon tubes 243, 244f
   bubbler tubes 243–244, 245f
   centralized computer control 251
   control systems 243, 249–251
   control terminal units (CTUs) 248
   D'Arsonval meters 246
   diaphragm elements 243, 245f
   digital multimeters 246f
   direct manual control systems 249
   disinfectant residual monitors 247
   distributed computer control 251
   duplexing 249
   electrical sensors 246, 246f
   electronic probes as level sensors 244
   equipment status monitors 246–247
   feedback proportional (closed-loop) control 250–251, 250f
   feedforward proportional control 250, 250f
   float mechanisms 243, 245f
   flow sensors 243
   helical sensors 243, 244f
   level sensors 243–244, 245f
   monitoring sensors (primary instruments) 243–247
   multiplexing 248–249
   on–off differential control 250
   pH monitors 247
   polling 248
   pressure sensors 243, 244f
   process analyzers 246–247
   proportional control 250
   receivers and indicators 247–248
   remote manual control systems 249
   remote terminal units (RTUs) 248
   scanning equipment 248
   secondary instrumentation 243, 247–249
   semiautomatic control systems 249
   signal transmission instruments 247
   strain gauges 244f
   supervisory control and data acquisition (SCADA) 251–252, 251f
   telemetric equipment 248
   temperature sensors 244–245, 245f
   thermistors (resistance temperature devices) 245, 245f
   thermocouples 244–245, 245f
   tone-frequency multiplexing 248
   totalizing wattmeters 246, 246f
   transducers as level sensors 244
   transmission channels 249
   turbidity monitors 247
   ultrasonic level sensors 244
   variable-resistance level sensors 244
Insurance Services Office 85
intensity 339
Interim Enhanced Surface Water Treatment Rule 14
internal pressure 102
iron 289

J

jar tests
   equipment for 337
joints 139–142
   ball-and-socket 110
   concrete pipe 120
   ductile-iron pipe 109–111
   fiberglass pressure pipe 117
   flanged 109
   grooved 111
   mechanical 110, 142
   polyvinyl chloride (PVC) pipe 116
   push-on 110, 139–140
   restrained 110
   shouldered 111
   steel pipe 113–114
   valves 196–197

L

laboratory certification 333–334
lead 10
Lead and Copper Rule 16–17, 324
   action level 16
lead service lines 172
   leaks 179
leakage 156–158
   failed pressure tests 158
   makeup water 157–158
   testing procedure 156–157
Liquid-Epoxy Coating Systems for the Interior and Exterior of Steel Water Pipelines 113
Long-Term 1 Enhanced Surface Water Treatment Rule 15
Long-Term 2 Enhanced Surface Water Treatment Rule 15–16

M

magnetic meters 287
mains 104
   air vents 149
   bacteriological testing 161–162
   booster disinfection 310
   chemical safety 166
   chlorine dosage 159–160
   chlorine gas for disinfection 161
   continuous feed method for disinfection 160

coupons 145
dead ends 82
disinfection 159–161
domestic use 85
excavation 127–138
failed pressure tests 158
final inspection 162
fire flow requirements 85
flushing 159
installation 123–150
installation safety 164–167
laying pipe 138–150
leakage 156–158
makeup water 157–158
material-handling safety 164–165
personal protection equipment 165
pipe handling 124–127
portable power tool safety 166
pressure requirements 85–86
quantity requirements 85–86
site restoration 162–164
sizing 84–86
slug method for disinfection 161
tablet method for disinfection 161
tee connections 142–144
traffic control safety 165
trench safety 165
trenching 128–138
tunneling 146
vehicle safety 167
velocity requirements 86
makeup water 157–158
manganese 289
Manual on Uniform Traffic Control Devices for Streets and Highways 165, 378
maps
 distribution system design 83
maximum contaminant levels (MCLs) 339
media relations 387
metric system 42
 abbreviations 42
 notations 42t
 prefixes 42
microbiological regulations 10
microorganisms 290, 291, 338
mixed-flow pumps 263
MMO-MUG (minimal medium) technique 338
monitoring
 See water quality monitoring
motors 242
multijet meters 285

N

National Primary Drinking Water Regulations (NPDWRs) 3
National Secondary Drinking Water Regulations 18, 18t–19t
natural organic matter (NOM) 309
needle valves 191–192
neutrons 239
nitrites 338–339
nitrogen 289

NOM
 See natural organic matter
NSF International Standard 61 115
nutating-disk meters 282

O

Occupational Safety and Health Act 373
Occupational Safety and Health Administration (OSHA)
 on confined-space safety 375–376
Ohm's law 241f
ohms
 See resistance
operations
 surface water systems 77–78
operator certification regulations 9
organic material
 chlorine and 289
orifice meters 286
oxidation 289
oxidation 337
ozone 309

P

permeation 115
pH 290
 chlorination and 290
 testing for 339
pinch valves 192
pipes and pipelines
 air vents 149
 asbestos-cement 105–106
 bar-wrapped concrete 119
 beam breakage 102
 bell holes 138, 139
 C value 103
 characteristics 101–103
 cleaning 138
 concrete 117–120
  prestressed 117–118
  reinforced 119–120
 copper 172
 corrosion resistance 102
 ductile-iron 108–111
 durability 102
 ease of repair 103
 ease of tapping 103
 economic considerations 103–104
 external load 101
 fiberglass pressure 116–117
 flexural strength 102
 galvanized iron 172
 gray cast-iron 106
 handling 124–127
 in-plant 104
 inspection 124, 138
 installation conditions 104
 installation cost 103
 internal pressure 102
 jointing 139–142
 laying 138–150
 lead 172
 mains 104

material selection   99–104
mechanical joints   142
placement   138–139
plastic   114–117, 172–173
polyethylene   116
pressure rating   102
pressure taps   144–146
push-on joints   139–140
restraint fittings   148
service line material   172–173
service lines   104
shear breakage   102
shipment   123–124
smoothness of inner surface   103
stacking   125–126
standards   100–101, 106
steel   112–114
strength   101–102
stringing   126–127
tee connections   142–144
tensile strength   102
thrust anchors   148
thrust blocks   148
tie rods   148
transmission lines   104
unloading   124–125
water hammer   102
water quality maintenance   103
wooden   105
See also mains
piston meters   282
plastic pipe   114–117
  fiberglass pressure   116–117
  materials   115
  permeation   115
  polybutylene (PB)   116
  polyethylene (PE)   116
  polyvinyl chloride (PVC)   115–116
  properties   114
  standards   115
plastic service lines   172–173
  leaks   179
polybutylene pipe   116
Polyethylene (PE) Pressure Pipe and Fittings, 4 In. (100 mm) Through 63 In. (1,600 mm), for Water Distribution and Transmission   116
Polyethylene Encasement for Ductile-Iron Pipe Systems   109
polyethylene pipe   116
polyvinyl chloride pipe   115–116
  advantages and disadvantages   115–116
  joints and fittings   116
  taps   175
positive-displacement meters   282
  nutating-disk   282
  piston   282
positive-displacement pumps   266–267
  reciprocating   267
  rotary   267
pressure   90–94
  absolute   93, 93f
  dynamic   89–90
  failed tests   158
  gauge   92, 93f
  head   92
  hydrostatic   89
  internal   102
  requirements in mains   85
  static   89
  testing procedure   156–157
  water height   90–93
pressure rating   102
pressure taps   144–146
pressure-relief valves   185, 192
Prestressed Concrete Pressure Pipe, Steel-Cylinder Type   118
prior appropriation doctrine   74–75
  beneficial use   75
  legal complications   75
  priority in time   75
propeller meters   285–286
proportional meters   286
protons   239
public relations
  caring   384
  and company vehicles   387
  courtesy   384
  customer service card   386, 386f
  good communications   384
  maintenance and repair crews   385–387
  media relations   387
  meter readers   384–385
  property damage   387
  role of   383–384
  and water distribution personnel   384–387
public water systems   2–3, 3f
  chlorine and   293
  community   2
  nontransient, noncommunity   3
  transient, noncommunity   3
pumps and pumping
  axial-flow   262–263
  booster   265–266
  capacitor-start, single-phase motors   256
  centrifugal   261, 262, 263–264, 268–278
  centrifugal-jet combination   266
  cycling   270
  deep-well   265
  diesel engines   258
  engine operation and maintenance   259
  gas engines   258
  gasoline engines   258
  internal-combustion engines   258–259
  mechanical protection of motors   256–257
  mixed-flow   263
  motor control equipment   257–258
  motor control systems   257–258, 258f
  motor starters   257, 257f
  motor temperature   256
  motors   255–257, 255f
  positive-displacement   266–267
  pump, motor, and engine records   259
  radial-flow   262
  reciprocating   267
  reduced-voltage controllers for motors   257

repulsion-induction, single-phase
    motors 256
rotary 267
single-phase motors 255–256
split-phase, single-phase motors 256
squirrel-cage, three-phase motors 256
steam engines 258
submersible 265
synchronous, three-phase motors 256
three-phase motors 256
throttling 270
types 266–267
velocity 261–266
vertical turbine 264–266
volute 261
wound-rotor induction, three-phase
    motors 256
PVC Pipe—Design and Installation 115
PWS
    See public water systems

R

radial-flow pumps 262
    See also centrifugal pumps
record keeping 4, 5t
records
    pump, motor, and engine 259
reduced pressure zone backflow
    preventers 348–350
    installation 350
regulations 1, 338
    consumer confidence reports 8–9
    containers 329
    contaminants 3, 324
    cross-connection control 9
    customer faucets 328–329
    dead-end sampling points 328
    Disinfectants and Disinfection By-products
        Rule 12–14
    disinfection by-products (DBPs) 10
    dissolved oxygen (DO) 321
    and EPA 2
    federal 1–9
    Ground Water Rule 17
    holding times 334–335
    laboratories 333
    lead 10
    Lead and Copper Rule 16–17
    microbiological safety 10
    monitoring, reporting, and record
        keeping 4, 5t
    National Primary Drinking Water
        Regulations 3
    National Secondary Drinking Water
        Regulations 18, 18t–19t
    operator certification 9
    preservation 334–335
    problems 335
    public notification 3–4
    public water systems 2–3, 3f
    raw-water sample collection
        procedures 329–330
    record keeping 334

Safe Drinking Water Act 1–4
sample
    labeling 334
    points 324
sample faucets 328–329, 332
sample-point locations 325
sample-preservation techniques 335
sampling frequency 324
sampling point selection 323–329
    distribution system 324–328
    groundwater sources 323
specific to distribution systems 10–19
state 9–10
state or USEPA enforcement 10
state primacy 2
state technical assistance to operators 10
storage 334–335
Surface Water Treatment Rule 14–16
time of sampling 335
Total Coliform Rule 10–12
total dissolved solids (TDS) 322
transportation 336
    shipment 336
turbidity sample 327
types of samples 321–323
    composite samples 322
    continuous samples 323
    grab samples 321–322
water quality monitoring 4–7
*Reinforced Concrete Pressure Pipe,
    Noncylinder Type* 120
*Reinforced Concrete Pressure Pipe,
    Steel-Cylinder Type* 119
reporting 4, 5t
reservoirs 221–222
    concrete 224
    earth-embankment 222–223
    ground-level 184–185
resistance (electricity) 241
Revised Total Coliform Rule 10–12
rights
    See water rights
riparian doctrine 74
rotary valves 193

S

saddles
    See service saddles
Safe Drinking Water Act (SDWA) 1–4
    on consumer confidence reports 8–9
    contaminant regulations 3
    and EPA 2
    on monitoring, reporting, and record
        keeping 4, 5t
    and National Primary Drinking Water
        Regulations 3
    on operator certification 9
    on public notification 3–4
    and public water systems 2–3
    rules 1–2
    on state or USEPA enforcement 10
    and state primacy 2
    on water quality monitoring 4–7

safety   373
    and causes of accidents   373
    chemicals   165, 379
    in confined spaces   375–376, 377f
    hand tools   377
    hydrants   214–215
    main installation   164–167
    material handling   164–165, 374–375, 375f
    and OSHA   373
    personal protection equipment   165, 373–374
    portable power tools   166, 377–378
    regulatory requirements   373
    traffic control   165, 378–379, 378f
    in trench work   165, 375
    vehicle   167, 379
    water storage facilities   236–237
Safety Practices for Water Utilities   164–165
sampling
    bacteriological sample points   327
    chlorine residual samples   325
    coliform analysis   327
    collection   329–331
    collection procedures
        distribution system   330–331
        special testing purposes   331–332
        treatment plant   330
    See also water quality monitoring
sanitary surveys   14–15, 17
SCADA
    See supervisory control and data acquisition systems
SDWA
    See Safe Drinking Water Act
service clamp   177
service lines   104, 171–181
    adapters and connectors   173
    copper tubing   172, 179
    corporation stops   173
    curb stops and boxes   174–175
    electrical thawing   179
    galvanized iron pipe   172, 179
    hot-water thawing   180
    lead pipe   172, 179
    leaks and breaks   179
    plastic tubing   172–173, 179
    records   181
    responsibility for   180–181
    size   171
    taps   175–178
    types   172–173
    warm-air thawing   180
service saddles   177
service taps   175–178
    direct insertion   175–176
    drilling-and-tapping machines   176, 177
    dry   175
    location   177–178
    procedure   176
    saddles   177
    self-contained   177
    sleeves   175–176
    small drilling machines   177
    wet   175

shear breakage   102
short circuits   242
site restoration   162–164
    backfilling trenches   162
    curbs, gutters, and sidewalks   163
    ditches and culverts   163
    grass replacement   163
    machinery and construction sheds   163
    pavement repair   162
    private property   164
    roadway cleanup   164
    traffic restoration   164
    trees and shrubs   163
    utilities   163
    watercourses and slopes   164
Sizing Water Service Lines and Meters   171
slimes   289
sodium hypochlorite   290, 292–293, 379
softening   339
specific gravity   293
Stage 1 Disinfectants and Disinfection By-products Rule   12
Stage 2 Disinfectants and Disinfection By-products Rule   13–14
Standard for Fire Hose Connections   207–208
standpipes   221
static electricity   239
static pressure   89
steel pipe   112–114
    advantages and disadvantages   112
    cement-mortar lining   113
    collapsed   112
    epoxy lining   113
    in-plant   114
    joints and fittings   113–114
    use   112
    wrinkle bending   114
Steel Pipe—A Guide for Design and Installation   112
steel tanks   223–224
Steel Water Storage Tanks   236
sterilization
    autoclave for   338f
storage battery   242
stringing   126–127
supervisory control and data acquisition (SCADA) systems   251–252, 251f
surface water   321
Surface Water Treatment Rule (SWTR)   14–16, 290, 309
    Interim Enhanced   14
    Long-Term 1 Enhanced   15
    Long-Term 2 Enhanced   15–16
    and sanitary surveys   14–15, 17
SWTR
    See Surface Water Treatment Rule
system hydraulics   89
    dynamic pressure   89–90
    hydraulic head and column height   89, 90f
    hydrostatic pressure   89
    static pressure   89
    velocity   90

system mapping 363
  and automatic mapping/facility management/geographic in-formation systems (AM/FM/GIS) 370–371
  comprehensive maps (wall maps) 363–364, 364f
  engineers' abbreviations 367
  plan and profile drawings 367
  sectional maps (plats) 364–366, 365f
  sources for sectional maps 365–366
  symbols 367, 368f
  valve and hydrant maps 366, 366f
  See also Equipment records
system security 379–380
  biological threats 380
  chemical threats 380
  cyber threats 380
  physical threats 380
  water supply threats 379–380

T

tanks
  storage 310
taste-and-odor control 294
tee connections 142–144
tensile strength 102
THMs
  See trihalomethanes
thrust 147–148
  anchors 148
  blocks 148
  control 147–148
  locations 147
  restraint fittings 148
  tie rods 148
tie rods 148
total coliform (TC) 338
Total Coliform Rule 10–12
  Revised 10–12
total trihalomethanes (TTHMs) 10, 309, 310
tracer studies 309
transformers 240
  AC 242
transmission lines 104
treatment 339
tree system 82
trihalomethanes (THMs) 12
TTHMs
  See total trihalomethanes
tunneling 146
turbine meters 285

U

ultrasonic meters 287–288
unaccounted-for water 73, 284
US Environmental Protection Agency (USEPA) 1–19, 309
  information on regulations 9

V

vacuum systems 294
valves 183–198
  actuator operating speed 196
  air-and-vacuum relief 84, 192
  air-relief 186
  altitude 184–185
  blowoff 84
  boxes 197–198
  butterfly 193–194
  bypass 189
  check 194
  classifications 186–194
  corporation stops 173, 184
  curb stops 184
  curb stops and boxes 174–175
  cutting-in 190
  diaphragm 192
  discharge 184
  distribution system isolation 183
  dry-barrel hydrants 205
  electric actuators 196
  elevated tanks 228–229
  gate 187–190
  globe 191
  ground-level reservoirs 184–185
  handwheel operators 195
  horizontal gate 188–189
  horizontal thrust 189
  hydrant auxiliary 183
  hydrants 208–209, 211
  hydraulic actuators 196
  inserting 190
  joints 196–197
  key 195
  manual operation 195
  needle 191–192
  nonrising-gate 188
  operation 195–196
  pinch 192
  pneumatic actuators 196
  power actuators 195–196
  pressure-reducing 184
  pressure-relief 185, 192
  preventing backflow 185
  pump control 184
  records 198
  regulating pressure 184–185
  resilient-seated gate 190
  rising-stem gate valves 188
  rotary 193
  slide 190
  for starting and stopping flow 183
  storage 196
  tapping 189–190
  throttling flow 184–185
  uses 183
  vaults 197–198
  water service 184
velocity 90
velocity head 96–97, 97f
velocity pumps 261–266
  axial-flow 262–263
  mixed-flow 263
  radial-flow 262
  slip 262
Venturi meters 286
vertical turbine pumps 264–266
  advantages and disadvantages 265
  booster 265–266

centrifugal-jet combination 266
deep-well 265
submersible 265
viruses 290
voltage
drop 241
volts 241
volume 21–28
circles 22
combinations of representative surface area 22
cones 26
cubes 21
definition 21
formulas 21, 23
rectangles 22
representative surface area 21
spheres 26
triangles 22
trough 23

W

wastewater 339
water age 309
water hammer 102
water meters 281–288
in basements 171
in boxes 169–170
in buildings 171
compound 283
current (for mainlines) 285–286
current 283
customer water meters 281–284
desirable characteristics 281
detector-check 284
exposed 169
large-customer meters 283–284
locations 169
magnetic 287
mainline 284–288
multijet 285
nutating-disk 282
orifice 286
piston 282
positive-displacement 282–283
propeller 285–286
proportional 286
standards 281
turbine 285
types 281
ultrasonic 287–288
Venturi 286
water quality 313–318, 321
ANSI/NSF Standard 61 100–101
blending 317
booster disinfection 310
breakpoint chlorination 310
and construction activities 315
and customer complaint investigations 315
dead ends 317
degradation in water storage 221
distribution system design 316–318
emergency monitoring 315–316
and finished water storage facilities 315, 317–318

flow requirements 317
groundwater 78
hydraulic modeling 316–317
monitoring 314–316
multiple-barrier approach 313
nonroutine monitoring 315–316
and pipeline design 317
pipeline materials 317
in pipes 103
planning considerations 316
pressure zone adjustments 317
requirements in mains 86
routine monitoring 314
sampling plan 314–315
water quality monitoring 4–7, 5t, 321
chain of custody 336–337
chemical contaminants 332
composite samples 5
drinking water 339
flow-proportional composite samples 5
grab samples 5
groundwater 321
laboratories 333–334
laboratory certification 8
record keeping 334
sample collection 5, 7–8
sample faucets 6–7, 7f
sample labeling 334
sample point selection 6
sample storage and equipment 6
sampling 321–332
special-purpose samples 8
time composite samples 5
water rights 73–76
absolute ownership 75–76
appropriation–permit system 76
correlative 76
groundwater 75–76
prior appropriation doctrine 74–75
reasonable use 76
riparian doctrine 74
surface water 74–75
water storage 217–237
blending water sources 220
capacity requirements 220
cold-weather operation 233–234
concrete reservoirs 224
concrete tanks 224
construction materials 222–224
decreasing power costs 219
detention times 220
disinfection 234–236
earth-embankment reservoirs 222–223
elevated tanks 217, 219, 221, 225–226, 227–232, 234
emergency storage 221
facility safety 236–237
fire demand 219
ground-level 226–227, 232–233
ground-level tanks 234
hydropneumatic systems 222
inspection 236
location 225–227
maintenance 233–236
operating convenience 218

operation and maintenance 233–236
  peak-hour demand 217–218
  pumping requirements 218–219
  purposes 217–220
  records 236
  reservoirs 221–222
  service types 220–221
  source or pump failure 219
  standpipes 221
  steel tanks 223–224
  supply and demand 217–218
  surge relief 219
  types of structures 217
  water quality degradation 221
water systems 77–79
  groundwater 78
  purchased water 78–79, 218
  surface water 77–78
  water storage 218
  See also distribution systems
water use 69–71
  community 72
  condition of system 73
  domestic 69
  industrial 69–70
  metering 72–73
  per capita 52–53
  public 71
  reasonable 76
  season 71–72
  time of day 71
  variations 71–73
Welded Carbon Steel Tanks for Water Storage 223
Wet-Barrel Fire Hydrants 206
wet-barrel hydrants 206
  inspection 212, 213
Wire- and Strand-Wound, Circular, Prestressed Concrete Water Tanks 224